THE BALDWIN LOCOMOTIVE WORKS

Studies in Industry and Society

PHILIP B. SCRANTON, SERIES EDITOR

Published with the assistance of the Hagley Museum and Library

1. BURTON W. FOLSOM, JR.
   *Urban Capitalists: Entrepreneurs and City Growth
   in Pennsylvania's Lackawanna and Lehigh Regions, 1800–1920*

2. JOHN BODNAR
   *Workers' World: Kinship, Community, and Protest
   in an Industrial Society, 1900–1940*

3. PAUL F. PASKOFF
   *Industrial Evolution: Organization, Structure, and Growth
   of the Pennsylvania Iron Industry, 1750–1860*

4. DAVID A. HOUNSHELL
   *From the American System to Mass Production, 1800–1932: The
   Development of Manufacturing Technology in the United States*

5. CYNTHIA J. SHELTON
   *The Mills of Manayunk: Industrialization and Social Conflict
   in the Philadelphia Region, 1787–1837*

6. JOANNE YATES
   *Control through Communication: The Rise of System
   in American Management*

7. CHARLES W. CHEAPE
   *Strictly Business: Walter Carpenter at Du Pont
   and General Motors*

8. JOHN K. BROWN
   *The Baldwin Locomotive Works, 1831–1915:
   A Study in American Industrial Practice*

9. JAMES P. KRAFT
   Stage to Studio: Musicians and the Sound Revolution, 1890–1950

10. EDWARD S. COOKE, JR.
    *Making Furniture in Preindustrial America: The Social
    Economy of Newtown and Woodbury, Connecticut*

11. LINDY BIGGS
    *The Rational Factory: Architecture, Technology, and Work in
    America's Age of Mass Production*

12. THOMAS R. HEINRICH
    *Ships for the Seven Seas: Philadelphia Shipbuilding in the Age
    of Industrial Capitalism*

13. MARK ALDRICH
    *Safety First: Technology, Labor, and Business in the Building of
    American Work Safety, 1870–1939*

A STUDY IN AMERICAN INDUSTRIAL PRACTICE

The Baldwin Locomotive Works

1831–1915

JOHN K. BROWN

THE JOHNS HOPKINS UNIVERSITY PRESS • BALTIMORE AND LONDON

The Johns Hopkins University Press
2715 North Charles Street
Baltimore, Maryland 21218-4363
The Johns Hopkins Press Ltd., London
www.press.jhu.edu

LIBRARY OF CONGRESS CATALOGING-IN-PUBLICATION DATA

Brown, John K.
  The Baldwin Locomotive Works, 1831–1915 : a study in American industrial practice /
John K. Brown.
    p.   cm. — (Studies in industry and society ; 8)
  Includes bibliographical references and index.
  ISBN 0-8018-5047-9
  1. Baldwin Locomotive Works—History.   I. Title.   II. Series.
TJ625.B2B76   1995
338.7′62526′0973—dc20   94-44850

A catalog record for this book is available from the British Library.

CONTENTS

List of Tables and Charts   vii

Preface   ix

Acknowledgments   xxi

Introduction   xxv

1 Establishing the Baldwin Works, 1831–1866   1

2 The Locomotive Industry, 1860–1901   28

3 The Character of Innovation in Locomotive Design   57

4 Management at Baldwin, 1850 1909   92

5 The Baldwin Workforce, 1860–1900   127

6 Building Locomotives, 1850–1900   163

7 Triumph and Eclipse, 1900–1915   198

Conclusion: Baldwin, the Capital Equipment Sector, and the
Nineteenth-Century Economy   234

APPENDIX A: Baldwin's Annual Output and Employment, 1832–1932   241

APPENDIX B: Data on Major American Locomotive Builders   243

Abbreviations and Original Sources   247

Notes   251

Bibliography   309

Index   323

TABLES AND CHARTS

TABLES

2.1 U.S. Railway Mileage, 1830–1910   30

3.1 Decennial Breakdown of Baldwin Production by Market
Segment, 1850–1900   77

4.1 The Baldwin Partnerships, 1831–1909   97

5.1 Baldwin Workers' Trades and Levels of Skill, 1876   132

5.2 Baldwin Employment Peaks by Decades, 1860 1910   133

5.3 Trades of Apprentices, 1854–68   141

5.4 Disposition of Apprentices, 1854 73   142

6.1 Measures of Baldwin's Production, 1850–1900   166

6.2 Indices of Production and Productivity at Baldwin: 1850, 1860,
and 1870   181

6.3 Indices of Production and Productivity at Baldwin: 1870, 1880,
and 1890   186

6.4 Indices of Production and Productivity at Baldwin: 1890 and 1900   194

CHARTS

2.1 Baldwin's Output for Domestic and Foreign
Customers, 1870–1900   45

4.1 The Baldwin Partners, 1831–1909   98

4.2 Organizational Chart of the Baldwin Locomotive Works Management,
December 1859   104

5.1 Baldwin's Annual Employment and Output, 1880–97   134

7.1 Baldwin's Annual Employment and Output, 1890–1907   203

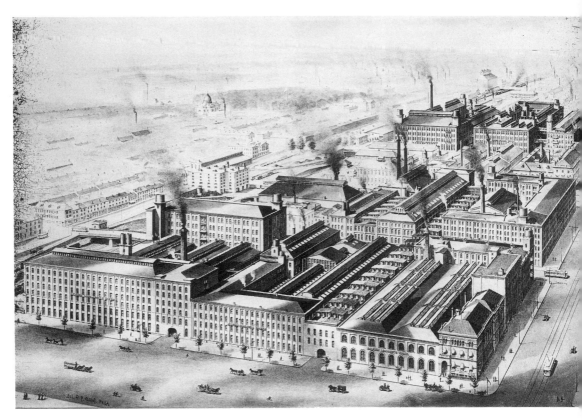

When Matthias Baldwin opened his first factory building on Broad Street in 1835, the neighborhood still had a rural quality. Over the next thirty years a throbbing industrial district grew up around the expanding locomotive works. This view circa 1905 shows the plant at the height of Baldwin's success. The shops filled seven city blocks and occupied portions of five more. AUTHOR'S COLLECTION

PREFACE

As the industrial age gathered force and momentum during the nineteenth century, the steam locomotive came to symbolize the new agencies of technology, commerce, speed, and power that reordered Western society and marked the most fundamental changes ever in humanity's lot on earth. Countless authors have traced the social and cultural impact of steam railways, but we remain only dimly aware of how these engines of change were actually made. That is the concern of this book, a history of the Baldwin Locomotive Works of Philadelphia, the largest and most successful locomotive builder in America or the world in the nineteenth century.

The chapters that follow detail aspects of the company's history and its place in the larger American industrial scene. My concern here at the outset is to build a locomotive. Compared with the bedlam of an erecting shop, the mind's eye is far quieter and much more orderly. Yet one must first see the process of building an engine to situate all that follows in this book. So join the throng on the way to work in the Baldwin plant early one morning in October 1905.

Just before the starting whistles blow out at seven A.M., Baldwin's day shift of over ten thousand men hurries through the teeming streets of Bush Hill, an industrial district rendered in brick and soot, crowded with factories, railway lines, coal yards, and worker tenements. The men split up into separate groups, heading for the different shops of a factory spread out over twelve city blocks where the six thousand workers of the night shift are ready to lay down their tools. Upon arriving, the day men will begin work on any of the approximately 450 new locomotives that on this morning are at varying stages of completion in Baldwin's regular production schedule. From the foundries to the erecting shop, workers in eight skilled trades and scores of specialties apply their talents to engines ranging from little 4-ton electrics to massive 150-ton steamers. Baldwin

In 1905 draftsmen occupied two floors of Baldwin's office building at 500 North Broad Street. Like production workers, the designers specialized in particular components or aspects of design. The flat-file tables in the middle contained thousands of drawings for standard parts. Draftsmen used these standards whenever possible—even in custom engines. The boy leaning on one table (*center*) was a runner who took finished drawings down to the shops. H. L. BROADBELT

did not make a single standard steam locomotive, but it largely followed common steps in building these custom products.[1]

It all begins in the drawing room, actually two large floors of the office building at Broad and Spring Garden Streets, where 150 draftsmen work up new designs. Under Baldwin's eight-week production schedule, the draftsmen have three weeks on average to get out the plans for a new type of engine—and much of the firm's output at this time is custom designs. Referring to the contract specifications, they build a locomotive on paper, using established standard parts where possible and drawing new component designs as needed. After plans are drawn and checked, they first go to the purchasing department, which orders materials from outside vendors. Great quantities of costly semifinished materials, like boiler flues, steel tires, forged axles, and patented appliances, come from specialist firms, the expense reaching over 50 percent of the final price of a Baldwin engine.[2]

On this morning many orders have advanced to weeks four and five in the production schedule. This fortnight sees the receipt of materials from outside makers and the first work inside the factory. Highly skilled blacksmiths in the smith shops swelter despite the cool autumn weather, forging iron and steel parts ranging from leaf springs to engine frames. Individual smiths work up springs and a hundred other small parts on their own anvils, judging how far and fast to work the glowing plastic iron before returning it to the fire. Large pieces like a locomotive frame require a crew of four or five, using a jib crane to swing the work from the fire to the massive steam hammer needed for large forging jobs. While the smiths test their skills against iron in a pitched battle of ringing hammer blows, raging heat, and choking smoke, skilled molders and their assistants in the foundry begin making sand molds for cylinders, wheel centers, and other castings, using wooden patterns made up by the patternmakers. With this painstaking work under way, the foundry cupolas are charged with coke fuel and pig iron. By midafternoon the molten iron is ready for pouring. A single cylinder casting can require over 10,000 pounds of the bubbling liquid; the founders judge when the heat is right for the pour by eyeing the iron's brilliant glow in the ladle.

We leave the foundry behind for our next stop in the boiler shop, although much of the heat and smell of liquid iron accompany us. Crowded conditions have forced Baldwin to place one of its boiler shops on a floor above the iron

A team of Baldwin forge men stand by their steam hammer sometime in the late nineteenth century. They are working a wrought-iron axle under the hammer, with the axle suspended from a jib crane so that it could be swung back into the fire for reheating as needed. Such work required coordinating the skills and efforts of a team of workers. CARL HOOPES

Located above the iron foundry, this boiler shop ran the length of a city block. Flat boilerplate was sheared, drilled, rolled, and stamped into separate components in other shops. Boilers were assembled here; their courses (or sections) were riveted together and fireboxes, tube sheets, staybolts, and steam domes installed. Without any workers to indicate the scale of this 400-foot-long room, it appears deceptively small. AL GIANNANTONIO

foundry. Boilermaking has the potential to be a real bottleneck in the production schedule. All parts entering into a boiler come from outside suppliers, so most of the steel plates, flues, staybolts, piping, gauges, and insulation do not arrive at Baldwin until the fifth week of the production schedule. The firm allots weeks five and six to the massively labor-intensive process of fabricating boilers, which can reach 40 feet in length and weigh up to 43,000 pounds even before the flues are added. To make each boiler in this two-week period, Baldwin allots four shops and over three thousand men on two shifts to the task. They employ powered tooling wherever possible: plate planers, drillers, punchers, and rollers that accommodate sheets over 20 feet long, as well as overhead cranes, hydraulic presses, and power riveters. Notwithstanding these tools, the work remains notably labor-intensive, and the firm must rigorously subdivide tasks to stay on schedule. Each stage has its own specialists: markers, drillers, rollers, flangers, riveters, chippers, and caulkers, many requiring helpers. Working in small groups that are paid piecework rates, the men hurry from plate to plate and machine to machine. The work is rushed but cannot be slapdash. Boilers under pressure are notoriously lethal, and Baldwin's customers often send inspectors into the plant to ensure that construction of their orders complies with railroad standards. The boiler shops are not as noisy as one would think. Huge hydraulic presses do most of the flanging work—making steam domes, for instance— while much of the riveting is accomplished by large hydraulic riveters rather than the brute-force hammering of the past.

While the boilermakers shear, roll, punch, flange, press, and rivet steel sheets as if they were paper, over five thousand machinists turn to precision machining operations. Baldwin's machine shops are scattered across seven city blocks, with the two major shops on Broad Street occupying four- and six-story buildings. Since the machinists must await the products of the foundries and smith shops, today they are working on orders in weeks six and seven of the production schedule. Machining work runs the gamut from crankpins the size of a forearm to driving wheels 6½ feet in diameter to 5-ton cylinder castings. Most of the workers run machine tools driven by electric motors and using the new high-speed steel cutting tools. But operations like the final fitting of cylinder heads and journal bearings still call for the old ways: a fine file, a scraper, and practiced eyes and hands. In both hand and machine work, the men must utilize special fixtures, jigs, and templates that, when used with care, help ensure that parts are made with sufficient precision to be interchangeable among all engines in a batch order. Railroad inspectors are also intrusive in the machining departments, checking that Baldwin's work meets the railways' standards of precision. The machinists generally labor alone, although they need the help of crane operators and laborers to set up cylinders for boring or engine frames for planing. Here too the work is paid by the piece and highly specialized, but it calls for fine skill and judgment to bore a 10,000-pound cylinder casting on a true axis and to the correct diameters. A machinist's error at this point could ruin the casting, already the product of countless hours of labor by patternmakers, founders, and casting cleaners.

This hydraulic bull riveter greatly speeded boilermaking while improving the quality of work over hand-riveting. An overhead crane picked up the boiler (*left*) and lowered it over the riveter. A boy placed the glowing-hot rivets in holes in the seam to be riveted. The hydraulic ram pressed the rivet against the anvil reaching up inside the boiler, setting the head. Then the crane operator repositioned the boiler for the next hole in the seam. Machine-riveting produced uniformly tight joints, improving boiler safety. HAGLEY MUSEUM AND LIBRARY

xiv   PREFACE

A group of boring mills, with driving-wheel centers mounted for boring the axle holes and facing the hubs. Judging from his attire, the man shown was probably a foreman, not an operator. Workers on these machines used special measuring tools, known as gauges, to bore wheels to the required dimensions. It was skilled but highly specialized work. One look at this shop suggests some of the dangers in locomotive building—the loose plank floor and absence of safety guards on the machines. AL GIANNANTONIO

The work of draftsmen, patternmakers, founders, blacksmiths, boilermakers, pipe fitters, and machinists all converges in the erecting shop—a single room filling most of a city block on Broad Street, only a short distance from City Hall and the heart of downtown Philadelphia. Erecting occupies the eighth week in the production schedule. Today the shop is filled to its capacity of seventy-five engines. In previous decades this final stage of construction had been a notable bottleneck to improving output. But now frames and other machined parts are made with sufficient precision to come together with only minimal fitting, while electric overhead cranes speed the handling of large parts like boilers.[3] Today the army of thirty-two hundred workers in the erecting shop requires only a week on average to complete a locomotive. While all the men here are involved in erecting, a great range of skills and tasks come under that heading. Most are machinists and fitters, but large numbers of boilermakers, pipe fitters, insulation men, sheet-metal workers, painters, and laborers also work in the shop.

Most visitors to the erecting shop find it a confusing bedlam of parts and men, a chaos of designs in varying stages of completion. The first stage in erecting generally involves setting up the cylinders, precisely aligning the bare machined frames on jacks, and bolting the cylinders to the frames.[4] Next a flatcar arrives from the boiler shop across Buttonwood Street carrying a boiler that the 100-ton overhead crane silently places on its frames. Knots of busy men then descend upon the developing engine, each group specializing in a different task and proceeding quickly from one locomotive to the next. While erectors bolt the boiler to the cylinder saddle, boilermakers place hundreds of flues inside the boiler shell, each tube requiring ten separate steps to make a tight joint in the flue sheets.[5] Then the engine is wheeled—plucked off its jacks by an overhead crane and placed on its driving wheels, which have been laid out on an adjacent track.

Now it looks like a real locomotive, and the pace of construction increases.

Specialist gangs drill, tap, and mount the remaining boiler fittings, the boiler is tested with hot water at 200 pounds per square inch, boilermakers caulk leaky flues and fittings, and the engine is prepared for a live steam test. The most skilled machinists in the plant set up the connecting rods, reversing gear, and valve motion. The bearings of these moving parts must be precisely aligned, so the machinists make final adjustments with files and scrapers. Other workers cover the boiler with its magnesia insulation and the sheet steel jacketing that protects the lagging, and a crane operator lowers the complete painted cab onto the engine. The locomotive is filled with 180-pound steam from a shop boiler, and a machinist cautiously opens the throttle. With the engine jacked up, the main drivers revolve in the air, showing that all is working properly.

After this crucial test, the remaining work is hurried forward. Erectors add the brakes, lubricators, air pump, injectors, handrails, cab fittings, running boards, fire grates, ash pan, leading truck, and dozens of other parts. The engine is nearly finished, and the overhead crane places it on the ready track leading out of the building. A shop switcher will soon haul the new locomotive uptown on the tracks of the Philadelphia and Reading to Baldwin's finishing

In the early years of the twentieth century Baldwin's output reached its all-time peak. Here in the erecting shop circa 1905 dozens of designs stand in varying stages of completion in one of the shop's two main bays. The overhead crane in the background delivered heavy materials and lifted finished engines over those still building. AUTHOR'S COLLECTION

Two engines stand in the early stages of erecting. The design and buyer cannot be identified, but that shown in the foreground is probably a 2-8-0-type freight locomotive. The wheels, cylinders, and frames await the boiler—not the normal erecting sequence, but procedures were often varied to compensate for delays in parts reaching the erecting shop. Note how the walls were whitewashed to reflect all the natural light possible. AL GIANNANTONIO

shop at Twenty-sixth Street. There it will join its tender from the Seventeenth Street shops, painters will finish lettering and detailing, and a Baldwin delivery engineer will take charge of the new locomotive. Under his care it will travel across country to its purchaser.

In a mere eight weeks Baldwin's thousands of workers take a novel design and translate it from words on a specification sheet into drawings and then into a finished mechanism in steel and iron, weighing upwards of 150 tons and capable of hauling 6,000-ton trains. While customers around the globe order uniquely different locomotives, the methods used in production reduce the constant variety and potential chaos to routine and order. During the month of our imaginary visit, October 1905, more than seven engines rolled from the erecting shop every working day. For the year as a whole Baldwin built 2,250 locomotives in hundreds of designs for customers around the nation and the world.

In October 1905 a new 2-8-0 freight locomotive for the Danville and Western Railroad stands on the turntable at Baldwin's uptown finishing shop. Most engines for American buyers were delivered without power in slow freight trains. Nonetheless, Baldwin sent along its own delivery engineers to ensure that the trip went safely, to set up the engine for the purchaser, and to iron out any last-minute difficulties that might otherwise hold up payment. LIBRARY OF CONGRESS

The company had reduced engine construction to a routine process, but that scarcely detracted from the aura of mystery surrounding these machines that shouldered the world's work. Locomotives provided the animated focal point of railroading, in turn the dominant force behind the economic and social transformations of industrial society.[6] From Baldwin's seven managing partners to its thousands of laborers, all involved took great pride in creating the leading technology to symbolize power, speed, and the progress of civilization. Locomotives so fascinated the general public that Baldwin needed admission tickets for visitors, and Baedeker's tourist guide cited the factory as a "highly interesting" attraction.[7]

People of the nineteenth century had a passionate faith in this technology that conquered time, tamed vast distances, and tapped the wealth of continents.[8] To modern eyes theirs was a naively romantic view, but bare facts suggest how it originated. Baldwin engines climbed Pikes Peak and hauled the Trans-Siberian Express. They roamed the American West on the nation's five major

Here five new Baldwin engines pose in 1893 on the tracks of the Philadelphia and Reading just outside the factory. Before going into service they were destined for display at the World's Columbian Exposition in Chicago. The lead engine, a 4-4-0 for the Baltimore and Ohio, represented a high point for this standard American-type wheel arrangement, used since the 1850s. Soon thereafter heavier cars and faster schedules made this type obsolete for most mainline services. BEN KLINE

The Pennsylvania Railroad had owned this Baldwin 4-4-0 for less than a month when it was pressed into service to haul a special train carrying Abraham Lincoln's body westward for burial in Illinois. Thousands of Americans turned out to see the train as it slowly passed, the engine and cars draped in mourning. RAILROAD MUSEUM OF PENNSYLVANIA (PMHC)

transcontinental lines. A Baldwin took the first train of the Jaffa and Jerusalem Railroad into Jerusalem, while others ran in England, birthplace of the railway age. They also played supporting roles in life's human dramas. In 1865 a Baldwin engine decked in black mourning crepe pulled the train carrying the body of the murdered Lincoln westward toward Springfield. Forty years later a wealthy California eccentric and former goldminer, Death Valley Scotty, paid the Santa Fe to carry him on a record-breaking forty-five-hour dash from Los Angeles to Chicago.[9] Like Pony Express relays, a succession of nineteen engines, seventeen built by Baldwin, propelled the old prospector's special express at lightning speed across the continent while the nation's newspapers chronicled the feat.

Partly because of such incidents, a highly romantic symbolic burden has attached itself to the steam locomotive from its origins to the present. Of all the icons of nineteenth-century technology, this machine has exerted the most enduring fascination. The subject of Whitman's poetry and Hawthorne's prose, gist for artists from George Inness to Charles Sheeler, and endlessly recorded in thousands of photographs, steam locomotives still haul an infinite freight of mental imagery—as totems of power and speed, as noble animated iron steeds, and as evocations of a mythic national past of confidence and simplicity when every boy dreamed of being an engineer.

This book offers an entirely different perspective on the locomotive, however. The first history of a major American locomotive builder in the nineteenth century, it seeks to describe how the Baldwin Works sold and made these complex and expensive capital goods. Equally important, it details how the firm's owner-managers ran such a high-technology company. Baldwin made locomotives for over a century and stood at the top rank of American industry for sixty years. Beyond the firm's own merit, tracing Baldwin's development provides a new perspective for understanding the era when the forces of industry, management, skilled labor, and technology first combined to remake the Western world.

ACKNOWLEDGMENTS

THAT ONLY ONE NAME APPEARS ON THE TITLE PAGE IS SOMEWHAT DECEPTIVE, for as I am thankfully aware, this book drew upon the talents and interests of scores of people whose generous contributions have immeasurably improved the final product. My debt to others began long before the project itself. In the mid-1950s the Baldwin company destroyed most of its internal business archive, records spanning over a century of operations. Management took this drastic step deliberately, seeking a decisive break with the past as Baldwin exited the locomotive market and embarked on new ventures building other lines of heavy capital goods. Such wholesale destruction could have prevented any historian from reconstructing the company's past. But steam locomotives exert an enduring fascination on human imagination, and countless people stepped in to preserve Baldwin's history, knowing the firm's leading role in creating America's railway age. Rail fans, archivists, former workers, and managers' descendants pulled records from the trash, squirreled papers away in attic trunks, collected engineering drawings, and preserved thousands of photos. Once I began my inquiries, a rich array of records came to light, enough to reconstruct the past with reasonable certainty. If instead of locomotives Baldwin had made steel heddles, hardware cloth, or some other anonymous industrial product, then its history would likely have been forever lost.

I cannot hope to note all who provided essential aid in researching and writing this account, but a reckoning is due. Throughout the project I received unstinting assistance from the Hagley Museum and Library, and thanks go particularly to Glenn Porter, Susan Hengel, Jon Williams, Chris Baer, Mike Nash, and Rob Howard. At the Historical Society of Pennsylvania, Linda Stanley was an unfailing guide through collections rich in Baldwin material. John Van Horne of the Library Company of Philadelphia took a continuing interest in Baldwin. At the Smithsonian's National Museum of American

History, Jack White generously shared his research notes and his unparalleled knowledge of American locomotives, while Steve Lubar provided leads on sources and invaluable criticism of my inchoate ideas. The late Ben Kline of the Railroad Museum of Pennsylvania gave unflagging aid and encouragement that belied his seemingly gruff nature. I conducted further research at a dozen libraries listed in "Abbreviations and Original Sources," to which I am also indebted.

Many private individuals also shared archival material and knowledge about Baldwin. Descendants of the firm's seventeen partners provided access to their records, including John H. Converse, William Devoe, John du Pont, Mrs. Joseph C. Hoopes, William P. Scott, and Mrs. William Vauclain. Former Baldwin employees offered valuable assistance, such as Henry Rentschler, H. L. Broadbelt, Al Giannantonio, Matt Gray, John Kirkland, Chester Ludwig, Frank Moore, and Barry Whitehair. Others interested in Baldwin's history who contributed to this work include C. A. Brown, Vernon Gotwals, Carl Hoopes, Trudy Kuehner of Morgan Lewis and Bockius, Walt Myers of the Adwin Company, and Bill Withuhn. I received invaluable financial support from the Hagley Museum and Library, the Smithsonian Institution, the Historical Society–Library Company Summer Fellowship Program, and the Business History Society's Rovensky Fellowship.

The people and institutions listed above gave essential aid in my research, but history involves far more than compiling facts. In writing this account I again received valuable advice and criticism from a number of people. My interest in Baldwin began as a dissertation topic many years ago in the history department at the University of Virginia. From that beginning to the present, I have enjoyed a stalwart critic, an unflagging advocate, and an insightful editor in my thesis adviser, Olivier Zunz. He has responded to my half-baked, overweight drafts and heated, overwrought arguments with sound criticism that consistently sought to sharpen my ideas without imposing his own views. His friendship and professionalism as a mentor became a guiding influence throughout this work and my entire graduate career. As a graduate student I was lucky to serve two masters, a fortunate position providing me with twice the advice as well as the tie-breaking vote in cases of disagreement. My second adviser, Bernie Carlson, unstintingly gave of his expertise in American technological and business history. His close reading and detailed criticism of each chapter tactfully revealed when I was becalmed or fogged in while helping to steady my course from start to finish.

Beyond these two advisers, my ideas were challenged and my prose improved by a number of other readers. At Virginia, Mark Thomas and Nelson Lichtenstein generously helped to ground my work where it touched upon their respective specialties in economic and labor history. Steve Lubar at the

Smithsonian read the dissertation, and his penetrating criticism improves every section here. David Hounshell graciously provided a vital sounding board for ideas that often differ from his own published work. Walter Licht gave key encouragement and help as I began the project, following this up with a close reading of the dissertation. The leading scholar of Philadelphia's industrial history, Philip Scranton, has also given generously of his time on repeated occasions, responding to conference papers and giving a detailed evaluation of the entire manuscript. Thus I am especially pleased that this book appears in the Johns Hopkins series first edited by Glenn Porter and now under Phil Scranton's direction. At the Johns Hopkins University Press, Bob Brugger has expertly guided this manuscript and its author through the unimagined complexities of publishing—complexities eased by Mary Yates's careful copy editing. Others who have read all or part of these chapters include Locke Brown, Jack White, Daniel Nelson, and Wendy Brown. I am grateful for their counsel. The work is dedicated to Wendy, in what can only be the most meager acknowledgment of her immeasurable aid over the many years that the Baldwin Works has figured so prominently in both our lives.

The generosity of all those listed here has improved this book immeasurably, yet the usual stipulation applies. I alone am responsible for those errors of fact or interpretation that may remain.

INTRODUCTION

IN THE SPRING OF 1872 A DELEGATION OF HIGH OFFICIALS FROM THE JAPANESE
government toured the United States to observe firsthand the character of a
commercial, industrializing society. After centuries of isolation from the West,
Japan's leaders had decided to embark on a program of national industrial
development, copying Western industry and technology to become a modern,
powerful nation-state. The Japanese particularly wanted to observe America's
capital equipment makers, creators of the transportation and production
machinery central to an industrial economy.[1] Accordingly, the Japanese mission's first stop after official Washington, D.C., was Philadelphia, the nation's
leading center of heavy industry.[2] Here the visitors watched artisans constructing the machinery that in turn created industrial America: textile machines,
ships, machine tools, railway cars, stationary and marine steam engines, and
mill machinery. They began their tour at the Baldwin Locomotive Works, the
nation's leading capital equipment company and the largest locomotive builder
in the world. There they saw twenty-five hundred men toiling in an 8½ acre
plant, turning out engines weighing up to 43 tons at the rate of eight finished
locomotives a week.[3] The Japanese mission sought to discover the foundations
of Western economic dynamism, and America offered no better exemplar of
that strength than the Baldwin company.

Despite the vital developmental role played by Baldwin and other machinery
builders, historians of nineteenth-century American industry have largely
ignored the capital equipment sector. Our industrial history concentrates
instead on American System volume manufacturers of such consumer goods as
guns and integrated corporations making bulk commodities like oil and steel.[4]
As America's unique and vastly important contribution to industrialization and
economic growth, mass production certainly deserves its biographers. But U.S.
historians' exhaustive focus on such high-volume producers has led many to

assume incorrectly that firms that pursued other production strategies were small, insignificant, or justifiably ignored.[5] Thus capital equipment builders, an extensive sector of large firms, are generally overlooked. Only such economic historians as Thomas Cochran and Nathan Rosenberg have noted the general importance of the machinery makers.[6] As Rosenberg writes of these firms: "Their growing skill in solving problems in specialized machine production ought to be regarded as the basic learning process underlying nineteenth-century industrialization."[7] Yet only a few business, labor, and technological historians have written accounts incorporating the nineteenth-century history of these companies.[8]

This study of the origins, growth, and operations of America's largest capital equipment company, the Baldwin Works, seeks to redress that oversight and open the capital equipment sector to the inquiry of other historians. Using Baldwin as a case study, I argue that the capital equipment builders pursued a related set of policies in innovation, management, production, and labor relations that constituted a distinctive business strategy. Common in the sector, these policies clearly distinguish capital equipment builders from high-volume manufacturers. Once they are examined and acknowledged, these defining attributes of capital equipment firms will underscore the breadth of nineteenth-century business practice and greatly revise our present understanding of the character of industrial development.[9]

My study of the Baldwin company originated from a suspicion that capital equipment builders must have faced unique operational issues compared with those confronting the integrated corporations and American System manufacturers that have dominated recent industrial historiography. Put simply, I thought locomotives had to evoke managerial, marketing, and production challenges different from those posed by oil and steel or sewing machines and bicycles. As my research progressed, I learned that Baldwin's operations diverged entirely from the practices of high-volume manufacturers.

While the locomotive builder was exceptional in comparison with mass producers, many other large and influential firms in the nineteenth-century industrial economy also engaged in the batch production of expensive, custom capital equipment. It is difficult or impossible to find detailed measures of the heavy machinery sector, but in Daniel Nelson's listing of the largest American industrial employers of 1900, over 25 percent built capital equipment.[10] The sector included such world-renowned firms as Pullman Car, Hoe Printing Press, Cramps Shipyard, and the Niles-Bement-Pond machine tool works. These shipyards and railway car builders, machinery firms and locomotive works, all faced similar challenges in product innovation, production management, and labor relations. The traits shared by the machinery builders were so extensive as to suggest the existence of two fundamental and divergent formats among

American metalworking firms in the nineteenth century: the capital equipment builders and the American System consumer product manufacturers.

Delineating this distinction between *building* and *manufacturing* is my analytical goal here, although I did not originate these terms. The two formats were well known and widely discussed at the turn of the century.[11] Joseph Wickham Roe, an early technological historian, opens his treatise *Factory Equipment* by analyzing these industrial systems: "Two well-defined methods of [metalworking] production are used in industry, and the principles which differentiate them run through all stages of factory production, even to marketing[;] . . . we may call them *building* and *manufacturing*. The predominate use of one or the other affects the nature of the whole plant, its equipment and the methods used."[12] By 1900 building techniques predominated among most lines of capital equipment, while light machinery such as guns, sewing machines, typewriters, and bicycles were generally manufactured articles. The two metalworking formats practiced by American System manufacturers and capital equipment builders varied fundamentally in almost all respects. Confronting very different concerns in the character of demand, customer relations, and product innovation, the manufacturers and the builders developed contrasting management systems and marketing strategies, and they employed different production machinery and workforce skills.

In the eighteenth century all metalworking production followed the building pattern: mechanisms from guns to early textile machines were individually made to order to suit customer preferences and built by skilled artisans in labor-intensive production. Early in the nineteenth century the Armories of the U.S. Army pioneered in manufacturing standard products, using a succession of special-purpose machines to make rifles with interchangeable parts. A number of entrepreneurs then seized on Armory practice to produce consumer products to standard designs, including clocks, guns, watches, sewing machines, bicycles, typewriters, and eventually automobiles. The great size of the American consumer market enabled and impelled manufacturers to adopt American System production techniques. These companies built upon the Armories' precedents, using a sequence of special-purpose machines and specialized semiskilled operatives to turn out prodigious quantities of the required parts to interchange standards. After assembly the manufacturers sold their finished products at wholesale prices to national distributors for eventual retail sale. To promote demand, they used intensive advertising campaigns based on brand names. Manufacturers also developed bureaucratic management structures to control their operations, meshing raw materials purchasing with established production targets and coordinating distribution with national advertising. All these developments aimed at a single goal: to produce uniform products in bulk, seeking ever higher volume and declining unit costs.

As American System manufacturers embarked on this developmental path during the antebellum era, the capital equipment builders grew to meet the particular demands of their own sector. The markets for heavy machinery were relatively narrow, so firms seeking growth utilized America's developing transportation network to cultivate extensive contacts with customers across the nation. Despite the breadth of these markets, builders and buyers were limited in numbers and generally relied on direct sales without any intervening distributors. The close relationship between the two parties also derived from the character of technical change in heavy engineering products. Buyers of capital equipment demanded a strong voice in design because they sought machinery that was particularly well adapted to their own operations.[13] As a result, most capital goods were custom designs, or at least customized versions of standard products. For example, in just one year, 1890, Baldwin built 946 locomotives to 316 different designs. While American System manufacturers based their operations on rigorously standardizing product technology, capital equipment builders thrived by producing custom designs, following the technological lead established by their customers' needs or directives.

Firms in the capital equipment sector existed in an environment marked by fluid and continuous technological change. American System manufacturers lessened their risks by controlling innovation in their own product lines. Marketing standard products, these firms launched new or improved models only when internal considerations suggested that the rewards to innovation—increased sales, profits, or market share—would outweigh the costs of reworking production and rendering older models obsolete. By contrast, capital equipment builders constantly updated and modified their products. Their customers continuously demanded such changes as they sought better operational efficiencies or broader capabilities through improved capital goods.[14] Since capital equipment markets were defined by technical flux rather than inflexible standards, competitive pressures among firms also impelled the builders to innovate continuously. Those who did not soon stumbled in the marketplace. Such constant innovation suggests why this sector, more than high-volume or repetitive manufacturing, was a mainspring in driving productivity improvements in the nineteenth-century economy.[15]

The high cost of capital equipment and its relatively narrow and specialized markets caused far greater fluctuations in demand than was true in consumer product manufacturing. Manufacturers first made their products and then sold them; the reverse was generally true for the machinery builders, and they secured most contracts by competitive bidding. Since they mostly created custom products to special order, builders could not smooth out variations in sales by producing for inventory, nor could they rely on brand names or advertising campaigns to stoke demand. Direct competition between builders waxed as sales fell and waned as they rose. When demand declined, makers of expensive capital goods competed on price and credit terms. In strong markets those issues receded while early delivery dates became important to selling these made-to-order products. In the long run a reputation for quality products, good workmanship, and close technical support provided crucial competitive advantages.

Because the builders' ability to influence demand and product technology was circumscribed, they had neither the need nor the means to fund extensive managerial bureaucracies. Management systems, as opposed to managerial personnel, could be quite important, however. Builders required extensive records to bid accurately on jobs, to work up new designs, to order materials in the right sequence and quantities, and to ensure rapid and efficient construction of their complex products. High-volume manufacturers relied on their standard products and special-purpose production machinery to impose a measure of system and regularity; capital equipment builders needed paperwork to order otherwise chaotic operations.

Production management in repetitive manufacturing centered on efficiency, in contrast to the builders' efforts to optimize their facilities and workers. Efficient and optimal production differed fundamentally, the two approaches involving contrasting tooling requirements and workforce skills. Manufacturers secured their productive advantage by deriving efficiencies from the volume output of standard products. They based production on routinized operations by semiskilled labor working special-purpose tooling. On the other hand, builders sought to optimize their knowledge, experience, plant, and workers by turning out a range of complementary products. In meeting customers' varied design needs, productive flexibility was essential. To that end, the builders mostly used general-purpose machine tools, such as lathes, planers, and slotters, adaptable to a variety of work. Such tooling required highly skilled workers possessing the versatility to turn out parts in various sizes and designs. Essential partners in production, workers and their skills lay at the core of the productive flexibility and technical virtuosity that constituted builders' essential marketing advantages. But work in the capital equipment sector had its drawbacks. The volatility in demand for heavy machinery caused frequent layoffs. The labor-intensity of constructing capital equipment also gave many employers the incentive to experiment with piece rates and other schemes that restructured work to improve productivity.[16] Nonetheless, the degradation of work and workers was limited by the need to maintain optimal production capacities, which derived fundamentally from workers' skills.

This summary of the differences between building and manufacturing can only touch on the main points. These themes and further details are fully described in the following chapters. It is worth emphasizing that I see the two formats as descriptive terms rather than as strict definitional boundaries. Some large manufacturers such as Westinghouse and General Electric also made such capital equipment as generators and turbines, often to standard designs. Builders like Baldwin utilized a limited number of manufacturing techniques when feasible, particularly in making standard detail parts for use across a number of differing designs. By the 1880s and 1890s some capital equipment firms making machine tools or stationary steam engines had adopted standard designs, which they marketed through distributors.[17] Makers of light capital goods like office equipment often emulated the high-volume machinery makers. No doubt other caveats could be noted. Despite such exceptions, the distinction between American System manufacturing and capital equipment building is still pronounced and useful when we consider the range of issues involved: the character of demand, product innovation and customer relations, management structures and systems, production requirements and labor relations, and marketing efforts. When studies of other American capital equipment firms follow this history of the Baldwin Works, these distinctions may be qualified, but the main outline will remain.[18]

This book details the history of the Baldwin Works from 1831 to 1915.[19] In this account I have attempted to advance beyond a conventional, single-firm business history by combining traditional narrative chapters with others taking an analytical approach. Baldwin's importance required that its story stand alone at points. But the five core chapters are thematic and use the firm's history to outline the range of issues that uniquely defined the capital goods sector. To analyze the particular challenges of building capital equipment, this study draws on concepts and issues of economic, business, labor, and technological history. In combining these fields and approaches, my goal is to provide a comprehensive industrial history that focuses on Baldwin while illuminating the range of issues involved in building heavy machinery. Such a multifaceted analysis is predicated on the belief that the issues confronting Baldwin—and other builders—in technological innovation, customer relations, management, labor, production processes, and the general economy arose and evolved in dynamic interaction; none can be understood absent the others.

Chapter 1 is a narrative account of the Baldwin company's founding in 1831 and its early history through 1860. Here the various challenges of building heavy machinery are introduced. The next five chapters are thematically organized, each covering an individual topic over the last half of the nineteenth century.

The locomotive industry as a whole is considered in chapter 2. Concerns include the character of demand for engines and the nature of competition between builders. For forty years the industry veered between competitive periods and episodes of collusion and price fixing. Over the long run Baldwin grew to dominate the field by using the full extent of its resources to cultivate a range of markets, including a vigorous export trade.

Baldwin's dominance derived largely from its willingness and ability to meet the design requirements of its customers. Their needs varied substantially at any given moment while also evolving over time. Chapter 3 looks at the causes and consequences of continuous technical change in locomotives, an issue of vital importance to the locomotive builder. Much of the focus here is on the relationship of railway operations to locomotive design, since American mainline carriers directed the overall pace and direction of innovation in motive power.

In meeting the major carriers' demands for custom locomotives, Baldwin pioneered in new internal systems of coordination and control, described in chapter 4. These managerial systems presaged aspects of the systematic and scientific management movements by fifteen to forty years. Chapter 4 also examines premodern, noncorporate, nonbureaucratic aspects of business management such as Baldwin's partnerships and its use of inside contractors.

Labor relations and the character of the workforce are the focus of chapter 5. Baldwin employed legions of skilled workers from eight craft trades in building large and complex custom locomotives. Although the company could never replace men with production machinery, the labor-intensity of engine building impelled management to take a highly directive role over the character of the work.[20] Thanks to Baldwin's workplace culture and a general trend of increases in real wages, workers did not resist this encroachment on their skills, and the company enjoyed fifty years of tranquil labor relations.

On the factory floor, workers' skills and management's capital and oversight came to bear on the challenges of building locomotives in all the varieties required by customers. Chapter 6 describes the construction process.[21] Although never achieving any revolution in production, by the 1860s Baldwin attained parts interchangeability within its system of batch output. Thirty years later it became the first large American company to harness electric power to industrial use. These and other changes in construction techniques occurred in a discernible pattern that is the particular focus of this chapter.

Chapter 7 combines the themes of the preceding five chapters in giving a narrative account of Baldwin's operations between 1900 and 1915. By this time the locomotive builder had grown to become an industrial giant; its 18,500 men in 1907 placed the firm among America's four largest industrial employers.[22] Baldwin achieved this commanding size by fully refining its custom-building techniques. But 1907 proved to be the high point for the company; thereafter American mainline carriers' demand for motive power began a long-term decline. Baldwin's officers did not recognize this fundamental reversal for many years, and they were largely unprepared to deal with it. For fifty years their company's development had centered on effectively responding to the railways' demands. This reactive stance was the basis of Baldwin's nineteenth-century strength.[23] After 1907 it would prove a fatal weakness, hindering Baldwin's ability to take control of its own destiny.

Although Baldwin's decline dated specifically to the reversal of its major market, the company's waning fortunes mirrored broad trends in the American economy. The industrializing society of the nineteenth century was transformed during the years from 1900 to 1920 into the modern consumer-driven economy, where standardized products, hierarchical managerial structures, and market control strategies dominated.[24] As a result, the whole capital equipment sector declined in strength and importance. Many builders failed entirely in America's post–World War II deindustrialization, and most surviving capital equipment firms are no longer builders in the old sense. Today's makers of locomotives, airplanes, and mainframe computers generally control their own product technologies, and most mirror mass-production manufacturers in making only standard designs. The skilled construction of custom engineering products, tailor-made to suit buyers' needs, is largely a thing of the past. The following chapters describe this lost segment of the industrial economy.

I

ESTABLISHING THE BALDWIN WORKS

1831–1866

IN A NUMBER OF SMALL WORKSHOPS LOCATED IN THE SHADOWS OF THE BELL tower of Philadelphia's Independence Hall, a group of remarkable men labored with tools and metals during the decade of the 1820s, creating the foundations of a second American revolution.[1] Heirs to the Quaker City's heritage in technological innovation, a lineage that included Benjamin Franklin, David Rittenhouse, and Oliver Evans, these younger technologists are generally not well known to history, nor are their separate inventions considered epochal. Their importance lay in their collective strength and in mutual relations that coalesced to form a community of mechanicians. Nathan Sellers, Isaiah Lukens, Patrick Lyon, David Mason, Matthias Baldwin, and other Philadelphians set the American Industrial Revolution on a self-sustaining, irreversible course by working in concert to create machines that in turn were used to build other machines or products.[2]

Although these men often had particular technical specialties, as a rule they were generalists, interested in a range of metalworking techniques and mechanical applications. This was partially dictated by economic necessity, since the novelty of their work and the limits on their markets precluded specialization. But their interest in machinery and its improvement had deeper motivations than simply the profit motive. These mechanicians were driven by a passion for the mechanical arts and by a common faith that new technologies served as the symbols and instruments of social progress. They manifested this faith in 1824 in creating the Franklin Institute, which quickly became a vital agency for the promotion and diffusion of technical knowledge.[3] Soon thereafter Philadelphia's technical ferment would make the city the nation's leading center for the construction of heavy engineering products, including textile machinery, steam engines, machine tools, iron ships, railway cars, and locomotives.

Matthias Baldwin, shown late in life. By this time he ranked among the wealthiest men in Philadelphia. But he never retreated from active involvement in social affairs, remaining an influential elder of the Presbyterian Church, serving on the boards of the Philadelphia County Prison and the Franklin Institute, speaking out as an abolitionist, and assisting the work of the Civil War Sanitary Commission, which gave aid to Union soldiers. HAGLEY MUSEUM AND LIBRARY

Heading the list of those firms creating the machines that built industrial America would be Matthias Baldwin's company, the Baldwin Locomotive Works, which became the largest locomotive builder in the world. Baldwin, a former jeweler's apprentice, turned to locomotive construction in 1831, at the dawn of the railway age. His young firm drew on Philadelphia's resources of technical knowledge, capital, and skilled artisans to create a factory system of locomotive building. That accomplishment proved as novel and important as the locomotive itself. By 1837 Baldwin employed three hundred men in a $200,000 factory, a scale of industrial production with few precedents or parallels in America beyond the textile manufacturers of Lowell, Massachusetts.

Aside from the issue of size, locomotive building presented entirely new and different challenges compared with those confronting the textile magnates. Matthias Baldwin had to marshal artisans from eight craft trades to build products of unprecedented complexity and cost, using a combination of extensive hand skills and expensive machine tools. Locomotive construction required a range of raw materials and semifinished components, all of the highest possible quality. In an age of slow freight and communications, Baldwin had to develop a network of suppliers from the coal and iron fields of interior Pennsylvania to New York, Connecticut, Massachusetts, and England. Because of the inadequacies of antebellum financial markets and institutions, Baldwin looked to his suppliers for economic as well as technical assistance. While surmounting the challenges of locomotive production, his young company soon encountered the roller-coaster demand cycle that would come to rule the capital equipment sector. Even when sales were successfully concluded, Baldwin remained bound to his customers by performance guarantees, installment payments, and demands for design improvements.

Successfully managing any sizable business enterprise was a notable feat in the antebellum era, but Baldwin pioneered in a particularly difficult field.

Crucial challenges in production, sales, and customer relations arose from the high cost and technical complexity of capital equipment and its narrow and specialized markets. Such characteristics sharply differentiated the nation's heavy machinery builders, many of which clustered in Philadelphia, from the American System consumer product manufacturers of New England. As Baldwin successfully navigated his young enterprise through these conditions, his firm also mirrored the genesis of an entire order of industrial production.

MATTHIAS BALDWIN

Matthias Baldwin did not set out to become a pathfinder in new styles of industrial organization. His interests focused on technology, on designing and building mechanisms, and this remained his passion throughout his life. Baldwin was born in Elizabethtown, New Jersey, in 1795, the son of a successful carriage maker who died when the boy was four years old. As a teenager, Matthias moved to Philadelphia to indenture at the jewelry firm of Woolworth Brothers. After completing his apprenticeship in 1817, he became a journeyman jeweler with Fletcher and Gardener, also of Philadelphia.[4] This early training had more relevance to his subsequent career than might seem apparent. Fletcher and Gardener were noted silversmiths, giving Baldwin experience in metalworking and in the kind of visual thought essential to designing three-dimensional artifacts like machines.[5] The machinist's trade hardly existed at the time, but Baldwin's training as a whitesmith provided valuable skills for the day when mechanisms in metal would become more widespread.

That day was not long in coming. In 1824 Matthias Baldwin entered a partnership with another Philadelphia mechanician, David Mason, to make tools for wood engravers and bookbinders. From hand tools their business soon expanded to include the first hydrostatic presses made in America, high-precision lathe slide rests, and engraved cylinders used in printing calico textiles.[6] These three product lines are suggestive of important general trends in the American mechanical arts of the period. The first two were borrowings from contemporary English designs, the slide rests incorporating a number of improvements over the European originals. In turning to textile machinery, the partners entered the major market for all early American machinery builders, for the textile industry pioneered in machine-based production.[7]

If these products indicate broad expertise in the Baldwin-Mason partnership, their next effort suggests a technically venturesome spirit, particularly in Matthias Baldwin. As their general machinery business grew, the partners moved to a larger shop on Minor Street, in the heart of the old city. Needing power for the new shop, Baldwin designed and built a new stationary steam engine. Despite his lack of experience in steam technology, he created an efficient, powerful engine that incorporated novel design features and remained

Baldwin built his first stationary steam engine to power his Minor Street shop circa 1825. A 6 horsepower vertical design, its chief feature was a yoked connecting rod that cut the engine's overall height by one-third compared with conventional vertical engines. This engine continued to power various Baldwin shops for many decades until donated to the Smithsonian Institution, where it remains on display today. RAILROAD MUSEUM OF PENNSYLVANIA (PMHC)

in use for over forty years. Although the success of this engine soon brought orders for more, Mason felt that his partner was straying too far from established markets, and he dissolved their partnership in 1829.[8] Baldwin successfully continued on alone, running a general jobbing machine shop that turned out a range of products.[9]

Beyond his activities in mechanical innovation, Baldwin's character was marked by interesting complexities that affected his subsequent business career. Caught up in the religious revivals of the Second Great Awakening, he converted to evangelical Presbyterianism in 1827.[10] For the next thirty-five years he conducted his own Sunday school class while also contributing substantial funds to erect five new Presbyterian churches in the city.[11] Once his company was well established, Baldwin's philanthropy became so generous that it suggests he believed worldly success should serve as an instrument for Christian salvation.[12] Baldwin took the responsibilities and beliefs of his faith quite seriously, serving in the 1830s with the Union Benevolent Association, a poor-relief agency, and donating 10 percent of his company's income to the Civil War Christian Commission. When he encountered pious believers down on their luck or out of a job, he habitually offered them employment at the works. Although a teetotaler, he was not a dour man, enjoying humor in others and possessing a dry wit himself.[13] Like many successful industrialists, Baldwin was a member of the Whig party. He supported the abolitionist movement and argued for equal treatment of Northern blacks.[14] His profound religious convictions probably motivated such beliefs, but Baldwin welcomed change in many aspects of life, equating it with social and moral progress. In keeping with the expansive spirit of his times, he wrote to one customer in 1852 that "not to go ahead now[a]days is to go behind very fast."[15] At that time Baldwin had twenty years of experience in promoting a technology that would transform the nation.

THE DAWNING OF THE RAILWAY AGE

In October 1829 Robert Stephenson's *Rocket* lived up to its name at the Rainhill Trials in England, demonstrating the speed and strength of steam power on iron rails. Americans eagerly followed English developments in this new technology, immediately sensing the potential that steam railroads offered to a nation determined to span a continent. In 1829, even before the run of the *Rocket,* the Delaware and Hudson Canal Company imported four British engines.[16] This first American venture failed, largely because the wooden tracks used by the D&H as an economy measure crushed under the engines' weight. But this reversal scarcely dimmed American interest in the new technology.

Merchants, manufacturers, and mechanicians in Philadelphia became particularly interested in railway development as a means to promote the city's trade with the interior. By 1830 construction had begun on the Philadelphia and Columbia Railroad, an 81-mile segment in a network of state-sponsored canals and railways designed to link Philadelphia with Pittsburgh.[17] The Philadelphia and Columbia was something of a gamble, given that the first American-built locomotive was completed only in 1830.[18] The novelty of the technology only whetted public curiosity, however, and a Philadelphia mechanician named Franklin Peale decided to gratify this interest and spur technical development by commissioning a model locomotive to operate on public display at his Philadelphia Museum. He turned to Matthias Baldwin, who constructed a successful engine in early 1831 despite having never seen an actual locomotive.[19] Although only a model, the engine was no toy. Throughout the summer of 1831

Baldwin's Lodge Alley shop was not much to look at, but it provided the space he needed to finish his first locomotive. In the heart of the old city, the shop was one of many small metalworking establishments in the neighborhood. Baldwin shared ideas, techniques, and workers with this network of fellow mechanicians—all working as industrial revolutionaries.
SMITHSONIAN INSTITUTION

Matthias Baldwin's first locomotive, *Old Ironsides,* closely followed an English "Planet"-class design by Robert Stephenson and Company, which was imported to America by the Newcastle and Frenchtown Railroad in 1831. Baldwin helped assemble that import, the *Delaware,* learning a great deal about this novel technology. Although he soon improved his products, *Old Ironsides* proved a fine design, remaining in active service for over twenty years.
RAILROAD MUSEUM OF PENNSYLVANIA (PMHC)

it ran on a circular track at the Philadelphia Museum, hauling two cars that carried eight people.

Baldwin's well-known technical ability, his work in stationary steam power, and his experience with this little engine soon brought him an order from the Philadelphia, Germantown, and Norristown Railroad for a full-size locomotive. Drawing on the design of a contemporary English engine that had been imported to Delaware, Baldwin built his first standard-gauge locomotive over a number of months in 1831 and 1832 at his new Lodge Alley shop.[20] Named *Old Ironsides,* the machine first ran on November 24, 1832, and was acclaimed a success from the start. This in turn launched an enterprise that would build roughly one-third of all steam locomotives ever constructed in America.

After these tentative first steps, two developments firmly established the Baldwin Locomotive Works. Matthias Baldwin knew that early American railways saved on construction costs by having lighter track and more curves in their routes than did their English counterparts. He decided to alter the basic English locomotive design to suit conditions in this country. To that end he adopted an innovation, called the leading or pony truck, developed by an American, John B. Jervis. This set of four unpowered wheels, ahead of the drivers, helped guide an engine through the many curves that were common on early American railroads, thereby lessening the chance of derailments.[21] Starting with the Jervis truck, Baldwin created a completely new design with many improvements over *Old Ironsides*. His design and his company received a major boost in 1834 by securing a large contract for engines for the now-completed Philadelphia and Columbia. Of his first ten locomotives, seven went to that line.[22] With the security of a major customer and confidence in the future of railroading, Baldwin decided to expand his facilities. In 1835–36 he built a substantial brick factory, surmounted by a cupola, fronting on Broad and Hamilton Streets in a district known as Bush Hill, then outside the Philadelphia city limits. The new factory and its tooling cost roughly $200,000, a huge sum for the era.[23]

Originally Bush Hill had been the picturesque site of an eighteenth-century country estate of that name. By 1835 the area was rapidly becoming an industrial center, focusing particularly on the production of capital equipment. Baldwin's new neighbors included the Norris Locomotive Works, the Bush Hill Ironworks, and Rush and Muhlenberg, makers of stationary steam engines and the successor to Oliver Evans's pioneering work in that field. These firms were soon joined by others, including two of America's leading machine tool builders, William Sellers and Bement and Dougherty. Such companies moved

This was the basic configuration of Baldwin's standard locomotive design of the 1830s, although most early models had a wooden frame, unlike the iron one shown here. The swiveling Jervis truck (*right*) helped the engine stay on the track. Only the main driving wheels (*left*) were powered, limiting the design's hauling capacity. Nonetheless, Baldwin enjoyed great early success with this type of engine, partly because he had few competitors in the American locomotive market of the 1830s. RAILROAD MUSEUM OF PENNSYLVANIA (PMHC)

Although large factory structures were rare in America before the Civil War, their design quickly settled along certain conventional lines. This 1860 view of Baldwin's 1835–36 plant at Broad and Hamilton Streets includes most of the common elements: a multistory long and narrow brick building—ell-shaped, in this case—pierced with many windows to maximize natural light. The long interior spaces suited power transmission by overhead shafting while giving foremen easy oversight of workers. The white wooden cupola contained a bell to call workers to their jobs. Blacksmiths occupied a single-story wing to the left, and machinists filled the rest of the building. BROWN UNIVERSITY

to the area because of its open land and the good rail connections provided by the Philadelphia and Columbia. Both were essential to capital equipment builders, which required large factories for the production of machinery, ready access to raw materials like iron and coal, and the ability to ship their products to customers across the nation. These builders also needed the city's skilled artisans, and soon Philadelphia's characteristic brick rowhouses filled in the land around the factories. Baldwin's plant would remain in Bush Hill for the next ninety-three years.

The first American railway boom was well under way by 1835, amply justifying Baldwin's decision to build the new plant. In the following year the young company made forty engines, with forty more rolling out of the factory in 1837. Customers from Massachusetts to Georgia and as far west as Michigan purchased these locomotives.[24] Baldwin employed three hundred men in 1837, up from the thirty he had worked with on *Old Ironsides* (see appendix A for Baldwin's annual output and employment figures). In just six years Matthias Baldwin had transformed himself from a successful master craftsman in metalworking into one of the leading industrial magnates in Philadelphia. There is no record of Baldwin's profits in this period, but evidence from another builder suggests that he could have made more than $60,000 in 1836 and again in 1837.[25] If Baldwin did garner such huge rewards from his new status and the booming locomotive market, he also learned some of the difficulties of managing such a large operation. Among them were the demands of his workers, who saw the flush of orders and knew the power they held in their skilled hands. In March 1836 they convened a meeting to pressure their boss into raising wages. Rather than face a "turnout" at a time of full order books, Baldwin raised the price of his engines from $6,700 to $7,000 and granted the wage increases.[26] At this early point in the industry, America had few locomotive builders, and demand for engines far outstripped their capacity to supply the railways' needs. This seller's market soon ended, however, and Matthias Baldwin learned how volatile the demand for expensive capital equipment could be.

SURVIVING THE PANIC OF 1837

In the late winter and early spring of 1837 the American economy entered one of the periodic liquidity crises that would recur throughout the century. As the circulation of funds slowed, the Panic of 1837 gripped the nation's commerce. The timing could not have been worse for Matthias Baldwin, who had based his 1835–36 plant expansion on the strong market for motive power. During 1837 he maintained the production of the previous year, but the lack of ready money quickly became a problem.[27] His men had to receive cash wages every week, and bills for raw materials also came due regularly at a time when payments for engines were beginning to lag. By February Baldwin had completed three locomotives for the state-owned Philadelphia and Columbia, yet the legislature failed to appropriate payments. His resort to short-term bank loans was a temporary solution at best.[28] Baldwin tried to rely on patchwork financing while delaying payments to his suppliers, but his situation deteriorated steadily as orders, credit, and cash all grew scarce. From a high of forty engines in 1837, output fell to only nine locomotives in 1840. By then debt had overwhelmed the young enterprise. At some point in 1838 or 1839 Baldwin was forced to call in his creditors, mostly suppliers and commission merchants holding protested notes. He showed them his own valuation of his assets, offering to surrender all that he had but admitting that "a forced sale would not pay them twenty-five cents on the dollar." [29] Rather than follow this mutually disastrous course, Baldwin's creditors agreed to extend his notes, hoping that an eventual recovery in the locomotive market would allow him to pay off his obligations.

While this arrangement took care of past-due debts, Baldwin turned in 1839 to restructuring the company to meet future needs. To secure the working capital necessary to continue operations, he looked to his longtime supplier of iron forgings, Stephen Vail of the Speedwell Iron Works. In return for a total

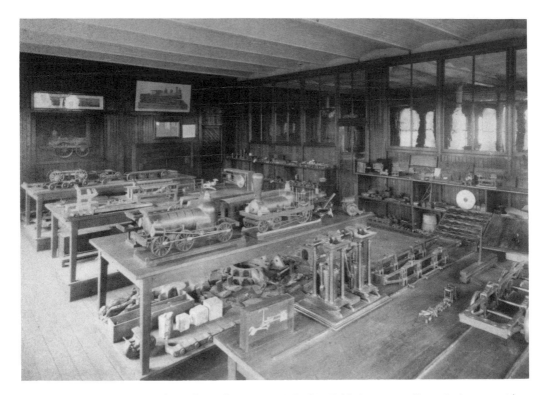

Like most early mechanical innovators, Matthias Baldwin customarily worked out new ideas in three dimensions, building scale models to test his concepts and develop improvements. When this view of Baldwin's model room was taken in the 1890s, the practice was largely obsolete, but the firm preserved his models as a tribute to the founder. A visitor to the room in 1882 saw "engine trucks, motors, self-acting canal engines, smoke stacks, tools, and in fact everything a practical mechanic could think of in a short space of a life time." The two engine models prominently displayed on the middle table were flexible-beam designs.
HISTORICAL SOCIETY OF PENNSYLVANIA

investment of roughly $20,000, Vail's son George became the owner of a one-third interest in Baldwin's reorganized firm, and he moved to Philadelphia to join its management.[30]

Baldwin's next move amounted to a rather gutsy bet. Given time, he believed he could develop a new locomotive design with far more power than his existing models. Success in this endeavor might reverse the flagging fortunes of his company—if he could find the time needed for experimentation and development. To that end he enlisted a second partner, George Hufty, to take charge of administration and production. Hufty had been a machinist in Baldwin's factory, and he evidently made little or no capital contribution in return for his one-third interest in the new firm of Baldwin, Vail, and Hufty.[31] Hufty provided something as valuable as cash: he had the expertise to relieve

Matthias Baldwin of day-to-day management. Under the prevailing beliefs of the era, such a capable manager deserved the recompense of an ownership interest.

With Hufty and Vail running the factory, Matthias Baldwin worked ceaselessly throughout 1839 and 1840, developing his new plan of engines.[32] All his locomotives to date had only a single powered axle with the unpowered Jervis pony truck up front. Though successful, this design lacked the strength to haul long freight trains. Other builders had developed a design with two powered axles and the Jervis truck, but Baldwin thought he could do better.[33] After an abortive geared engine, Baldwin developed what he called his flexible-beam design. Working out his ideas with scale models, he created an engine that transmitted power to three axles. Because the two axles in front could pivot, the engine passed through tight curves easily. The design also had the whole weight of the engine carried on powered wheels, thereby increasing both adhesion to the rails and hauling capacity.[34] This blend of good tracking ability and high power would eventually prove a winning combination. When the first new flexible-beam engine went into production in 1841, Baldwin returned to daily administration and George Hufty was dropped from the partnership. Baldwin and George Vail continued on as co-owners.[35]

That partnership would also have a short life. The locomotive market

Baldwin's first flexible-beam engine had six powered wheels, but he soon developed an eight-wheeler for even heavier trains. The front truck pivoted like a Jervis leading truck, but its wheels also received power, transmitted by connecting rods from the cylinders. With eight driving wheels, such engines were well suited to hauling slow and heavy freight trains.
RAILROAD MUSEUM OF PENNSYLVANIA (PMHC)

remained in the doldrums in 1841, when Baldwin built only eight engines. The firm still made a range of non-locomotive products, and these, coupled with replacement-parts sales and repair jobs, kept the enterprise barely afloat. The nation's commerce was still depressed in 1841, as was railway construction, and the new flexible-beam design suffered a number of teething problems. Given these difficulties, George Vail grew doubtful that his investment would ever pay off, so he withdrew from the company in July 1842.[36]

Baldwin lacked the cash to pay off Vail's interest, so a deal was struck. Baldwin would take on a new partner, Asa Whitney, a former superintendent of the Mohawk and Hudson, and the new firm would assume the debts of its predecessor. How much money Whitney paid into the locomotive builder's capital is unknown, but his widespread contacts in the railway industry were an important contribution to the new firm of Baldwin and Whitney. This factor, coupled with general economic recovery and the new flexible-beam design, finally brought a healthy volume of orders, and in 1846 the firm made forty-two locomotives. By that time Baldwin had paid off all his 1830s debts with interest as well as discharging his obligations to the Vails.[37] In turn, Whitney withdrew from the partnership in 1846 to establish his own railway car wheel factory across the street from Baldwin's plant. By this time the Baldwin company had achieved some stability, having passed through the difficult stages of startup, rapid early growth, and then a severe market reversal. For the next eight years it reverted to the sole ownership of Matthias Baldwin.

THE LOCOMOTIVE INDUSTRY IN THE 1840S

The evolutions in Baldwin's partnerships following the 1837 panic arose from the great volatility in demand for engines. These swings would recur frequently throughout the century. Other characteristics of the locomotive industry also became apparent during the 1840s. These challenges and constraints confronted all the major firms in the industry, and they defined much of Baldwin's operations. Chief among them was a shift from a seller's market to one generally favoring buyers as new firms entered the industry. Matthias Baldwin had been one of the first American locomotive builders, but by 1848 he faced at least seven major competitors. During the railway boom of the mid-1850s the industry grew to number over forty firms.[38] Even though many of these new entrants were small shops or general machinery makers with only a sideline in locomotives, their presence fundamentally reordered the industry and vitiated Baldwin's first-mover advantages. With so many firms competing for business, the railroads quickly sensed their advantage and pressed it home. They began to force the builders to compete on a variety of levels, including price, credit terms, quality, and delivery dates.

One indication of the newly competitive market after the mid-1840s was the carriers' growing practice of soliciting bids for engines.[39] Ten years earlier, builders had simply named their price. In the 1830s Baldwin had required payment in cash, 50 percent when an engine was half completed and the remainder due on delivery. By 1844 the locomotive builder was agreeing to a half payment in cash on completion, with the balance due six months after delivery.[40] In the early 1850s Baldwin even accepted total payment in the stock of certain lines.[41]

It could be argued that Baldwin's ability to accommodate such terms reflected the strength of his company, but he had to meet these demands or lose the business to others who would. Credit sales posed substantial risks, since few railroads of the 1840s and 1850s were established businesses with steady cash flow. New carriers were notably hazardous in this regard. They almost always required credit terms to purchase engines while they completed their routes and developed new traffic. With so many new lines emerging in the 1840s and 1850s, Baldwin had to take the risk to secure business. Even if these customers eventually paid their bills on time, accounts locked up in extensions of credit offered little help when the locomotive builder had a pressing need for cash. Here Baldwin derived some benefit from his location in Philadelphia, where information on customers' creditworthiness was available and banks might discount the carriers' commercial paper. So long as the national economy did not seize up in one of its periodic liquidity crises, Baldwin could generally raise the cash he needed to pay his hands and to meet his other obligations, but securing adequate funds was often difficult.

Issues of quality also rose in importance during the 1840s with the heightened competition in the industry. Quality involved both the workmanship of the product and its technical suitability to the demands it would face in service. Since the 1830s Baldwin had warranted his locomotives against defects in construction by extending a thirty-day performance guarantee. For the most part the company's own delivery engineers went out with new engines to set up and run them during that trial period.

As Baldwin knew quite well, his designs had to fulfill the technical and operational needs of his railway customers. Put another way, he realized that his products were only one element, albeit a crucial one, of the entire economic and technical system of railroading. Through the 1830s Baldwin also possessed more knowledge of the engines' performance capabilities than did most neophyte railroaders. For example, he compiled data on the speeds and hauling capacities of his designs on both level track and a number of gradients.[42] Customers used this information to gauge the amount of traffic their lines could carry. Baldwin also expanded his product line to meet different carriers' varying needs. From the one standard design of the mid-1830s, Baldwin offered three engine classes in 1840.[43] As the competitive pace in the locomotive industry quickened in the mid-1840s, even three models became insufficient. Railroads increasingly sought locomotives with particular characteristics to suit the terrain of their routes and the types of service they ran, such as heavy freights or fast passenger trains. To meet this demand Baldwin's standard product line grew to fifteen different sizes or types of engines by 1846, including various classes of his flexible-beam design.[44] This design variety greatly taxed Baldwin's production processes, but he had to meet his customers' needs or see them go elsewhere.

Resolving not to carry these new competitive burdens alone, Matthias Baldwin passed some of them along to his own suppliers. Since the earliest days of locomotive production Baldwin had relied on a large network of firms to supply raw materials, semifinished components, parts, and fittings. With engines of the 1840s surpassing 40,000 pounds and containing upwards of four thousand parts, it is easy to imagine Baldwin's reliance on these vendors. By that decade he was ordering Pennsylvania coal and iron plate; boiler flues from a Philadelphia manufacturer; files from Pittsburgh and England; nuts, bolts, and copper sheet from New York; forgings from Baltimore; brass from Boston; and wrought-iron tires from Connecticut and England. Sometimes he bought directly from these firms; in other cases he used commission merchants as intermediaries. Baldwin's ties to this constellation of companies had both a technical basis, dependent on the quality of their products, and a financial aspect, and the two became intertwined.

Like most classic entrepreneurs, Matthias Baldwin used his suppliers to finance his own firm's growth. Even when he received cash-on-delivery payments for engines in the 1830s, he established credit terms whenever possible with his own suppliers, often giving them notes of up to four months.[45] As his railway customers began to demand credit, Baldwin in turn asked for longer payment schedules for his own purchases. When the carriers started to pay for engines with railway stocks, the locomotive builder made suppliers take these equities in partial payment for orders.[46] Finally, Baldwin took out short-term loans on occasion from friendly vendors when his own till was empty and he faced such pressing obligations as payments to his workers.[47] This variety of transactions transformed Baldwin's technical network with suppliers into a financial one, helping to finance the growth of his company and the railroads' needs as well. Such improvisations were a necessity in an age without commercial banks making industrial loans. Although the system was shaky, it worked to everyone's benefit most of the time.

Baldwin's technical demands on his suppliers also illustrate some of the challenges of running such a complex enterprise in the antebellum era. Loco-motives required the highest-quality materials obtainable; cracks in a forged axle or flaws in boilerplate spelled disaster, death, and unwelcome publicity. Once Baldwin found vendors capable of meeting his quality requirements, he

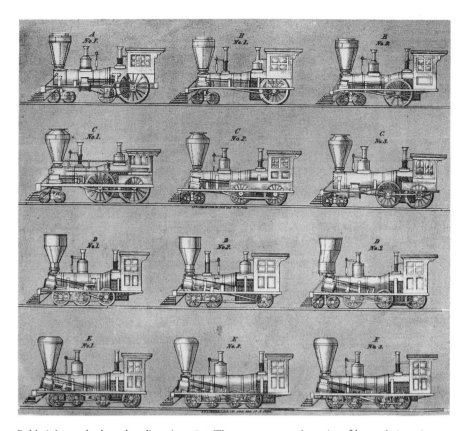

Baldwin's standard product line circa 1855. The company used a series of letter designations to differentiate its locomotives by the number of powered axles. Class A were experimental designs, and B types mirrored Baldwin's 1830s products. The engines shown on the top line were largely obsolete by 1855, having only a single powered axle. The class C types on the second line had two driven axles. Called the American or 4-4-0 type, this configuration became quite popular in passenger and freight service. The D types on the third line had three powered axles to haul freight trains. Finally, the four-axle E-class engines spread the engines' weight and power over eight points on the rail, improving hauling capacity while lessening damage to the tracks. Each class was available in different weights to suit railways' differing needs. HAGLEY MUSEUM AND LIBRARY

tended to continue those relationships for many years. Close relationships with suppliers were also dictated by the nature of Baldwin's production processes, particularly the need for speed in locomotive construction.

Most locomotives were made-to-order products. Once they had ordered an engine, however, railroads generally wanted them quickly. Baldwin tried to accommodate his customers with a sixty-day production schedule, which in turn placed demands on his suppliers. The locomotive builder operated with a type of rudimentary just-in-time inventory system for raw materials. For example, after receiving an engine order, Baldwin notified his primary boilerplate supplier, Steele and Worth of Coatesville, Pennsylvania. That firm kept the plate dimensions for Baldwin's boilers on file and used the information to roll and cut the requisite sheets. Concurrently, Abbott and Ferguson of Baltimore worked up forged axles to drawings on file, and a number of other suppliers turned to do their part. Once completed, these materials were shipped to Philadelphia by coastal vessels, canal boats, or railroads, with bills of lading arriving in advance of the shipments, which allowed Baldwin's foremen to mesh production plans with the flow of materials.

This supply system required close synchronization of suppliers' work to Baldwin's needs, both in scheduling and in the dimensions of parts and materials. By the mid-1840s this had become a complex problem, since Baldwin's standard product line included fifteen classes of engines. While some detail parts were used across a number of classes, the variety in the product line heightened the challenges of coordinating suppliers' output with Baldwin's production demands. Even within a given design, this posed difficulties. For example, a single boiler called for sixteen variously sized sheets of boilerplate, all furnished by the supplier to finished dimensions. Baldwin needed to rely on his vendors for this work, but ensuring coordinated effort was a daunting task. Locomotives of the mid-1840s contained thousands of parts, and Baldwin made up to forty-two locomotives a year at that time in a variety of classes.

Although the system had its challenges, it also offered a number of benefits. With just-in-time deliveries of most parts, Baldwin avoided tying up scarce working capital in inventory for lengthy periods. By ordering boilerplate and other parts made near or at finished dimensions, he saved on freight charges and on his own labor costs. In an age of limited managerial oversight and scarce capital, this type of subcontracting to specialist firms made a great deal of sense. These technical and financial ties to a network of suppliers greatly aided Baldwin's survival, operations, and growth.

THE FACTORY SYSTEM OF LOCOMOTIVE PRODUCTION IN THE 1840S

The increasingly competitive state of the locomotive industry in the 1840s and the mounting variety in Baldwin's product line made Matthias Baldwin's primary task of the decade even more difficult. That task was to establish a factory system to build locomotives. This effort, begun in 1835 with the move to the Broad Street factory, was well under way by 1837, when three hundred men toiled in the $200,000 plant. But the sales plunge following that year's panic caused massive layoffs, and the plant ran at a fraction of its capacity until the mid-1840s.[48] By then the whole competitive structure of the industry had changed, placing a new premium on efficient production. Measures to improve efficiency also became more important in this decade because Baldwin no longer had the productive advantage of building only a single, standard design. Industrial management was still in its infancy in the 1840s, however, and Baldwin first had to reestablish his factory system before he could improve it.

After Asa Whitney's withdrawal in 1846, Matthias Baldwin ran his company as a simple proprietorship until 1854. For help in managing the enterprise he relied on a few trusted associates. After his short stint as a partner, George Hufty became general superintendent and oversaw production. George Burnham handled the company's finances and most of its correspondence, while a draftsman named William Pettit assisted Matthias Baldwin in design matters.[49] These four men were the only central managers at the firm until the mid-1850s. In this regard Baldwin mirrored most industrial companies of the era, which had very small managerial staffs.

Given this organizational structure, the shop foremen necessarily exercised great authority and discretion. Circa 1850 Baldwin probably had seven foremen: one for each of the three machine shops; one each for the boiler, foundry, and smith shops; and a foreman of painters.[50] These men handled a whole range of matters. They scheduled the work in their departments, dickered over prices with suppliers, and kept their own accounts on the costs of materials used in their shops. In personnel matters the foremen hired and fired workers, set wages, oversaw apprentices, and selected men for layoffs when necessary. They oversaw all work in their departments, ensuring that parts were made on time to the required dimensions and coordinating output among the shops. These considerable powers made foremen the absolute masters of their own departments. In establishing this structure, Matthias Baldwin drew on his own artisan background, giving foremen most of the authority held by master craftsmen of the early nineteenth century.

Under this factory system Baldwin necessarily relied on the exertions and skills of artisans in a variety of crafts. Locomotive building required sheet-iron- and coppersmiths, boilermakers, patternmakers, molders, carpenters, blacksmiths, machinists, and ornamental painters. The firm also employed unskilled carters and laborers in large numbers to muscle materials and parts around the shops. By the late 1840s roughly four hundred men worked in the Broad Street factory. Given the infancy of machine building and steam technology, some of the skilled trades, such as machinists and boilermakers, were relatively new occupations. Despite this novelty and Baldwin's large need for such craftsmen, foremen had little difficulty finding skilled men to staff the shops.[51] Philadelphia's large and growing pool of craftsmen, coupled with Baldwin's own apprenticeship program, seems to have provided adequately for the company's needs.[52]

At times Baldwin's skilled men only grudgingly accepted the factory regime of their employer. Despite the novelty of their trades, they knew their skills gave them power, which they were willing to wield if necessary. Baldwin had averted a strike in 1836 only by granting wage increases. In 1844 Asa Whitney had to bargain with the men when the company needed extra production beyond the normal ten-hour day.[53] Lathe hands decided that they would work four overtime hours a day, in return for time-and-a-half wages. They preferred this to seeing Baldwin hire extra lathe men for a second shift, which would have increased his supply of hands in the long run and thereby decreased their leverage. Baldwin's workers also took their Saint Monday devotions from time to time, avoiding work when it suited them.[54] Finally, the men knew that the volume of Baldwin's output largely depended on their own exertions. Since they received daily wages rather than piecework pay, they possessed little incentive to exert themselves for their employer's benefit.

If these examples suggest a gulf between management and labor, the two sides also came together at points. In 1844 a committee of workers invited Matthias Baldwin to a banquet with his men at the Columbus Hotel.[55] In the same year, many of Baldwin's workers marched in a great Whig party processional through Philadelphia in support of Henry Clay, whose American System promised tariff protections and internal improvements. Sharing their party affiliation, Matthias Baldwin helped finance this parade of "manufacturers and mechanics."[56] Inside the factory Baldwin had a consistent policy of favoring his workers' sons for places as apprentices. Finally, when the demand for engines fell off, as it so often did, Baldwin tried to mitigate or avoid layoffs by lowering locomotive prices. As he told one customer, he offered such price cuts "to keep a larger number of hands employed during the winter."[57]

These and other efforts partially bridged the gap between Baldwin and the men, but the Broad Street plant was a large-scale factory, not one of the small craft shops of an earlier era. The old close relationship between journeymen and master craftsmen increasingly faded as the company's growing scale of operations and its vulnerability to national economic trends and cycles simply made the old ways irrelevant. For example, in 1848 Baldwin's output fell 48 percent compared with the preceding year. The decline proved calamitous to the company's cash flow, since the whole structure of its revolving credits to customers fell apart during these market plunges. With no cash coming in, the workers went unpaid for a month during the fall of 1848. During those weeks Baldwin tried desperately and unsuccessfully to raise the necessary funds.[58] Unaware of this and knowing real hardship themselves, his men convened a meeting and sent him a petition, detailing their privations and asking that he take out a loan to pay them. They argued that "Mr. Baldwin . . . is generally known to the community as an extensive manufacturer and a wealthy man [therefore] we cannot but believe his credit is good for any amount of money necessary."[59] Although his credit was in fact shaky at best, Baldwin eventually found money to pay the men, resolving the immediate problem. But the incident undoubtedly created bitterness among workers. It underscored the vulnerability that came with exchanging time and effort for wages in a large and increasingly impersonal factory.

A number of factors had contributed to this incident, including the short-term fall in sales, mounting competition in the locomotive industry, and the patchwork character of commercial credit in the antebellum economy. Internally the chief problem was the labor-intensity of locomotive production. At this time Baldwin's 215 workers needed roughly two months to build a locomotive.[60] Put differently, each engine required roughly eleven man-years of effort, a figure that did not include all the preliminary work done by suppliers. Locomotive building was so consuming of time and skills that almost half of Baldwin's total revenues went into payroll costs.[61]

By 1850 building these complex and unwieldy products required coordinating the efforts of hundreds of men. They often worked in groups, as in smithing and boilermaking, while skilled machinists and erectors needed laborers' help in setting up heavy components. This need for a multitude of workers had originally impelled Matthias Baldwin to organize production by the factory system. Although machine tools were certainly essential to production, operations at the Broad Street factory in the 1840s centered primarily on deploying concentrated labor, rather than capital. With so much of his revenues going into payrolls, some time would pass before Baldwin could build up his base of fixed capital. In the meantime, cash shortfalls like the one that triggered the November 1848 incident over wages were bound to recur unless something changed.

COMPETITIVE CHALLENGE, LIQUIDITY, AND PRODUCTIVITY IN THE 1850s

During the 1850s the competitive pressure in the locomotive industry grew even more troublesome for Baldwin's operations. But the decade did have its positive developments, at least until the Panic of 1857. Railroad building became a national obsession in these generally prosperous years, with many new carriers starting up, particularly in the South and Midwest. As a result, locomotive demand increased, as did Baldwin's sales. On the other hand, railway development and the general prosperity brought more new entrants into locomotive building. New railway companies turned this heightened competition to their advantage, using commission merchants to solicit competitive bids for engines on terms of long credits or partial payments in stocks. Older and more established carriers generally paid in cash or in high-quality commercial paper, but they placed an additional demand on engine builders.

During the 1850s such lines as the Baltimore and Ohio began to establish their own specifications and designs for new motive power, rather than simply accepting the builders' standard products. Baldwin had to accept the inefficiencies of custom production or lose the business to competitors. For the most part Baldwin acquiesced in this loss of design control, which placed new strains on production. Finally, the surges in demand and in Baldwin's business during most of the decade presented their own challenges. By 1857 annual output had grown to sixty-six engines, but spiraling labor and materials costs accompanied the increase. With many accounts tied up in credits, such growth may have even exacerbated Baldwin's recurring shortfalls of working capital. In sum, the 1850s were fraught with difficulties for locomotive builders. Matthias Baldwin was not without power or resources, however, and throughout the decade he instituted a number of changes inside the factory that promised to improve productivity and ease his chronic liquidity problems.

During the spring and summer of 1850 Baldwin enjoyed the happy circumstance of full order books. Even this was troubling, however, since output failed to keep pace with either promised delivery dates or overall demand for engines—despite increases in the workforce to four hundred men. The price of reliance on workers' exertions had become clear to Baldwin, as he wrote to his financial manager, George Burnham:

> Now this state of things is by no means gratifying and if we fail as much in future as we have the last month . . . the wages will consume all your money. . . . We ought to be able to make one [locomotive] a week with this force if all would do anything like their duty. I would suggest that the work should at once be given out by the piece in all cases, even though the cost may be increased, for if we can't make anything [i.e., profit] let us have some reputation for filling our engagements. I am satisfied all the trouble is owing to the fact that the completion of the work is not the object of the men but [they want] only to kill time, and do as little as they can. . . . This is the moth that is destroying us and with the loss of credit arising from the constant borrowing, we are in no pleasant condition.[62]

During the early 1850s the company switched many tasks over from day to piece-rate pay, even the most skilled and demanding work, and productivity improved.[63] The institution of piecework was Baldwin's first real advance in industrial management since adopting factory production in the first place. Baldwin's craft workers did not object to the change, largely because the firm set the rates quite high to foster the desired productivity. In acquiescing, however, they started down a road where their work was increasingly defined by the company and its needs, rather than by their craft training.

After installing piecework the firm established a number of new managerial practices during the 1850s. In 1854 Matthias Baldwin took on a new partner, his longtime boiler shop foreman, Matthew Baird. Probably because of Baird's influence, the firm soon dropped such products as stationary engines and boilers and adopted a policy of building only locomotives.[64] That decision to specialize allowed the firm to build on the foundation of piecework with a rigorous program to standardize locomotive parts. Now standard components could be used across a number of Baldwin-standard engines or even in custom designs. Such policies had an immediate impact on efficiency and profitability. In 1850 the company made thirty-seven engines, a total near its maximum capacity, and Matthias Baldwin took $15,000 in profits from the firm; four years later the firm had financed an expansion in facilities, it built sixty-two engines, and Baldwin paid himself over $71,000.[65] The company was not yet out of the woods. Short and sharp recessions in 1855 and 1858 caused renewed liquidity problems and austerity, but the new attention to management provided immediate benefits. This focus on internal controls helped to ameliorate the effects of problems outside the firm that Baldwin could not influence: sales fluctuations, severe competition, and the resultant leverage of the carriers over prices, credit terms, and design issues. The emphasis on management also had other consequences, recasting the character of work and the relative power of managers and workers over production.

THE RIGHT TO MANAGE AND WORKERS' RESISTANCE

By the late 1850s Baldwin's workforce straddled a transition from a heritage as craft workers to a future of industrial wage labor. Workers' skills themselves were not at issue in this transformation; locomotive building would require highly skilled personnel down to the end of steam power in the 1940s. Well-trained and capable workers were so essential in the 1850s that Baldwin strengthened its craft apprenticeship program in 1853 by requiring formal indentures for apprentices. The real issue for Matthias Baldwin and Matthew Baird was harnessing workers' skills and exertions to the task of producing profits as well as engines. With employment reaching four hundred to six hundred men during the 1850s, Baldwin's managers had created a huge industrial workforce that towered above the small shops typical of the period. With the adoption of piecework, the partners had begun to exert industrial discipline over this large force. Baldwin's skilled men went along with this change because it paid well, but they soon protested other signs of their transformation into dependent wage labor.

In 1857 Baldwin enjoyed its best sales year ever, with six hundred men making sixty-six locomotives. Then the boom collapsed in the Panic of 1857. The partners tried to stem the decline by cutting prices and by traveling to Cuba and the Southern states to drum up business.[66] They were only partially successful, however, and had to order large layoffs. With output off by 50 percent in 1858, discharges of two hundred men cut the workforce by one-third. To avoid deeper cuts, the company switched back temporarily to day pay to keep more men employed, although at lower wages.[67] This provided scant consolation to the scores of laid-off Baldwin workers left jobless in the depressed city. Hardship and resentment built up into action that winter as many of the unemployed men formed the first trade union of machinists in America, the Machinists and Blacksmiths International Union (MBIU). A longtime Baldwin machinist, Job R. Barry, was a founder and key leader of the union, and many of the firm's current and laid-off employees followed him into the MBIU, providing the bulk of the fledgling organization's membership.[68]

The recession provoked by the panic turned out to be relatively brief. By March 1859 orders for engines had picked up again. While on a sales trip in the South, Matthew Baird wrote to the company's general superintendent, urging him to increase the workforce to six hundred men. Baird added, "It is best not to let the hands know any more of the business than is necessary (I mean the amount of work) as it will have a tendency to make them difficult to manage."[69] Baird was probably unaware of the MBIU, which kept its existence a secret from employers while building its membership, but he did know that skilled men held latent power over production.

In March 1860 a flood of orders crowded Baldwin's factory with work, and those men decided the time had come for action. Announcing the presence of their union in the factory, they presented a petition to management. The workers' immediate issue was a cut in overtime wages from time-and-a-half to time-and-a-quarter.[70] The company had taken this action immediately after the 1857 panic as sales and prices fell and competition mounted. Given the labor-intensity of locomotive building, the company had "found all overwork a loss."[71]

To the workers the overtime issue was a real grievance. From the company's point of view, however, the mere presence of the union ensured that the dispute would center on who exercised the right to manage production—the partners alone, or management and workers. It was the classic question of industrial capitalism, and Matthias Baldwin answered it decisively, if undiplomatically. After receiving the petition Baldwin went into the erecting shop, the center of union support, and stood on a workbench to address the men. He said that he might pay time-and-a-half for special rush jobs, whereupon a man interrupted to demand that rate for all overtime work. As an onlooker later recounted, "Mr. Baldwin looked at the man for a minute and did not speak. He then said, 'men this shop is as much to you as it is to me.' He got very red in the face. 'Now before I pay time-and-a-half for all overtime, I'll see this shop burn to the ground.'"[72] With that the strike was on.

The MBIU had surprising strength for a union that had largely recruited in secret. Almost all of Baldwin's highly trained machinists and blacksmiths,

This photo dates roughly to 1860 and shows the interior yard behind the factory. The erecting shop (*left*) was known to workers as the "Crystal Palace," for its many windows. Skilled machinists and erectors from this shop accounted for most of the strikers in 1860. Strewn about the yard are an unfinished locomotive, forged engine frames, axles, and finished wheel sets. SMITHSONIAN INSTITUTION

approximately 170 men, walked off the job early in March. Although this represented only one-third of the total force, the strike threatened to suspend production entirely. The strikers held mass meetings, picketed the shop, and tried to win over public opinion.[73] There were also ugly incidents like assaults involving picketers and workers. Union sympathizers who remained on the job attempted to sabotage Baldwin's business by putting scrap iron into the cylinders of newly completed engines to cause breakdowns.[74]

The company took a clear and adamant position. It described the strikers as "a secret society of mechanics whose object is to regulate matters for employers," whereas Baldwin and Baird were determined to maintain "the liberty of making laws for our own shop."[75] Given all their commercial uncertainties, the Baldwin partners were grimly determined to retain control inside the factory. The firm announced that the overtime issue was non-negotiable, and it gave few strikers a chance to return. It also began immediately to recruit replacements, drawing men from New England, New York, Baltimore, and the Pennsylvania Railroad shops in Altoona. While awaiting the new recruits, experienced hands used apprentices and laborers to erect engines.[76] By the end of March the company had six hundred men on the payroll, more than the prestrike total, and Baldwin built a record eighty-three locomotives in the year as a whole. The MBIU had failed absolutely, the nascent union was largely crushed, and Baldwin would remain a nonunion shop for the next fifty years.

Twelve years after the strike a union leader, Jon. C. Fincher, wrote up a description of the MBIU's formation and the strikers' grievances. Labor historians have used this account to argue that employers' efforts to undermine craft skills were the leading factor in this labor-management contest.[77] This is incorrect. In fact Baldwin's overwhelming reliance on artisan labor led to its antiunion stance. The company depended so entirely on craft skills in building engines that it refused to tolerate any effort by workers to translate those skills into a countervailing power in the workplace.[78] Although the strikers' ostensible issue was overtime wage rates, Fincher's account suggests that in fact labor's main grievances centered on broad economic and social trends the union men found particularly threatening. These included the growing scale of factory operations, the end of face-to-face dealings between men and masters, and the peril in these large factories of massive layoffs as a result of business depressions. The MBIU's failure at Baldwin had its ironies: a new union in a young trade fighting to preserve old-style craft traditions at a large factory employer whose industrially based organization of work represented the wave of the future.

THE COMPANY ESTABLISHED

By 1860 Matthias Baldwin had secured his factory system of locomotive production, as the strike itself demonstrated. In that contest the partners showed they possessed the power and will to defend their conception of the firm and their right to manage it. Their ability to influence events outside the factory that affected their business remained circumscribed, however. Vigorous competition, credit sales, and commission merchants who played the builders off against each other continued to bedevil the company. But the focus on internal efficiency during the 1850s had eased Baldwin's perennial shortfall of working capital, allowing the firm to reinvest profits in plant and equipment. By 1860 Baldwin had $350,000 in fixed capital, and it ranked second in the industry in employment and units produced.[79] With its internal controls and substantial capital base, the company seemed poised for future growth. First, though, it had to survive another reversal.

The Civil War had far-reaching ramifications for Baldwin's business, first plunging the company down and then lifting it up to a position of some security. The initial fall came from Baldwin's great dependence on Southern railways as its primary market. In 1860, 79 percent of Baldwin's output went to carriers in states that would soon leave the Union. The company failed to see the secession crisis coming, and when it finally happened, Baldwin found itself holding over $100,000 in uncollectible commercial paper.[80] Concurrently, the opening of war caused a general slowdown in the Northern economy, and Baldwin's 1861 output fell by 52 percent compared with the year before.

These events quickly gave way to better tidings for the firm as the Union shifted to a wartime footing. Locomotive sales to the U.S. Military Railroads partially counterbalanced the loss of Southern customers. Even more important were Baldwin's ties to the Pennsylvania Railroad. During the war years the Pennsylvania's traffic soared, thanks to its route through the nation's industrial belt. Baldwin took almost all of that carrier's orders for locomotives, producing over one hundred engines for it.[81]

Aside from renewed orders, the years after 1861 also brought fundamental changes in business transactions in the locomotive industry and throughout the Northern industrial economy. With the booming economy, the supply-demand relationship in locomotives finally shifted in Baldwin's favor. Wartime inflation presented a new threat to financial stability, but the new national currency, the greenback, provided the means to counter that threat. Seizing on the demand for engines, in the summer of 1862 Baldwin began to require cash for all sales.[82] A year later the company instituted a new policy "not to receive orders at fixed prices" at all. For the duration of the war Baldwin sold engines only on a cost-plus basis. This policy arose from high inflation rates for both labor and materials.[83] Thanks to the greenback and strong demand, Baldwin, like many companies, turned inflation to its own advantage by ending credit sales altogether. The cash economy lessened the company's vulnerability in business cycle downturns and freed up working capital, finally ending Baldwin's need for short-term loans to bridge working-capital deficits.[84] The hand-to-mouth financing of the past was over, and strong profits allowed expansion in fixed capital to achieve further growth. The sales boom also gave Matthias Baldwin and Matthew Baird extravagant personal incomes of roughly $210,000 each in 1864 alone, the highest in Philadelphia.[85] The wartime economy finally provided the Baldwin company with the financial strength it had lacked for thirty years. It soon used that strength to dominate its industry.

The Baldwin Works enjoyed a secure, but hardly dominant, position in the locomotive industry when Matthias Baldwin died in September 1866. Most of Philadelphia noted his passing. Few of his contemporaries had surpassed his contributions to Christian and charitable causes and to Philadelphia's developing reputation as a center of heavy industry. Leading politicians, all the city's Presbyterian clergy, businessmen, and many employees of the works

Baldwin made this 4-6-0, or Ten-Wheeler, for its best customer, the Pennsylvania Railroad, in December 1866, three months after Matthias Baldwin's death. It stands at the corner of Broad Street and Pennsylvania Avenue, with the 1863 machine and erecting shops — built to cope with the wartime boom — in the background. Such a wheel arrangement became popular in the 1860s, this being Baldwin's 166th model built to this general configuration, which offered more power than 4-4-0 types. The engine burned soft coal and had a steel firebox, a relatively new material in locomotive construction. Designed for freight service and part of a batch order of twelve, it cost $17,000 plus war tax. Soon postwar deflation brought such high prices down. RAILROAD MUSEUM OF PENNSYLVANIA (PMHC)

attended his funeral service in center city. Then a vast procession, including most of Baldwin's one thousand workers, began the funeral cortege to Laurel Hill Cemetery, on the city's outskirts. On their way, the marchers walked a circuit around the factory, as its cupola bell tolled in mourning.[86] The founder had arranged for the firm to survive his death and to continue its operations without interruption.[87] In fact his company's greatest successes still lay ahead, but he had laid the groundwork for that future.

Matthias Baldwin's personal character and abilities contributed greatly to the firm's survival and early growth. His willingness to experiment in an entirely novel technology suggests real foresight and inventive talent, although a number of other mechanicians also worked on locomotive development. What truly set Baldwin apart was the extent of his gamble in this untried field. Because he moved so quickly to create a factory system of locomotive production, he consolidated his technological innovations with the business and organizational capacity necessary for market success. Few nineteenth-century inventors or innovators possessed such a combination of technical ability and commercial acumen. Baldwin gained substantial first-mover advantages from his large factory and extensive production of the 1830s, although the depressed market of 1840–44 almost transformed him from a prescient industrialist into a man before his time. During those depressed years Baldwin drew on his essential strengths in technology and design to create the novel flexible-beam engine, which ultimately helped to retrieve the fortunes of his company.

The highly pitched competition of the 1850s and locomotive builders' loss of design control gave Matthias Baldwin a strong impetus to focus on factory management and internal productivity. Such concerns also rose in importance with the company's overall growth in employment, its four hundred workers of 1850 placing it among the largest American industrial companies. While the company founder had little interest in issues of daily management, he did originate the policy of piecework pay, also known as "payment by results." By this time Baldwin had also selected a new partner and trusted lieutenants who devoted themselves to developing policies to improve productivity and achieve further growth. Here he accomplished the feat of transforming an entrepreneurial firm, dependent on himself, into a managed company with stable operations and a promising future.

In the locomotive industry circa 1866, Baldwin vied for first place with the Rogers Locomotive and Machine Works of Paterson, New Jersey. Soon thereafter the postwar prosperity, America's burgeoning industrial economy, and the railway booms of the Gilded Age pushed the demand for motive power to new heights. The Baldwin Works took over one-third of this expanding market, dominating its industry and becoming the largest builder of capital equipment in America. As detailed in subsequent chapters, much of this great expansion developed from the company's accurate assessments of the opportunities available in dynamically changing markets. The firm also made astute internal adaptations to minimize risks. Highly fluctuating demand, customers' design influences, and labor-intensive production methods forever ruled the industry, but Baldwin's Gilded Age managers developed new methods to ride out these challenges and prosper.

2

THE LOCOMOTIVE INDUSTRY

1860–1901

As the decade of the 1860s opened, the Baldwin company had achieved a relatively secure and stable position in an industry fraught with risk. In 1860 the firm enjoyed its best year to date; its 675 workers made eighty-three engines, and sales topped $750,000. This highly credible performance placed Baldwin second in the industry, behind the Rogers Locomotive Works of Paterson, New Jersey. On the eve of the Civil War, Rogers, Baldwin, and third-ranking Norris stood among the largest factory employers in the country—exceeded only by the textile mills of Lowell, Massachusetts, and some Pennsylvania iron plants.[1] But size alone did not guarantee long-term success for locomotive builders in this era. Although Norris had led the industry in the mid-1850s, it ceased production in 1866. Rogers nearly shut down in the aftermath of the 1873 panic, delivering only eighteen engines in the following year.[2]

On the other hand, Baldwin pulled decisively ahead of its rivals in the 1860s and dominated the locomotive industry for the remainder of the century. How did the firm reach such a commanding position? Much of the company's success derived from its strategy of building very different engines for a variety of markets, in contrast to manufacturers' tactic of producing standard products in volume. In its growth Baldwin also surmounted or adapted to the range of challenges common to the capital equipment sector. Chief among them was the highly variable demand cycle for expensive capital goods. The boom-and-bust character of the locomotive market affected the nature of competition in the industry, as well as firm structure, managerial policies, and marketing efforts— topics also considered here. Swings in sales proved so frequent and perilously wide that prudent builders took long-term survival, rather than profits alone, as their primary concern. During the Gilded Age both goals were evident, for example, in the locomotive industry's recurring efforts to ameliorate demand and price fluctuations through price fixing and market share agreements.

Notwithstanding these collusive policies, Baldwin proved uniquely successful in adapting to the constraints of the sector while also seizing upon opportunities for growth. For those firms that survived the short-term perils of the market, growth beckoned in the half century after 1860 as expanding arteries of iron and steel trackage remapped the nation and reordered its economy and society.

AN OVERVIEW OF THE AMERICAN LOCOMOTIVE INDUSTRY

During the nineteenth century the railroad industry expanded to traverse the entire American continent, knitting the nation together and becoming the most potent agent for the economic and social transformations now described as the Industrial Revolution. Although of European origin, the new technology of steam power on iron rails ideally suited the demands of American geography and commercial needs. By 1840, just eleven years after the success of the English *Rocket,* the United States possessed more railway mileage than all the nations of Europe combined.[3] By the 1850s this network of iron rails bound regions of the country together as the first great trunk lines linked the Atlantic seaboard to the Midwest. In 1869 the first transcontinental line reached the Pacific, and by the 1890s an intricate web of rails traversed the central plains, allowing farmers throughout most of the region to travel no more than 15 miles by wagon to reach a rail line.

The impressive growth in railway mileage shown in table 2.1 augured well for the locomotive builders who supplied the motive power needed to animate that network. But the table only partially illustrates the dynamism of nineteenth-century railroading. Throughout the century the volume of traffic carried on the national network increased at a pace that far outstripped the mileage growth rate. Between 1870 and 1910 railway mileage grew roughly sevenfold. Over the same forty years the volume of freight and passenger traffic increased over sixteen times.[4] This dynamic improvement in railway productivity had many roots. As far as the locomotive industry was concerned, it involved sales of more engines and of progressively better engines, each capable of hauling more freight and passengers.

While historians have debated the extent to which railroads remade the U.S. economy, their impact on certain industries cannot be overstated. For example, railway demand dominated the iron and steel industries, with the carriers taking roughly half of all American steel production from 1870 to 1907 for rails alone.[5] Our present concern, the locomotive-building industry, grew directly out of the railroads' requirements for motive power. Commencing from such pioneers as Baldwin, Norris, and Rogers, the industry numbered forty firms by 1854. During the nineteenth century over one hundred companies entered the locomotive market, although relatively few succeeded for long.[6]

Although the first American railways imported English engines, the native

TABLE 2.1 U.S. Railway Mileage, 1830–1910

Year	Miles	Year	Miles	Year	Miles
1830	23	1860	30,626	1890	208,152
1840	2,818	1870	52,922	1900	258,784
1850	9,021	1880	115,647	1910	351,767

Source: Historical Statistics, pt. 2, pp. 728, 731.

locomotive-building industry quickly took over this market, largely because U.S. railroads differed greatly from their English counterparts, and the imports soon proved unsuitable. To reduce the initial expense in building, American lines curved sinuously to avoid hills or valleys. U.S. railroad builders also cut costs by using extremely light rails, often a composite of iron strapping over wooden stringers. While British locomotives suited conditions in their native land, in America they frequently crushed the light track or derailed at the first curve. In all, early U.S. carriers imported no more than 120 British locomotives. Aside from the technical shortcomings of the imports, the nascent American locomotive industry benefited from a substantial tariff barrier of 25 percent on the value of imported machinery.[7]

In another divergence from English practice, most railroads decided not to build their own motive power, preferring instead to rely on contract builders such as Baldwin. In the early years the national shortages of capital and skilled labor were powerful reasons for the carriers to contract for engines, particularly since so many independent shops stood ready to fulfill their needs.[8] Furthermore, early builders like Matthias Baldwin knew more about steam technology than did the first generation of railroaders. By the Gilded Age the leading contract shops had adopted specialized tooling and methods, allowing them to produce engines at a lower unit cost than most railroad shops could manage.[9] So the carriers generally looked to the builders to supply their needs.

The structure of the locomotive industry changed substantially between the 1830s and the 1860s. Initially locomotive building was a sideline for textile machinery builders or general jobbing machine shops like Baldwin's. Early locomotives were rudimentary machines, made largely by hand, and the chief difficulty in their construction lay in the unwieldiness of their components rather than in any need for great precision. Because production of such simple engines required skills and knowledge but little startup capital, small metal-working firms entered the market with relative ease, turning out engines to meet the needs of new lines nearby. This trend reached its peak during the railroad boom of the early 1850s, then the Panic of 1857 threw at least half of these small producers into bankruptcy.[10]

For the remainder of the century the size of the locomotive industry saw little change, although individual builders were buffeted by demand fluctuations and evolving markets, and many firms failed. Surviving data from the U.S. Manufacturing Census show the size and rank of most major producers in the industry between 1850 and 1880 (see appendix B). In 1850 and 1860 the first tier of locomotive builders included such leading firms as Baldwin, Norris, Rogers, and Hinkley.[11] By 1860 these companies largely specialized in locomotives, although Rogers also maintained a sizable general machinery business. The second tier in the antebellum period included Taunton, Mason, Cooke, Grant, and Manchester. Frequently these unspecialized firms were textile machinery makers with a profitable sideline in locomotives. Such companies often managed to continue in the locomotive business after the Civil War. The third tier of builders in 1860 were marginal outfits like Souther and Wilmarth, which despite past strengths had been mortally wounded by the 1857 panic.

Firm rankings and market shares changed dramatically after the Civil War, as seen in the 1870 and 1880 censuses. Baldwin had been climbing in the late-1850s and had taken a commanding lead by 1870, which it lengthened in the following decade. Previously first-tier Norris and Hinkley were either out of the business or soon to withdraw.[12] The second-tier firms noted above were joined by Pittsburgh, Dickson, Schenectady, and Rhode Island. After 1870 few new entrants succeeded in the mainline locomotive business, for the existing players invested heavily in new machinery and facilities, raising capital costs to daunting heights. By 1870 Baldwin dominated the industry, surrounded by a cluster of smaller firms that fell further and further behind during the remainder of the century. The Philadelphia giant accounted for 30 to 45 percent of all locomotive production in America in every decade from the 1870s to the 1930s, with the remainder split among roughly fifteen firms.[13]

THE BUSINESS CYCLE AND THE LOCOMOTIVE INDUSTRY

Throughout the nineteenth century the most crucial issue defining locomotive builders' operations was the highly cyclical nature of demand for engines. Between 1870 and 1915 Baldwin saw seven occasions when output leaped by more than 50 percent from one year to the next, and five cases when demand shrank by that much. The largest swing occurred between 1885 and 1886, when production jumped 127 percent. Such fluctuations placed great strains on the firm, which required highly flexible management policies to ride out the bucking market. Baldwin's experience reflected the entire locomotive industry, which saw far wider variations in demand than did consumer product manufacturers or such producer goods industries as pig iron and rails. Only shipbuilding, another capital goods industry, faced higher fluctuations in output.[14] Note that locomotive *demand* and *output* are used synonymously here, since by the 1870s engines for mainline American railways were largely custom-made to customers' specifications. Production for inventory to ease demand swings therefore became largely impossible.

The cyclical character of demand and production arose from the accelerator-multiplier effect—a characteristic of all capital equipment markets. *Fortune* magazine described this phenomenon in a 1930 feature on Baldwin: "Comparatively slight fluctuations in the business of the railroads cause very much greater fluctuations in the orders they place for new equipment. Thus a 10 per cent increase in traffic might well call for fifty locomotives instead of the normal twenty-five. This would of course be a 100 per cent increase for the builder. And declines in traffic react with equal exaggeration on the equipment makers."[15] Railway traffic in turn reflected the volume of economic activity for the nation as a whole, and locomotive sales provided a sensitive barometer of the business cycle itself, forecasting recession or booms with impressive accuracy. Consequently the nineteenth-century business community watched the locomotive industry closely, knowing that changes in its health would quickly be mirrored across the nation.[16]

The locomotive builders never complained directly about the volume fluctuations in their business. Complaint could only alienate customers and was unlikely to change their buying habits to more regular purchasing. At least feasts followed the famines. But other authorities noted the costs of such irregular buying to all parties involved. In 1851 Matthias Baldwin's old Philadelphia friend, Henry R. Campbell, was working on the Vermont Central, and he described the carrier's pressing need for new engines. But he also noted that "it is rather too much to suppose that a railroad direction [i.e., management] will order power in time. They have never been guilty of such a discrete act. The fact is that true economy in these matters is not very well understood."[17] The interval of fifty years saw little change in motive power purchasing, and in 1905 the *Railroad Gazette* noted the economies Campbell had in mind: "The railroads themselves are largely responsible for this constant unsettled state of the market and they are the heaviest losers by it. . . . When business promises to revive . . . nearly every railroad holds off [car and locomotive purchases] until the last moment and then gives large orders with the demand that they be filled in time to meet the needs of a heavy increase in traffic."[18] As the *Gazette* went on to note, bunched-up demand caused higher prices for equipment and delays in its delivery from builders who only months earlier had been taking the few proffered contracts at low bids in a struggle merely to keep their doors open. This dynamic remains endemic to capital equipment markets.

The roller-coaster business cycles of the last half of the nineteenth century greatly exacerbated the accelerator effect. Depressions shut down many locomotive builders temporarily while pushing others into bankruptcy. Among the leading firms in the industry, Wilmarth and Hinkley failed after the mid-

1850s depression, the Panic of 1873 took down Grant, and Rhode Island fell into bankruptcy in 1896.[19] Unlike most industrial firms of the period, many locomotive builders—although not Baldwin—had incorporated by the 1870s. While it is unclear whether their primary motivation in incorporating was to raise capital or to protect it through limited liability, the latter was a substantial advantage given the acute threat of business failure in so volatile an industry.[20]

Baldwin and other builders responded to depressed markets with policies very different from those in high-volume manufacturing. Bulk producers of such goods as textiles, steel, and rails generally tried to bull through recessions by lowering prices, in an effort to maintain sales volume and generate sufficient income to meet their high fixed costs.[21] In dull periods American System manufacturers could at least produce for inventory. In contrast, the locomotive builders had to curtail production when their orders dried up. They laid off workers and cut pay rates while lowering engine prices in an effort to garner at least some new work. For example, after the Panic of 1873 Rogers discharged one thousand men, two-thirds of its total force. Grant laid off 700 hands out of 850, a dramatic step, yet it nonetheless fell into bankruptcy in March 1875. Like Rogers, Baldwin survived the panic, but only with dramatic cost cuts. In September 1873, before the depression began, Baldwin employed twenty-eight hundred men full time, with an average weekly wage of $14.57. Eight months later only fourteen hundred hands had jobs, they were working three-quarter time, and average pay per man was $9.47 a week.[22] Shortly thereafter the firm cut employment, pay rates, and hours even further. If possible, the locomotive makers soldiered on through these dry spells, hoping to maintain the nucleus of their organizations and a core of skilled men in anticipation of an eventual upturn.

After every depressed period in their business during the nineteenth century, the builders saw the market recover and reach new heights in demand before falling back again. These alternating booms and busts greatly affected the nature of competition in the industry and the terms of trade between builders and buyers. During recessions builders sought any business that surfaced, and buyers could dictate terms. At such times locomotive companies competed largely on the basis of engine prices and credit terms. With strong demand, power shifted from buyers to builders as general overcapacity in the industry gave way to full order books.[23] During booms builders could set their own prices within reason, and they spurned credit sales. They still competed among themselves, but now the contest narrowed to providing early delivery dates for these made-to-order products.[24]

As was noted chapter 1, these shifting terms of trade were evident when the overcompetition and credit sales of the 1850s gave way to cash and cost-plus terms during the Civil War. In the late 1860s and early 1870s the balance of power shifted to the railroads' advantage, as productive capacity in the locomotive industry grew faster than demand for engines.[25] Following the Panic of 1873 the industry was almost prostrate.[26] Then demand picked up, and the builders regained the upper hand at last. The trade press noted this change in 1881: "All the [locomotive] works in the country are over crowded with orders and are able to make their own terms. Prices are nearly one hundred per cent higher than they were three or four years ago. . . . Now the manufacturer gets cash on delivery, while in 1872 and 1873, he was glad often to take the bonds of the railway companies."[27] Business was so strong that one carrier seeking twelve engines in March 1881 could secure nothing earlier than a December 1882 delivery date.[28]

Under classical economic theory, such a situation should have caused startups in the industry, and indeed a number of new ventures were under consideration in 1881, particularly in the Midwest.[29] None became a reality, however, for the mercurial market for engines soon defeated such plans. The boom of 1881 gave way to the bust of 1884, when the industry's output fell 44 percent compared with a year earlier.[30] By this time locomotive building required substantial capital and stoical proprietors with the willingness to tolerate the market's instabilities. This was not an industry for new entrants seeking an easy and regular return on their investment. As matters stood in the 1880s, such long-established firms as Rogers and Grant found it difficult enough to compete with the giant that had grown in Philadelphia.

Beyond its effect on competition, the demand cycle for engines also had a major impact on the character of marketing in the locomotive industry. All the major firms advertised regularly in the railroad trade press, and special builders' exhibits became a staple of every international exhibition from the Centennial Exposition of 1876 to the New York World's Fair of 1939. Baldwin's marketing tactics included circulating lithographs of engines in the early 1850s and photographs after 1860. During the Gilded Age the company issued detailed illustrated catalogs, a practice it continued well into the next century.

Despite such efforts, locomotive builders had little ability to influence the volume of demand.[31] Mainline locomotives were costly items in the nineteenth-century economy, with prices ranging from $6,000 to $40,000, and demand for new motive power was highly elastic.[32] During sales slumps the builders attempted to elicit orders by cutting prices. But American railroads, confronting weak markets themselves, almost invariably deferred such purchases until traffic growth made the acquisition of new power imperative. No amount of advertising or salesmanship could counter this ineluctable reality, still true of the locomotive market today.[33] Although marketing per se was of limited import, the evolving structure of markets for locomotives presented perils to

Baldwin frequently displayed its products at national and international expositions. In 1903 the firm exhibited this 2-6-0 locomotive at the Osaka Exposition in Japan. Japanese carriers had been customers for decades, and this engine was destined for the 3-foot 6 inch-gauge Sanyo Railway. The exhibit included displays by Baldwin's subsidiary, the Standard Steel Works, makers of steel wheels and tires. SMITHSONIAN INSTITUTION

some builders, while others discovered avenues to growth. Here Baldwin led the industry.

BALDWIN'S GROWTH TO DOMINANCE, 1850–1880

Baldwin's growth arose from a range of related policies and advantages; no single factor was decisive. But over the long run much of the firm's expansion arose from its accurate assessments of changing opportunities in the markets for engines. In the 1860s and 1870s the company pioneered in a number of fields for locomotives, outside the American mainline railway market, and it garnered an increasing share of demand. In developing complementary product lines to achieve growth, Baldwin again diverged from the practice of American System consumer product manufacturers like Singer Sewing Machine and Remington Arms. While they sought intensive demand and economies of scale, Baldwin looked to extensive and varied markets. This strategy provided two benefits. In depressions the firm could draw business from a variety of sources — including

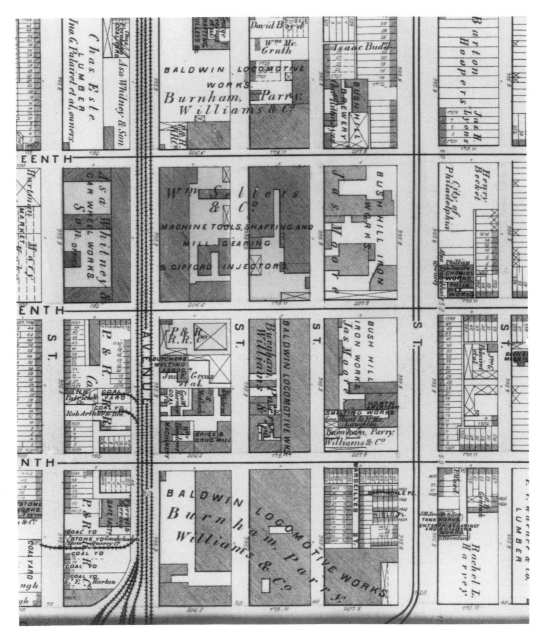

This 1875 map 9f that portion of Bush Hill surrounding the Baldwin factory shows how the firm derived substantial advantages from its location in this thriving industrial district. This twenty-block area included the Sellers machine tool company, a file works, Whitney's car wheel factory, the Bush Hill Iron Works, local trolley lines, coal and lumber yards, the Philadelphia and Reading's connection to the national railway system, and scores of houses for industrial workers. Baldwin drew strength from them all. HAGLEY MUSEUM AND LIBRARY

industrial, mass-transit, and export customers—rather than depending solely upon the mercurial purchasing of American mainline railways. During booms U.S. mainline carriers provided Baldwin's mainstay, but other product lines also helped fuel its growth and amortize overhead expenses. This strategy for growth depended on establishing flexible design and production techniques unknown in high-volume manufacturing.

The company's expansion circa 1870 built upon internal strengths dating to the 1850s. During that decade Baldwin dropped its non-locomotive product lines and began directing its energies to improvements in engine building.[34] Concurrently, Baldwin laid foundations for future growth by reinvesting profits in new facilities and equipment. By 1860 it apparently led the industry in capitalization. During the war decade the company established new production controls, achieved parts interchangeability within its system of batch production, and developed an embryonic staff of middle managers. These men followed Matthias Baldwin and Matthew Baird as partners, and they brought new ideas and capabilities with their rise to ownership.

Aside from these internal developments, the firm derived a range of competitive advantages from its location in Philadelphia, the nation's second-ranking industrial center during the nineteenth century, just behind New York City.[35] While other locomotive builders had also grown up in urban industrial districts, such as Boston, Paterson, and Pittsburgh, none of these cities offered the range of advantages Philadelphia provided to Baldwin.[36] The city possessed a remarkably broad industrial base, with nationally renowned products ranging from hats to iron ships. In particular, by 1860 the city had become a vast industrial district of skilled metalworking and a vital center of batch production. During the Gilded Age a score of factories in the Bush Hill neighborhood of Baldwin's plant turned out a range of complementary products; these included such firms as William Sellers and Dement and Dougherty (machine tools), Hoopes and Townsend (nuts and bolts), and A. Whitney & Sons (railway wheels). Elsewhere in the city companies like Midvale Steel, Cramp's Shipyard, J. G. Brill (trolleys and streetcars), and the Southwark Foundry and Machine Company contributed to Philadelphia's industrial strength. Baldwin derived many benefits from this constellation of nearby companies, purchasing their materials or tooling, and paying comparatively low freight charges on these bulky commodities.[37] The locomotive builder also drew at will from Philadelphia's reservoir of skilled workers, whose talents provided the chief foundation for the city's diverse metalworking sector. Given its great fluctuations in output and employment, Baldwin depended heavily on this pool of available labor possessing the range of skills required in locomotive building.

Philadelphia's status as an important rail center also provided Baldwin with noteworthy advantages over its rivals. In 1842 the Philadelphia and Reading

In the 1870s the Baldwin partners invested in the early construction of the Northern Pacific
—a transcontinental running from Chicago to the Pacific Northwest—cementing a close
relationship between the two companies that lasted for many decades. Here is an 1886
Baldwin engine built for the carrier, a 2-10-0 for heavy freight service in mountainous
territory. The Northern Pacific paid $13,225 in cash for this novel design. RAILROAD
MUSEUM OF PENNSYLVANIA (PMHC)

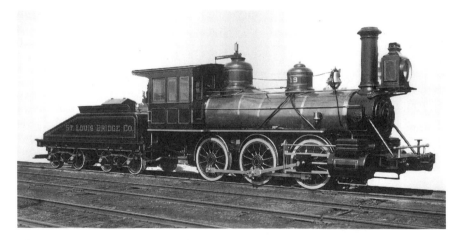

This 1884 engine illustrates some of the advantages Baldwin derived from its ties to the
"Philadelphia interests." The group played a leading role in constructing the Eads Bridge
over the Mississippi at St. Louis in the early 1870s. The St. Louis Bridge Company hauled
trains across the great steel-arch bridge, then through a downtown tunnel into St. Louis
Union Station. Built with Carnegie steel and financed by J. S. and J. P. Morgan, the
structure gave J. Edgar Thomson's Pennsylvania Railroad a connection to the West. Specially
adapted Baldwin products hauled trains over the bridge. This locomotive had a slant-back
tender to improve the backward visibility useful in frequent switching moves. Tunnel
operation required that it burn smokeless coke fuel. RAILROAD MUSEUM OF
PENNSYLVANIA (PMHC)

opened its line from the Pennsylvania coal fields to Philadelphia. Anthracite coal fueled the Industrial Revolution, as factories like Baldwin's burned the black diamonds under power boilers and used their concentrated heat for iron molding, boilermaking, and forging.[38] When the Philadelphia and Columbia line through Bush Hill came under the P&R's control, Baldwin gained an economical and advantageous supply route for this most essential commodity. The P&R also became a frequent buyer of Baldwin engines, purchasing them almost exclusively by the 1880s.

The nation's leading railroad, the Pennsylvania, had its headquarters in Philadelphia. The PRR gave Baldwin a crucial boost during the 1860s, and the two companies remained close allies for over a century. Baldwin cemented its relationship with the Pennsylvania quite early, even before its construction began. The carrier's chief engineer and later president, J. Edgar Thomson, had been one of Baldwin's "very best friends" since the 1840s.[39] During Thomson's presidency (1852–74) the Pennsylvania grew to become the largest carrier in the nation, with correspondingly large motive power needs. For the most part Thomson's friend, Matthias Baldwin, built these engines, greatly spurring his firm's growth.[40]

Philadelphia's strengths in finance and banking also aided Baldwin's expansion. As a partnership, Baldwin necessarily financed its substantial capital growth during the nineteenth century by reinvesting profits.[41] Given the frequency of credit sales in the locomotive industry, however, a builder's profitability often depended in turn upon accurate commercial information and ready agencies for discounting the commercial paper the carriers preferred to use in purchasing engines. Although New York was the nation's banking center, such Philadelphia firms as Jay Cooke & Co. and Drexel & Co. were also leading railway financiers. The evidence is scanty, but Baldwin seems to have enjoyed close ties with these bankers, who could provide information on the financial condition of carriers across the country.[42]

Bankers like Cooke, Drexel, and Anthony Drexel's New York ally, J. P. Morgan, became actively involved in promoting railway development in addition to their work as financial intermediaries for the carriers.[43] Here too Baldwin benefited from its ties to the bankers. For example, Cooke spearheaded construction of the Northern Pacific in the late 1860s, and Baldwin was involved in the line's development from the start.[44] Unlike Cooke, the Drexel and Morgan firms generally took a direct role in railway management only when carriers fell on hard times, imperiling bondholders' investments. On such occasions, however, the bankers had the ability to place equipment orders with favored firms. Thus the Morgans, acting as trustees for the bankrupt Cairo and Vincennes Railroad, purchased an engine named *J. S. Morgan* from Baldwin in 1875. When Drexel, Morgan & Co. intervened in the management of the failed Long Island

Railroad in 1877, that carrier ordered three Baldwin engines, although it had favored other builders in the past.[45] After J. P. Morgan began directing the bankrupt Philadelphia and Reading in 1886, that carrier stopped building its own motive power for many years and turned to Baldwin for new engines.[46] Other locomotive builders had their own ties to the banking community on which the railways depended, but Baldwin seems to have enjoyed uniquely close relations.

Baldwin's partners gained their entree with bankers and financiers in part through their own involvement in railway promotion, particularly in new western lines in the 1860s and 1870s. Matthias Baldwin and his successor, Matthew Baird, were members of a loosely organized group of wealthy investors known as the "Philadelphia interests." Led by J. Edgar Thomson, the group included Baldwin, Baird, PRR vice president Thomas Scott, Andrew Carnegie, Jay Cooke, Anthony Drexel, and Drexel's partner, J. P. Morgan.[47] In the 1860s these men financed a line, later known as the Kansas Pacific, that competed with the Union Pacific in an effort to become the Great Plains link in the nation's first transcontinental railway. Although that venture failed and the road was later absorbed into the UP, Baldwin sold over fifty engines to the carrier in 1870 alone. Other lines underwritten by the Philadelphia interests had financial troubles in the highly competitive environment of western railroading after 1873, yet they also served to boost Baldwin's sales.[48]

Another project involving the Philadelphia interests gave Baldwin the means to improve its product quality while lowering the cost of a crucial raw material: steel. Through the 1860s locomotive builders used steel only rarely, despite its superiority over wrought iron, largely because of its high cost. Once the Bessemer process had been developed in England in 1856, an age of cheap steel beckoned. Bringing that age to America required venture capital, however, which Baldwin helped supply. In 1863–64 the company evidently backed Alexander Holley's early work in Bessemer steel at Troy, New York.[49] In the same year Baldwin joined the Philadelphia interests in investing in the first large-scale Bessemer plant in America, the Pennsylvania Steel Company at Harrisburg. The locomotive builder acquired its own steelmaking subsidiary, the Standard Steel Works, circa 1873.[50] Baldwin ventured into steel to relieve its dependence on expensive, high-tariff English imports and to secure a quality substitute for wrought iron in making locomotive tires, wrist pins, and fireboxes. The firm's technical leadership here lowered its costs and risks while improving engine quality and marketability.

Consolidating all of these advantages, by 1870 Baldwin had become a powerhouse in locomotive sales and production. Annual output surpassed two hundred units for the first time in 1869. Four years later, 437 engines rolled from

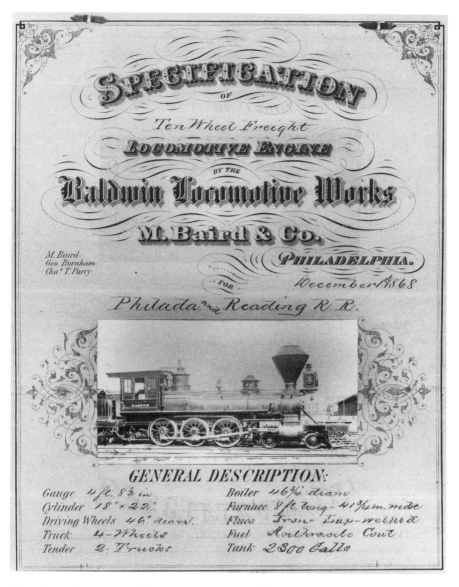

Cover sheet of an 1868 specification accompanying Baldwin's bid to build ten engines for the Philadelphia and Reading for $14,300 each. While proposals served as the basic financial contract, the accompanying specifications laid out the technical details of the bid or order. In this case the specification included only one other page, since Baldwin was bidding to build to the railroad's drawings. By 1900 specifications reached six highly detailed pages.

HAGLEY MUSEUM AND LIBRARY

The narrow-gauge *Ptarmigan* of 1880 looked like a standard American type, but the engine ran on 3-foot-gauge track, rather than the standard of 4 feet 8½ inches. Consequently the entire design was scaled down from normal mainline practice. This engine had 12- by 18-inch cylinders, weighed 42,000 pounds, and could haul 750 long tons on level track. Baldwin's heavy standard-gauge American types of the era had 18- by 24-inch cylinders, weighed 74,000 pounds, and hauled up to 1,400 long tons. While narrow-gauge lines cost less to build, their relatively low operating efficiencies doomed the concept. RAILROAD MUSEUM OF PENNSYLVANIA (PMHC)

the erecting shop. By this point Baldwin had developed its flexible production techniques, and it produced quite varied engines for a number of different markets. Each of these demand segments required unique marketing tactics.

As in the past, U.S. mainline railroads took the lion's share of Baldwin's output. In the period from 1865 to 1873 American railroading boomed ahead as track mileage doubled. Baldwin needed no marketing intermediaries, such as distributors or salesmen, to close sales in the mainline market. While there were hundreds of American railways during the Gilded Age, the locomotive builders were limited in numbers and easily identified. Before 1900 partners Matthew Baird, Charles Parry, Edward Williams, William Austin, and Samuel Vauclain provided a sufficient sales force, and they frequently traveled to customers to drum up business, to discuss design issues, and to close deals. For a given sale, the partners dealt with the railway president on financial matters, while technical issues were decided in conferences with the master mechanic or chief mechanical officer. Particularly when dealing with large carriers, the partners were prepared to negotiate over design, pricing, credit, and delivery aspects of a sale. Once a buyer had defined its needs through actual plans or specifications or informal discussions, Baldwin proffered its bid in a formal proposal accompanied by a detailed set of technical specifications.[51] If accepted, such proposals served as the basic contractual agreement, specifying price, delivery dates, and performance guarantees. Many Gilded Age carriers typically solicited proposals from a number of builders, but the Baldwin company's strong capital base provided a real marketing advantage, particularly when demand was weak, allowing the firm to extend credit to purchasers.[52] The mainline market presented an additional demand for builders in the postwar years as railway master mechanics exerted increasing sway over locomotive design. In response, Baldwin adapted its design and production capacities and announced its willingness "to modify or alter [designs] to suit the views of purchasers."[53] To meet the varying needs of less technically venturesome carriers, the company also broadened its standard product line to include forty classes of engines by 1871.[54]

Baldwin developed such productive flexibility primarily to serve the American mainline market, but this capacity also allowed the firm to seek out other market niches. In 1868 the company built its first narrow-gauge engine, and during the narrow-gauge fad of the 1870s it took roughly 45 percent of this business.[55] In another new venture, in 1870 Baldwin helped develop a market for specially designed small engines for use inside coal mines.[56] During the decade the firm also led the industry in creating a variety of special product lines: industrial engines for steelyards and other manufacturing applications, steam streetcars, and locomotives for elevated mass-transit lines. To develop

Before the coming of underground electric subways, cities like New York and Chicago provided for mass-transit needs with elevated steam railways built on steel supports over busy streets. The service required unique engine designs, small yet powerful, like New York Elevated's road number 24. Baldwin received this order, one of a batch of four, on July 22, 1878. The builder turned this little engine out in seven weeks. It sold for $2,750.
SMITHSONIAN INSTITUTION

This ad ran in Russian periodicals circa 1880. It touted coal-burning locomotives in particular and underscored Baldwin's ability to deliver engines "to all ports in the Russian empire on board ship." SMITHSONIAN INSTITUTION

such markets, Baldwin amassed comparative data, for example showing the cost savings of mine and industrial engines versus horse drayage; it produced catalogs showing standard types of these special-purpose models; and it advertised in specialist trade periodicals. While these markets were comparatively small, they offered noteworthy advantages.[57] The company lessened its dependence on U.S. mainline customers with these new product lines, which also provided more work to amortize tooling and other fixed costs. Most of these engines were Baldwin-standard designs, made relatively cheaply and efficiently with standard components.

The partners also worked concertedly to develop the export market after 1870, especially cultivating such sales during the Panic of 1873 and subsequent recessions. Here they competed primarily against British builders like Neilson, Dubs, Beyer Peacock, and North British Locomotive. Britain's colonial empire greatly aided its locomotive builders; Indian locomotives remained almost exclusively British for a century. British shops had to cultivate exports to their

colonies and elsewhere, since their home railways generally made their own
motive power. But the Americans made inroads abroad by offering pricing,
delivery, or technical advantages. During the Gilded Age Baldwin exported
primarily to Central and South America, Australia–New Zealand, Japan, and
Russia. Here American builders had a number of advantages. These countries
were in the midst of railway booms, and their terrain and rail lines often demanded U.S. locomotive design characteristics rather than British designs. With
this trade, traditional marketing techniques helped link Baldwin to potential
customers abroad. The firm advertised in foreign railway publications and
subscribed to London, Paris, and St. Petersburg newspaper clipping services.[58]
Commission sales agents in Havana, Rio de Janiero, Melbourne, and Yokohama
worked to secure contracts, and by the 1890s the company had its own London
sales office.[59] During domestic depressions or when notably large foreign orders
beckoned, partners Charles Parry and Edward Williams traveled abroad to
secure business.

This foreign trade helped Baldwin particularly by tempering the cyclical
downturns in the domestic locomotive market. As chart 2.1 suggests, Gilded
Age business cycles had international reach, with the U.S. market generally
declining concurrently with export sales. But in two instances, 1883–84 and
1890–91, Baldwin benefited from timing lags that allowed increased foreign sales
during declines in the home market. Overall, the chart suggests that the builder
used export production to fill excess capacity. Output for foreign and American
customers tended to decline concurrently, but even a declining export market
helped Baldwin survive in years of low domestic demand. In recessions exports

CHART 2.1 Baldwin's Output for Domestic and Foreign Customers, 1870–1900

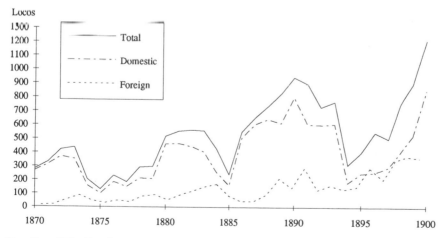

Data from "The Development of the American Locomotive," *Journal of the Franklin
Institute,* pp. 266, 269.

English engineers laid out the railways in Egypt, ensuring that those carriers turned first to British locomotive builders. But John Bull did not have a lock on the market, as this engine demonstrates. Shown new in the Baldwin erecting shop is construction number 15,857, a 2-6-0 type, built in April 1898 for the Egyptian State Railways. Part of an order of twenty, the engine sold for $9,000. While the locomotive was close to normal American practice, its tender betrayed strong British design traditions. AUTHOR'S COLLECTION

amounted to between 29 percent (1874) and 53 percent (1896) of total sales. During production booms this figure stood at roughly 20 percent. At times of domestic downturns Baldwin could offer foreign customers faster delivery and lower prices, since its labor and raw materials costs were also depressed. Finally, Baldwin simply sought foreign business more avidly during U.S. recessions. For example, in 1885 the company circulated broadsides, printed in Spanish, announcing price cuts in railway and sugar plantation locomotives. Aimed at the Caribbean and Central American markets, the flyers claimed that "these reductions bring the prices of all kinds of locomotives to a point which has never before been reached in the history of locomotive construction."[60] Such efforts bore fruit; in 1885 Baldwin shipped 34 percent of total output abroad.

Baldwin's foreign sales put the firm in direct competition with British builders, the international leader in locomotive exports. Initially the latter enjoyed a virtual monopoly in South America, Africa, Asia, and Australia, since British capital and engineers had built most of the railways in these territories, and many operated under British management. Baldwin succeeded in capturing some of Britain's trade by offering lower prices and faster deliveries, particularly during U.S. trade depressions.[61] In an 1899 sale of twenty locomotives to Burma, the Philadelphia company successfully bid at £1,750 apiece and six months to deliver, while a Glasgow builder wanted £2,000 and eighteen months. Baldwin's

success in such cases derived partly from particular factors that handicapped
British builders and partly from the intangible issue of entrepreneurial failure
in the late-nineteenth-century English economy.[62] Although British companies
tried to prevent American encroachment, the value of locomotive exports
from the United Kingdom declined from $9 million in 1890 to $7.3 million in
1900. Concurrently, U.S. builders' foreign sales grew from $1.3 million to $5.6
million.[63] Most of these sales went to Baldwin, the American locomotive export
leader throughout the Gilded Age.[64]

Baldwin developed the industrial, mass-transit, and export markets to
improve cash flow, maintain its core of skilled workers, and amortize an increasingly heavy burden of tooling costs during the depressions of 1874 to 1894. In
this period the partners knew two things with certainty: the U.S. mainline
locomotive market promised long-term growth, but that demand would be
very unstable. Handling each decade's rising peaks required more managers and
workers—and particularly more specialized production machinery—to keep
costs and delivery dates in line and to hold market share. The overall growth in
sales to American mainline carriers from 1865 to 1907 fueled most of Baldwin's
great expansion, but the other specialized niches served as the firm's safety net.

THE STRUCTURE OF THE INDUSTRY IN THE GILDED AGE

By 1870 Baldwin was far and away the leading firm in the locomotive industry,
with twice the output of its nearest competitor, the Rogers Locomotive Works.
The firm had surpassed its rivals by developing a range of markets, a tactic
based in turn upon its adept meshing of productive capacity with market needs.
By being the first to consolidate these policies and capacities, Baldwin redefined
the industry and regained substantial first-mover advantages. During the 1870s
the Philadelphia firm heightened its dominance, pushing its market share ahead
from one-quarter (1860s) to one-third (1870s) of all locomotive production in
America.[65] The company's ventures into new markets had created an industrial
giant; only the largest integrated steel mills employed more men.[66] Baldwin's
size gave it a strength possessed by no other builder, best seen in the industry's
struggles with the depressed market of the mid-1870s. While other builders
shut down entirely or fell into bankruptcy, Baldwin made engines and profits
throughout those lean years.[67]

Despite its dominance throughout the Gilded Age, Baldwin was unable,
and probably unwilling, to drive its remaining competitors out of business.
Until a 1901 merger of many competitors, Baldwin remained in the leading
role, followed by approximately ten much smaller firms making mainline
locomotives, with a cluster of other companies producing special-purpose
engines. The smaller firms survived through special geographical or technical
niches. For example, all builders could expect some loyalty from the railroad

By 1871 Baldwin's plant had grown to substantial size, occupying most of three blocks facing Broad Street, with additional buildings on three other blocks. The center structure shown here was the original factory; to its right is the erecting shop, built in 1869. HAGLEY MUSEUM AND LIBRARY

companies servicing their plants. In return for its freight business—both raw materials in and finished engines out—a Brooks or Schenectady received orders for power. Special product niches also supported a number of firms. Climax, Heisler, and Lima built engines for logging companies, while Vulcan and Porter enjoyed considerable success in light industrial locomotives. Baldwin competed with these builders but lacked the benefit of certain patents and specialization.

Builders of mainline road engines, such as Cooke, Grant, Manchester, and Richmond, used a number of techniques to survive and even prosper at times. All sought export work to cushion recessions, although Baldwin had secured the bulk of this business. To ease periods of soft locomotive demand, Manchester developed a profitable steam fire-engine business, Richmond built the engines for a U.S. Navy battleship, while Cooke went into bridge building in 1878.[68] In the driest spells the smaller builders simply shut down until orders returned.[69] During periods of high demand, with railroads all crowding into the market, competition among builders centered more on early delivery dates than on price. Even small firms could garner a share of this business. For example, when the Santa Fe wanted quick delivery of a hundred new custom engines in 1881, no single builder could handle such a large order in the midst of an already booming market. Therefore the carrier split its purchasing among Baldwin (40 units), Hinkley (35), Manchester (15), and Taunton (10).[70]

Such factors let Baldwin's smaller rivals survive. A steady third of the market was about all the industry leader could comfortably take on. To raise that share would have required much more capital investment to assure reasonable delivery dates in periods of high demand. This investment would have been difficult to service when the inevitable recession struck.[71] The Baldwin partners evidently understood that further expansion would not increase economies of scale. Manufacturers improved throughput, productivity, and market share by mechanizing production, but this strategy was not viable for most capital equipment builders.[72] Their growth came by filling a variety of market needs rather than exploiting a select few in depth. During strong demand, Baldwin and other builders' order books filled up, and projected delivery dates stretched out eighteen to twenty-four months. When this occurred in 1902, Baldwin was able to sell advance reservations in its books to the Baltimore and Ohio for $1,000 an engine.[73] Why overexpand when the company could make money from its lack of capacity? Given the highly unstable demand for their products, locomotive builders rightly perceived excess capacity as too great a threat to profitability to risk expanding facilities in anticipation of a market upturn. During the Gilded Age Baldwin and its competitors also tried repeatedly to find collective solutions to the problems of low demand, overcapacity, and poor profits during weak markets.

COMPETITION AND COLLUSION

Business historians have extensively researched industrial competition in the late nineteenth century, looking for the causes of the widespread merger movement of the 1890s.[74] Most of this analysis has centered on mass-production industries and sheds little light on competitive dynamics in the capital equipment sector. But the locomotive industry's response to fluctuations in demand has some parallels to the portrait by Naomi Lamoreaux of another specialty product industry, the makers of fine writing-paper. She describes how in periods of slack demand their "output declined in accordance with the drop in demand; prices fell, but not excessively . . . and the burdens of curtailment were spread throughout the industry."[75] Like these specialty manufacturers, locomotive builders sought to manage periods of narrow profit margins by colluding in an attempt to fix prices.[76] To modern eyes such actions seem unsavory at best, probably immoral, and deservedly illegal, but the practice was widespread and legal in the nineteenth century. The characteristics of the locomotive industry almost demanded collusion—a limited number of producers with high capital burdens in an industry regularly confronting periods of depressed sales and overcapacity. Collusion was simply another tactic to minimize the substantial risks of this sector. While the builders' attempts to fix prices failed to endure for long, they recurred frequently and deserve discussion.

The first known collusive effort in the industry dated to 1857, when twelve builders met in New York City.[77] At that time demand for engines was quite strong, but the builders wanted to establish an industrywide policy limiting credit sales. They may have also considered fixing prices, although the record is unclear.[78] Given that they soon separately reneged on the credit sales agreement, evidently an accord on prices, if it existed, also quickly came to naught. In 1862 the Rogers Locomotive Works tried again, calling for "a meeting of the principal Locomotive Builders of the country," noting that "we have all suffered more or less by many credits, competition in prices, and want of harmony in action."[79] The meeting took place, but little seems to have come of it. The builders tried again late in 1868; again they were ineffectual.[80] All of these efforts probably failed for the same reasons that defeated similar attempts by high-volume manufacturers in the nineteenth century. While agreements to fix prices were not illegal per se, such combinations in restraint of trade also had no legal force. If one signatory undermined the accord, attempting to secure more business by cutting prices, the other parties had little choice but to follow with their own price cuts—to the detriment of all.

Locomotive price fixing soon resurfaced. In 1872 the Locomotive Builders Association was created, with Baldwin, Brooks, Hinkley, and Cooke as members.[81] The LBA had officers, bylaws, and a constitution that spoke loftily about improving locomotive technology. It said nothing explicitly about prices. In acknowledgment of Baldwin's market power, voting in the LBA was based on plant capacity, and Baldwin partner Charles T. Parry presided over the association.[82]

The association's existence was well known in railroading circles, with its constitution appearing in the *Railroad Gazette*. Unlike the past efforts, the LBA seems to have succeeded in at least one of its goals. The members passed a resolution, later printed in the trade press, requiring that all engine orders be placed directly by railroads rather than by commission merchants ordering on their behalf. Railways had frequently used these middlemen to bid down prices among competing builders, with the merchants taking a percentage fee for the service.[83] The carriers could not have been pleased by the LBA's collusion, but they could hardly cry foul. They too had been meeting among themselves since the mid-1850s to set freight rates.

The builders' main purpose in the association, setting standard prices, succeeded only briefly, if at all. This attempt failed because of the custom nature of the products involved. Since locomotive specifications varied from road to road and order to order, it proved difficult to establish a standard pricing system that dealt adequately with all possible variations in a given size or type of engine. Compounding this problem was the 1873 panic and its harsh impact on the locomotive industry. At mid-decade, when two or three builders refused to collude any longer, the LBA evidently died.[84]

Undaunted, the builders tried again, circa 1891, with the American Locomotive Manufacturers Association (ALMA). Once more they had officers, bylaws, and a constitution. Members included Baldwin, Schenectady, Brooks, Pittsburgh, Rogers, Richmond, Cooke, Dickson, Porter, and Rhode Island.[85] This time the price fixers succeeded, since they met the technical-diversity issue head-on. Lists of standard prices survive and are exhaustive in their detail, with every locomotive type from 0-4-0 switchers to 4-8-0 mainline freight engines given a set price. Within each type, mandated prices also accounted for variations in size, cylinders, track gauge, and extra equipment. The pricing system even covered freight delivery charges for engines, and the lists applied to all sales in the United States, Canada, and Mexico. Notably, the ALMA accords did not extend to exports beyond North America, a Baldwin mainstay. In a further effort to promote stability and prevent secret undercutting, the association established output quotas for all members.[86] Those firms producing more than their quota allowance paid a penalty to the underachievers.

Evidence of the ALMA's success is inferential but persuasive. Surviving price lists cover the period from July 1891 to February 1897.[87] The successive lists indicate that the association continued right through a severe depression in the middle of the decade, changing prices in response to fluctuations in demand and in the costs of raw materials and labor. Prices for smaller engines generally held steady, while most larger types increased in price. The fact that successive lists were issued at all (i.e., that they were necessary) implies that no member broke ranks and that ALMA succeeded in holding prices up and advancing them whenever possible.[88] A comparison of the ALMA price lists with the selling prices for engines listed in Baldwin's order books also shows that the association succeeded in putting a floor under prices. Other evidence, however, suggests that collusion hardly resulted in exorbitant profits.[89]

This description of collusion and cooperation may seem to clash with the earlier emphasis on fierce competition in locomotive building. But the two portraits can be reconciled in that each was a response to the great gyrations in demand for engines. Some degree of competition always ruled the industry, in terms of quality and delivery dates, even during collusive periods. And actual price fixing occurred in no more than ten years of the industry's seventy-year existence in the nineteenth century. Most of the collusive efforts began at points of rising demand for engines, not during depressions or peak markets. Although it seems implausible that builders would collude during a period of improving demand, they had good reasons. First, they found it easier to act jointly at such points than in times of peak demand or steep depression. Also, as sales improved, the locomotive builders were often caught in a particular bind, described here during the 1880 revival from the late 1870s depression:

> As to profits in building, complaints are made that thus far they are not satisfactory, a fact mainly due to the increased cost of materials and labor since the period of greatest activity commenced. At the beginning all were greedy for orders, competition was sharp, and contracts were taken very low. At a later date the iron [price] boom brought new complexities. The trade therefore is not considered prosperous as regards net profits realized, but prospects are very hopeful.[90]

Although there seems to have been no collusion in 1880, this quotation well illustrates why narrow profits during other periods of improving demand drove the locomotive makers to combine. For any given engine, materials accounted for over half of a builder's final costs. Since the maker could purchase most materials only after its final bid was accepted, even modest increases in their prices could easily turn its contract bid into a financial loss. Anecdotal accounts confirm this bind. For example, in the flush demand around 1870, engine prices ranged from $6,000 to $12,000, but Baldwin's profit per unit ranged down to $245.[91]

Just what constituted a reasonable profit was an eye-of-the-beholder sort of question. Baldwin clearly enjoyed adequate to good profits over the long run, given its great expansion and its ventures in railroad promotion and steelmaking. But very few records survive to quantify its success. Its ledgers over the period 1853 to 1866 show a wide range in annual profit from approximately $113,000 in an off year (1858) to $1.1 million in 1866.[92] Figures from the manuscript schedules of the U.S. Manufacturing Census give an indication of profitability across the industry, although many of the data are problematic (for a discussion of their liabilities, see appendix B). At the time of the 1870 and 1880 enumerations, locomotive demand was strong and improving. But the 1870 census shows only seven firms as profitable, Baldwin and Rogers operating close to break-even points, and Schenectady running at a loss. In 1880 the situation was even worse, with only three firms—Baldwin, Cooke, and Manchester—making good profits, four at marginal levels, and Brooks and Rogers in the red.[93] Given such poor returns at a time of relatively strong demand, the builders had some justification in colluding to boost profits.

Proprietors' incomes and firms' survival depended upon a measure of long-term profitability. Capital equipment builders typically sought to build up cash surpluses during booms to tide them over when the next recession struck. Profits also provided the most important source of capital for investment in new facilities. In turn, the locomotive builders found that improving plant and equipment could be as difficult as it was important. Except during demand peaks, the industry frequently suffered from overcapacity. But locomotives steadily increased in size and weight throughout the nineteenth century, with average mainline engines growing from roughly 25 tons in 1860 to 72 tons in 1910.[94] Because of such evolution in their products, builders in this industry of recurring overcapacity had to reinvest continuously simply to survive, let alone grow. The funding required for capital improvements became increasingly burdensome in the 1880s and 1890s, as locomotives grew in size, weight, and complexity. The builders required larger shops, more and bigger machine tools, and heavier cranes. As the largest firm in the industry, Baldwin generally led its competitors in plant expansions. Some of the other builders also found the money for investment. Those firms that did not—notably Taunton, Mason, and Hinkley—soon left the business.

Wide fluctuations in demand, recurring years of overcapacity, and elevated capital requirements created formidable challenges for locomotive builders from 1880 to 1900. Those companies that surmounted these problems found sales aplenty as railway demand for motive power continued its long-term climb. The number of engines in service in America doubled between 1880 and 1900.[95] In the 1880s Baldwin held 31 percent of the market, and it grew to take a 39 percent slice in the following decade.[96] Baldwin's size and its first-mover advantages seem to have made the firm unassailable.

THE INDUSTRY AT ITS HEIGHT, 1890–1907

Despite Baldwin's successes, the 1890s proved a difficult decade for all locomotive builders. With a prolonged recession following the 1893 panic, the industry's total output for the decade declined slightly compared with the 1880s. This unsettled situation gave Baldwin and its smaller competitors an extra incentive to cooperate under the ALMA rather than face the mercurial market individually. When the ALMA collusion ceased is unclear; the last known price list is dated February 1897. But its success showed many builders the value of concerted action. With little conceivable chance of improving their lots separately, the largest of Baldwin's competitors decided that their main chance lay in combination. During the great Progressive Era merger movement, eight of the independents combined in 1901 to create Alco, the American Locomotive Company.[97]

A horizontal combination of locomotive builders had been under consideration since at least 1892.[98] The principals could not agree then on terms for an outright merger, so they resorted to the ALMA instead. Collusion was the next best alternative to combination in response to narrowing profit margins and mounting capital costs.[99] The consolidation movement started again in 1898, when a New England financier and industrialist, Joseph Hoadley, purchased the Rhode Island Locomotive Works.[100] Using a holding company,

the International Power Company, Hoadley then engineered Alco's formation
in 1901. His associates in this venture included a New York investment banker,
Pliny Fiske, and a millionaire speculator, Joseph Leiter, who three years earlier
had attempted to corner the Chicago wheat market.[101]

Alco was capitalized at $50 million, and by combining ten previously inde-
pendent builders, it achieved rough parity with Baldwin's productive capacity.[102]
Here was a potentially formidable competitor, but Baldwin's partners declined
an invitation to join the new combine.[103] Although their reasons are unknown,
the partners may have understood how the nature of their business and markets
would preclude a successful horizontal combination of the major locomotive
builders. Unlike contemporary mergers of continuous-process manufacturers
such as U.S. Steel and Pittsburgh Plate Glass, which achieved substantial
savings through vertical integration, a single great locomotive combine could
not achieve noteworthy economies in production, since demand was so variable
and engines were built in units or small batches to custom specifications.[104]
The Baldwin partners understood the limits of growth in their industry and
may have foreseen that one great locomotive combine would only go the
way of American Bicycle, American Shipbuilding, or any other horizontal
combination based solely on achieving price dominance. Mostly dismal failures,
these combines were unconcerned with, and unable to make, substantial
improvements in productivity. Soon new firms entered their markets and
undercut their artificially high prices.[105] One large locomotive combine would
still confront plunges in demand that would greatly tax its ability to service the
costs of consolidation.[106] Precisely why Baldwin's partners declined to join the
Alco merger is unknown, but these speculations do not seem unwarranted.

After 1901 Baldwin's partners confronted a very different competitive
situation. American locomotive production reached its all-time high in the
first decade of the twentieth century, but the two oligarchs competed rather
vigorously in this expanding market.[107] Their highly competitive bidding
seems surprising, given that the two firms essentially constituted the industry. Most likely the dynamic pace of technical change in locomotives of the
period prevented Baldwin and Alco from reaching an accord on pricing. With
the carriers buying more and larger engines, both builders made substantial
improvements in plant and equipment to meet the crescendo of demand in the
early 1900s. Thanks to these efforts Baldwin maintained its 39 percent market
share of the previous decade while Alco took 46 percent. In its best year ever,
1906, Baldwin's Broad Street factory turned out 2,666 engines—more than 8.5
locomotives each working day. To accomplish this feat the plant ran around the
clock, with seventeen thousand men working the two shifts.

The partners decided, however, that even this performance was insufficient
to meet Alco's competition and the railways' requirements for motive power. In

1906 Baldwin began construction of a new satellite factory in the Philadelphia suburb of Eddystone. With the new plant and its 1909 incorporation, the company increasingly took on the characteristics of a modern, twentieth-century industrial firm. Unbeknownst to Baldwin, these preparations for future growth were immediately undercut by the market it had served for almost eighty years. American railway demand for motive power had peaked in 1905.[108] Thereafter Baldwin and Alco began a long decline that accelerated in the 1920s and continued until the end of steam power. They were not alone in their troubles, as much of the capital equipment sector suffered through the 1920s and the years of national depression that followed.[109]

BALDWIN AND THE LOCOMOTIVE INDUSTRY
IN THE NINETEENTH CENTURY

If Baldwin increasingly entered on hard times during and after the 1920s, until then it had enjoyed a seventy-five-year ascendancy. Examining the roots of that durable success underscores particular attributes of the firm while also delineating key differences between capital equipment builders and high-volume manufacturers. The contrasts between these two formats in turn demarcate broad and important changes in American industry and society that took place between the builders' heyday in the nineteenth century and the manufacturers' supremacy in the twentieth.

The general environment for nineteenth-century capital equipment builders held so many risks and challenges that one marvels at their mere survival, let alone their impressive success in creating the technologies of industrial society. Chief among the perils was the bucking demand cycle for expensive capital goods. Every decade saw national recessions that imperiled all locomotive builders. Despite the variability in demand, Baldwin and its competitors had to accumulate sufficient profits to make the heavy investments in plant and equipment necessary for building such complex mechanisms. Before making engines, let alone profits, the builders also had to sell their products to customers across the nation and around the world. And by the 1870s the locomotive makers had lost the design initiative in their own product lines to their customers, yet they still competed among themselves on a variety of levels.

The industry responded to such challenges with a range of policies that allowed many firms to survive, and some to prosper at times. All locomotive builders sought to yoke themselves and their operations closely to the needs of American railways. When the carriers boomed, the builders at least did all right. And when the railroads deserted the market, locomotive makers learned to act collectively, holding up prices to allow profits sufficient to survive the dry spell. To meet the railways' need for custom locomotives, the builders drew upon the extensive skills of workforces that were the chief assets of such industrial districts as Bush Hill and Paterson. The ready availability of such skills allowed Baldwin in particular to broaden its product lines and markets. The Philadelphia builder survived market reversals and achieved its growth by fulfilling extensive demand, making engines in countless designs for customers in a variety of industries all over the world. Antithetical to the manufacturers' use of standard products to exploit intensive demand, this strategy nonetheless proved viable for almost a century, until America had evolved from an industrializing, producer-driven economy to a mature and consumer-oriented society.

Mass-production manufacturing, with its attendant revolutions in distribution and management, was taken up with fervor by American industry by the 1910s. Thereafter the capital equipment firms and their machinery-building techniques waned in importance. In the 1930s General Motors would even bring rigid standardization and manufacturing methods to the locomotive industry. That it would do so successfully would have astounded Baldwin's nineteenth-century partners. For eighty years their company's growth had been based on the strategy of building whatever the customer required. Baldwin geared its internal operations and marketing tactics entirely to that end. The character of technical change in steam locomotives explains why Baldwin was impelled to develop this capacity to meet its customers' design needs.

3

THE CHARACTER OF INNOVATION
IN LOCOMOTIVE DESIGN

NINETEENTH-CENTURY COMMENTATORS AND MODERN-DAY HISTORIANS HAVE written at length about the determined efforts of railroad managers to standardize the "terminology, measurements, parts, tools, and other equipment" used on their systems during and after the Gilded Age. Alfred Chandler asserts that cooperation "to promote standardization of equipment and operating procedures" was a "central theme" of American railroading during the 1860s and 1870s.[1] The period saw interfirm agreements on such technologies as standard track gauges, car couplers, car wheel patterns, and axle dimensions that unified the hundreds of individual railroads into a national system carrying through freight and passenger traffic across a number of different lines. A chief agency in this standardizing mission was the American Railway Master Mechanics Association. Created in 1868, the Master Mechanics Association devoted most of its consultative efforts to locomotive design, use, and repair.[2] In an 1874 speech to its annual convention, the noted mechanical engineer Coleman Sellers exclaimed that "the primary object of this Association may almost be said to [be to] introduce uniformity in all parts of the great railroad system of the United States."[3]

Despite Sellers's fervor for the standardizing mission of the association, and despite the general truth of Chandler's portrait of systematization, the master mechanics themselves largely resisted extending national standards to steam locomotives. Contrast Sellers's remarks with an 1899 address to the association by Joseph Bryan, the president of the Richmond Locomotive Works. Bryan opened his speech by quoting from an English newspaper that had asserted that U.S. locomotive makers built only standard designs. In response he exclaimed, "Think of that will you! Think of that, O you master mechanics, who cudgel your brains to find new designs and who for ten years have been changing your types of engines so rapidly as to make the poor manufacturer's head swim. . . .

It sounds like a roaring farce."[4] Bryan went on to plead for standard designs, noting "what a comfort it would be to the builder to . . . build something he had seen or heard of before. What an avoidance of useless expense and the assurance of better deliveries."

Although his remarks were somewhat hyperbolic—as were Sellers's—Bryan accurately characterized the master mechanics' influence in innovation. Throughout the history of steam locomotives, standard designs largely proved an unpopular and ill-fated concept. Railway master mechanics prevented the locomotive builders from establishing national standards for use on a number of lines, and within a given road standard designs were generally short-lived at best. One key factor that allowed for diversity in locomotives was that, unlike cars, they seldom traveled beyond the tracks of their own roads. But the causes for it are rooted in larger issues, and the reasons for this constant evolution and endless variety in locomotives are worth examining in detail.

Master mechanics had a variety of motives for opposing standards and promoting diversity in locomotive design, but their main concern is simply stated. Standards blocked progress. Over the nineteenth century, American railways saw astronomical growth in the volume of freight and passenger traffic moving through their lines. The railways carried that burden by achieving dynamic improvements in productivity, running more trains and especially better trains—faster, longer, and heavier than ever before.[5] Such trains required more powerful locomotives with every decade, so continuous evolution ruled locomotive design, rather than a standard, unchanging technology.[6] Design change was also fostered by railways competing among themselves for traffic, vying to offer better efficiency and productivity. Some carriers attempted to establish standard engine designs, "but in every instance these have given way to the urgency of keeping pace with other roads which have not attempted to bind themselves with the iron bands of standardization."[7]

The character of innovation in locomotive design was a crucial issue for Baldwin and the other builders and is examined here in detail. As Bryan's remarks indicate, the builders had little control over the design evolution of their major products, engines for mainline railroads. How and why did they lose the initiative in design to the master mechanics? These important questions have ramifications well beyond steam locomotives. Railway car makers, shipbuilders, machine tool firms, and most other makers of large capital goods also served customers who largely defined the pace and direction of technological change in their products. The custom character of most capital equipment marked the most crucial difference between builders and mass-production manufacturers of machinery. The continuous design evolution in heavy machinery meant that its builders had to utilize different marketing techniques, production tooling, and labor skills compared with the methods of high-volume manufacturers. Because the sources and character of innovation in locomotive design had such vital ramifications for Baldwin's operations, the subject merits detailed consideration.

Builders such as Baldwin, however, were far from powerless on matters of design. Unlike volume manufacturers, which worked up complete product designs and then anxiously hoped for their acceptance in the marketplace, capital goods builders like Baldwin often responded to their customers' specifications in a process of design mediation between buyer and seller. Such discussions were essential to both parties, since the locomotive builder had to give tangible reality to the master mechanics' design initiatives.[8] Even in cases where the customer provided complete plans—rather than general specifications—for a new locomotive, Baldwin retained some autonomy in deciding how to build the engine. Such mediation of ends and means was a crucial attribute in Baldwin's market and business success after the Civil War. But first we must turn to the prewar period to understand how and why the locomotive builders lost the initiative in the evolution of locomotive design.

EARLY AMERICAN LOCOMOTIVE TECHNOLOGY
AND THE QUESTION OF PATENTS

The earliest American locomotives were either English imports or American-built copies of English designs, such as Baldwin's *Old Ironsides*. But American engineers soon adapted these designs to suit the lightweight track, rough roadbeds, and many curves common to railways in this country. In 1832 John Jervis introduced his leading truck, which guided locomotives through curves and thereby helped prevent derailments. Jervis did not patent this important improvement, and other builders quickly copied his invention, including Matthias Baldwin.

The next innovation of importance to the long-term development of locomotives sought to increase the power the cylinders transmitted to the rails. Early engines, such as that shown on page 7, had only one set of driving wheels (i.e., wheels driven by the cylinders). With heavy trains or on steep grades, these wheels would lose their traction on the rail and spin uselessly. The remedy was to add another set of drivers and connect the two sets so that the cylinders would transmit power to four points on the rails instead of just two. Henry Campbell, a Philadelphia mechanical engineer and close friend of Matthias Baldwin, patented this 4-4-0 type in 1836.[9] Such an engine is shown on page 60.

The crucial invention here was not merely the second pair of drivers; successful 4-4-0 designs also incorporated a new form of suspension, the equalizer, patented in 1838 by another Philadelphia locomotive builder, Joseph Harrison Jr., an acquaintance of Campbell's and Baldwin's. One of his patent drawings for the equalizer is shown on page 61. Harrison's device proved very important; it allowed the transition from the early 4-2-0 type to the 4-4-0, "and few, if any, American locomotives were built without equalizers after 1840." [10]

Hindsight tells us that the American steam locomotive had acquired all its essential attributes by 1838, although further refinements continued for over a century. English contributions to the basic design included the horizontal fire-tube boiler, forced draft, semihorizontal cylinders, and power transmission from cylinders to wheels using direct connecting rods. The chief American additions were the Jervis truck and the Harrison equalizer, the latter allowing multiple driving axles. All these attributes remained in American steam locomotives until their demise in the 1950s. Significantly, of all the English and American inventors, only Campbell and Harrison patented their improvements, and they made their innovations available to all who would pay the royalties. [11] Therefore no locomotive builder controlled key patents in a way that could have prevented others from entering the field. This contributed to the competitive vitality of the antebellum locomotive industry. And the fact that no builder held an essential technological advantage in patents enhanced the master mechanics' ability to influence design matters during and after the 1850s.

This technically unfettered situation becomes apparent only with hindsight, however. The path locomotive development took during and after the 1850s was hardly discernible in the 1840s, when radically different designs competed for acceptance. Perhaps because so many basic attributes of locomotives were

Baldwin built this four-drivered engine for the Havana Bay and Matanzas Railroad early in 1861. Through driving rods the cylinders turn the front drivers, while connecting rods transmit some of this power from the front to the back drivers. This 4-4-0 was built for a Cuban line, and such exported products typically had a higher level of finish work than domestic locomotives—evident in the varnished wood cab and polished brass cylinder jackets. RAILROAD MUSEUM OF PENNSYLVANIA (PMHC)

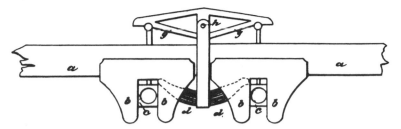

Detail drawing from Harrison's 1838 equalizer patent. The equalizing beam is marked *g*. The two circles, marked *c*, are the driving-wheel axles. When a wheel on one axle hit a bump in the track, the overhead beam passed some of the force over to the other axle, equalizing the shocks and strain and ensuring that all driving wheels stayed on the track. HAGLEY MUSEUM AND LIBRARY

unpatented, builders in the 1840s frequently sought a competitive advantage through patent protection of their own novel design contributions. Among leading builders, Ross Winans, Septimus Norris, and Matthias Baldwin in particular chose this means to enhance their competitive positions in the marketplace. Baldwin was at heart a mechanical engineer, not a businessman, and he took pride and interest chiefly in the technological and inventive aspects of his work. His technical creativity primarily accounted for the seventeen patents he received between 1833 and 1866. Following common practice at the time, Baldwin made most of these patents available to others in return for royalties. But in at least one case, and probably in a second as well, he balked at allowing others to use particular designs that he felt provided an important competitive advantage for his firm.[12] Ross Winans and Septimus Norris were far more adamant than Baldwin in attempting to use their patent monopolies to forestall competitors, either directly by refusal of rights or indirectly by setting high royalties.[13] By blocking general use in the industry of certain patented features, this practice in turn promoted diversity in the designs of locomotives made by the various builders.

THE BUILDERS IN CONTROL OF DESIGN, 1830S–1840S

These locomotive makers sought to protect their patents in the 1830s and 1840s because at that point they still had real value. Locomotives were a novel technology in a state of design flux, and many builders produced qualitatively different engines. Patented component designs like Baldwin's half crank, and patented engine types such as his flexible-beam locomotives, enjoyed considerable success in the nascent market of the 1830s and 1840s, despite their later irrelevance to the course of locomotive development. At this point Baldwin and other builders had a stronger influence on design evolution than did their railway customers.

The builders had control of design for a number of reasons. First, innovations such as Baldwin's half crank and flexible beam provided competent solutions to concrete problems. Such solutions grew out of the fact that the locomotive firms possessed far more of the technical knowledge required in locomotive design than did most of the early railways. For example, even before civil engineers made their initial surveys to locate rights-of-way for proposed lines, they contacted Baldwin to learn his engines' capacities in surmounting hills and passing curves.[14] Baldwin's answers would influence what route the line would take. After completion, early railroads continued to depend on the locomotive builders in a number of ways. In addition to supplying credit for purchases, firms like Baldwin were a primary source for the technically competent machinists, shop superintendents, firemen, and engineers the operating companies needed to run and maintain their engines.[15] When Baldwin sent out such trained personnel from its own factory, the firm's technological imperatives and design preferences often accompanied them.

Early locomotive builders also maintained control over the evolution of engine design by developing a variety of standard models to meet the individual needs of their customers. In 1840 Baldwin offered three sizes of engines, all in the same 4-2-0 configuration but differing in power and weight.[16] After the development of the flexible-beam design in the early 1840s, Baldwin's standard line included twelve models, divided into four classes, with two to eight drivers each. Within each class, three sizes of weight and power were available.[17] Other builders such as Norris also marketed their own standard lines in the 1840s.[18]

The builders turned to lines of standard products to achieve economies in production while meeting the various needs and service conditions of different railroads. Baldwin's two- and four-drivered engines of the 1840s were primarily for passenger trains, while the sixes and eights best suited slow freights. The variety of sizes available within a class—for example, 18-, 20-, or 24-ton eight-wheel freight locomotives—allowed railroads to select a size that would meet their power requirements while avoiding excessive wear and damage to tracks. These engines also came with rated performance guarantees; for example, Baldwin warranted its 18-ton freight 0-6-0 locomotive of 1845 to haul 840 tons on level track at 9.8 miles per hour.[19] The varied product line and performance guarantees were Baldwin's chief methods for defining and meeting its customers' requirements.

This system worked well for Baldwin and the railroads through the mid-1840s. Early railway mechanical officers had quite daunting managerial and technological challenges pressing from all sides. No doubt most were content to leave locomotive design questions to the builders. Firms like Baldwin showed an awareness of their particular needs, the locomotive companies frequently incorporated product improvements, and the rated hauling powers had the advantage of making the builders responsible for at least a minimum level of performance.

THE ORIGINS OF THE MASTER MECHANICS' INFLUENCE ON DESIGN, 1845–1850

While most mechanical superintendents in the late 1840s happily left design matters in the builders' hands, a few began insisting on innovations. Leading companies like the Baltimore and Ohio and technically competent officers such as J. Edgar Thomson started to push Baldwin and the other builders for improvements in design and performance. They had three basic reasons for seeking this influence. First, the operating officers saw firsthand a number of deficiencies in materials and design. Furthermore, the variety of engines on the market clearly indicated that the builders had yet to discover the "one best way" to improved performance. Finally, older roads were already experiencing the growth in business and traffic that continued into the next century, and they wanted more powerful engines to meet this increased load.[20] The ultimate power over design decisions had, of course, always resided with the market rather than with the builders, since customers soon rejected unsatisfactory products. But now some mechanical officers began to take direct initiatives in design evolution. While not a widespread phenomenon until the 1850s and later, its origins are important. Once Baldwin and the other builders lost the initiative in locomotive development to their customers, they had to devote much of their production to custom work, since each leading railroad tended to see its own needs as unique. In the postwar period Baldwin proved notably successful in adapting its management, labor force, and production process to this customizing dynamic. But the transition was difficult, and at first Matthias Baldwin strongly resisted losing design control.

The master mechanics' opening wedge to influencing locomotive design lay in their suggestions for improved components and materials. As long as these could be easily incorporated into Baldwin's standard line, the firm had no objection. For example, J. Edgar Thomson of the Georgia Railroad and Banking Company wrote in 1845 asking that wrought iron be substituted for cheaper and weaker cast iron "at points liable to fracture," and Baldwin acceded.[21] Baldwin also had no objection when the Little Miami Railroad sought to alter the standard 12-ton six-drivered engines it had on order by substituting larger connecting rods, wrought-copper boiler heads, and an iron tender frame.[22] Changes like wrought-iron parts or stronger connecting rods were relatively easy to make, and the firm probably welcomed the chance to improve its products in this incremental fashion.

Baldwin had an entirely different response in 1844 when the chief engineer of the Baltimore and Ohio, James Murray, asked the firm to bid on an order for six custom engines to be built to his specifications. The firm wrote back declining to bid because "it would require us to make a new set of patterns which would occupy nearly as much time as to construct the engines after the drawings and patterns are made." Baldwin suggested that the B&O instead take a standard design, adding somewhat disingenuously, "We do not pretend to prescribe to the purchaser the kind or class of engine best adapted to his railroad business."[23] Murray declined. It is likely that Baldwin had other reasons for refusing to bid on this contract, since the firm was still amortizing the costs for drawings and patterns of its own standard line of engines, developed in 1842, and these were selling relatively well.

In the fall of 1847 the Baltimore and Ohio was back in the locomotive market, soliciting proposals for four engines built to the company's detailed list of twenty-five technical specifications.[24] At first Baldwin again tried to persuade the railroad to take a standard model, but the B&O's chief engineer, Benjamin H. Latrobe, balked—noting that Baldwin's recommendation weighed 27 tons, 7 more than the company's specification. Fitful negotiations continued between Latrobe and Baldwin for four months, with neither side willing to give much ground. Baldwin sought release from seven specifications, but Latrobe remained adamant on the 20-ton weight limit—a limit Baldwin believed impossible to meet, given the other requirements. Finally, the specifications were revised slightly in January 1848, and instead of picking over the details, Baldwin submitted the winning bid of $9,000 for each of five engines.[25]

What had changed to allow the deal to go through? The B&O revised five specifications, but only slightly. The crucial factor was that Baldwin now needed the business. Sales plummeted from thirty-nine engines in 1847 to twenty in 1848, leaving the firm little choice but to take what the market offered. This shift to a buyer's market caused a corresponding change in the balance of power over issues in locomotive design. Even after receiving the B&O order, Matthias Baldwin confessed to a friend, "I am left quite destitute thus far this season."[26] Accepting a custom order was a difficult matter, since the drawings and patterns required for a new model could cost up to $2,000 at this time, equivalent to one-third or one-quarter of an engine's selling price.[27] Although Baldwin most likely skimped on expenses by using standard component drawings and patterns wherever possible, his developmental costs for this order were substantial nonetheless, and he tried to amortize them by interesting the Vermont Central and the Pennsylvania Railroad in the new design, illustrated on page 65.[28]

This incident is described in detail because it was one of the first cases where Matthias Baldwin had to submit to such extensive design direction from a customer. He sought to avoid this, in part shrinking from the high production costs of custom engines. He also took pride in his own engineering talents, believing his product line could adequately meet the railways' needs

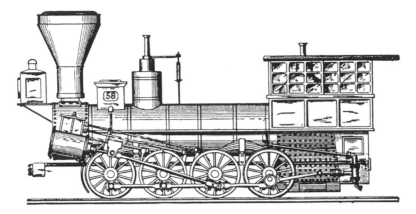

This engraving represents the custom-built freight locomotives produced by Baldwin under the Baltimore and Ohio's specifications. The three engines were named *Hector, Cossack,* and *Tartar.* The most obvious variation here from Baldwin's standard practice is the absence of his flexible-beam running gear. HAGLEY MUSEUM AND LIBRARY

without their meddling in his chosen field of expertise. In one sense the engines themselves proved him right; as he predicted, they exceeded the revised weight specification by 2 tons. But the B&O order was a forerunner of developments in the 1850s, when leading railroads would take the initiative and responsibility for furthering locomotive development. In this case the B&O specifications aimed at producing a powerful freight engine that would weigh less than standard models on the market, lessening wear and damage to the track. As built, the engines were almost 4 tons lighter than Baldwin's suggested alternative. The design also sought to burn bituminous coal efficiently, a particular concern for the coal-hauling B&O. By writing specifications to meet these goals, the railroad was trying for a better integration of its motive power into its whole technological system.

Given the high first cost and running expenses of locomotives and their essential contribution to railroad operations, it is not surprising that operating companies would sooner or later seek to take locomotive development into their own hands. Although most master mechanics were preoccupied in the years before the Civil War with more pressing operational issues, the period also saw more than a few turning to design issues. In 1848 the Vermont Central offered $10,000 to any builder who would develop an engine capable of running 60 miles per hour. Still desperate for business, Baldwin developed and built this custom design, which the firm also sold in a slightly modified form to the Pennsylvania.[29]

In these examples, railroads prescribed specifications for engines, leaving some latitude for builders over the specific design and components to be utilized. But even this limited discretion was lost to Baldwin in an 1848 order by the New York and Erie. This contract for eight engines was executed to complete drawings furnished by the railroad, most likely prepared by the Erie's highly competent master mechanic, John Brandt.[30] While such a complete assumption of design prerogatives was relatively rare for the period, it was a harbinger of change. As the noted railway authority Zerah Colburn wrote in 1856, "Purchasers, and not the builders of engines, now dictate the general direction of improvement in locomotives."[31]

A NOTE ON LOCOMOTIVE TECHNOLOGY

As indicated earlier, the 1850s saw a general design stabilization around the 4-4-0 type of engine for both freight and passenger service. This is not to say that all other types fell into total disfavor; they did not. But the widespread popularity of the 4-4-0 represented a general verdict by the market in favor of rugged simplicity and adaptability. Old customers pressured Matthias Baldwin to develop his own 4-4-0, which he finally did in 1845.[32] This was a defensive move, keeping those clients who desired this design from turning to other builders, and Baldwin continued to put great stock in his own designs. But increasingly in the 1850s the market rejected such builders' patented specialties

Matthias Baldwin's mechanical creativity often took form in highly original designs. This 1860 locomotive for the Mine Hill and Schuylkill Haven Railroad incorporated Baldwin's flexible beam and his variable valve motion, a complex mechanism designed to save steam (and fuel) when an engineer did not require full power. This engine also included features developed by other innovators, including a smokestack feedwater heater, paper boiler insulation, and driver brakes. Baldwin built two engines in this order, selling for $10,800 each, but none of these innovations found lasting favor. RAILROAD MUSEUM OF PENNSYLVANIA (PMHC)

as Baldwin's flexible-beam design and his variable valve motion, which came to be regarded as complications rather than improvements.

After design stabilized around the 4-4-0 type, much refinement and development remained undone, but it occurred within relatively well defined channels. The overarching pattern from the 1850s to the 1940s was a growth in boiler size (to create more steam) and in cylinder size (to increase power output). In turn, the growth in cylinders and boilers generally required adding more driving wheels, so that the heavier engines would not crush the track and would be able to utilize the higher power without traction losses. Within the 4-4-0 or American type, this transformation had the following results: between 1855 and 1908 tractive force—a measure of hauling power—grew from 7,000 to 22,000 pounds, and total engine weight went from 25 to 50 or 60 tons.[33] By 1900, however, the 4-4-0 was largely obsolete, having been superseded by engines with more drivers. The 4-6-0 grew in popularity after 1860, as did the 2-8-0 after 1875. In all, American designers created close to forty different types of wheel arrangements for steam locomotives.[34] These types need not be detailed here; suffice it to say that most were outgrowths of the development of larger boilers and cylinders.

Within this general pattern, mechanical engineers exhibited great diversity on questions of design, even within a specific type. Consider the general design issues in a 4-6-0, or Ten-Wheeler, type circa 1870. One built for freight service required high power at relatively low speed, and it would have smaller drivers, larger cylinder bores, and heavier frames than one for hauling fast passenger trains. Even within the freight-service category, the optimal dimensions for drivers, cylinders, boilers, and other parts varied greatly according to expected trainloads and the character of the terrain the engine would cross. Then there were questions of preferred materials, such as copper, wrought iron, or steel for fireboxes. A host of different patented specialties also had to be selected, including leading trucks, boiler tubes, tires, injectors, brakes, headlights, and smokestacks, to say nothing of such smaller fittings as oilers, pressure and water gauges, and water, steam, and safety valves. After the Civil War the number of such specialty items and fittings on the market exploded as the railway equipment business grew along with the railroad sector.

Another aspect of locomotive technology deserves mention: its thermo-dynamic characteristics. During the nineteenth century, engineers applied steam power to a range of applications, including ships, pumping stations, mill engines, road rollers, and threshing engines. Many aspects of the technology were common to all these devices. But as a rule, locomotives had to produce the most power per pound of mechanism, since they had to haul themselves, their fuel and water, and a substantial load over varying terrain at relatively high

Boiler backhead of a Philadelphia and Reading engine, built circa 1879. The boiler incorporated a patented wide firebox design developed by that line's inventive general manager, John Wootten, to burn waste anthracite coal. The culmination of forty years of experimentation by a number of inventors, Wootten's firebox finally provided the enduring answer to effective coal burners. This illustration also demonstrates how thermal stresses affected locomotive design. The use of two fire doors was unique, but they suggest the high firing rate, or forcing, typical of locomotives. The vertical links under each fire door provided a strong yet flexible method of attaching a boiler to the engine frames. As a cold boiler was fired up, it expanded considerably in length, requiring a strong but adjustable mounting at the firebox end. RAILROAD MUSEUM OF PENNSYLVANIA (PMHC)

speed. This power was achieved by "forcing" the boiler—making it burn an excessive amount of fuel.[35] This technique, necessary to achieve the required power, took a heavy toll on boilers, particularly before steel superseded iron in their construction. In the 1860s and 1870s a locomotive boiler's iron firebox typically needed replacing every three years.[36] Other parts were renewed less frequently but still often enough compared with other capital equipment. Running gear such as wheel bearings, valves, reverse motions, and the like also needed constant adjustment and servicing because of rough and jarring track. The frequency of repairs to running gear and boilers meant that most railways had ongoing maintenance programs and extensive repair shops, which incurred substantial labor expenses. The locomotive maintenance burden was heavy, and it reinforced the master mechanics' desire to influence locomotive design.

Beyond their maintenance expenses, steam locomotives incurred high running costs in fuel and labor charges. Over an engine's average service life of twenty-five years, these costs far exceeded its initial purchase price. Since maintenance and operating expenses weighed far more heavily than initial costs, master mechanics continuously sought improvements in power and reliability or reductions in fuel and labor requirements. The savings in these areas could far exceed the potential rewards offered by standard engine designs. Like other users of capital equipment, American railway master mechanics responded to these pressures with continuous innovation and custom designs, perceiving such efforts as technically rational and economically imperative. As we shall see, their experimentation also had more personal motivations.

THE MASTER MECHANICS TAKE THE INITIATIVE, 1865–1900

In the history of American railroading the period after the Civil War is noted for the systematizing, rationalizing, and standardizing efforts that created a national transportation network. The railways' chief mechanical officers or master mechanics undertook much of this work. But relatively little was done to create standard steam locomotives, either nationally or on a given road, largely because the master mechanics did not want them.[37] When they took over the initiative in locomotive design development from the builders after the Civil War, they sought an individualized product, one adapted to the particular needs of their roads. Unlike passenger and freight cars, which had to interchange with other carriers, making standards important, locomotives never traveled beyond their own roads. Therefore leading master mechanics could follow their own predilections in design. Indeed, they had powerful reasons to avoid consistent practice even within a given company, since locomotives were generally assigned to separate operating divisions—sections of the line ranging from 75 to 100 miles in length. Conditions varied greatly between divisions in such matters as the balance and intensity of freight versus passenger traffic, hilly or level terrain, curved versus straight track, right-of-way clearances, and weights of track.[38] These technological realities promoted locomotive design diversity on any given railway and blocked standardization, either on that carrier or nationally.

The long-term trend of increasing passenger and freight traffic promoted further variety in a road's locomotive fleet. As the volume of traffic grew, trains became longer and heavier, and in response the master mechanics sought more powerful engines. This basic factor drove the growth in locomotive weight and power after the Civil War. Typical heavy passenger locomotives made by Baldwin for the Lehigh Valley Railroad, for example, grew from a 39-ton 4-4-0 in 1874 to a 70-ton 4-6-0 in 1895 and a 119-ton 4-6-2 in 1905.[39] The engine of 1874 was probably still on the road when the 1905 Pacific type went into service, although the American type would have been relegated to branch-line service, and its days were numbered.

Another reason why diversity ruled locomotives lay in the empirical character of motive power development. Not until the 1890s did railroads begin to apply scientific testing to locomotive design and operations. Trial and error were the main routes to improved engines before 1900. The locomotive-building firms had little incentive to take the risks of producing a new design that might fail and face costly rejection in the market. But master mechanics had the

Alexander Mitchell's *Consolidation* or 2-8-0 type of 1866 provided good hauling power within a simple, even elegant design. Compared with Baldwin's ungainly flexible-beam models (see page 66), Mitchell had developed a clear winner. The *Consolidation* received Baldwin construction number 1,500 and sold for $19,000 cash, a relatively high figure reflecting its experimental character and the effects of wartime inflation. RAILROAD MUSEUM OF PENNSYLVANIA (PMHC)

working knowledge and the physical plants to try out new designs. The long-run trend of increasing traffic and heavier trains gave some of them sufficient incentive to experiment.

These developmental issues are seen in the origins of the Consolidation type, or 2-8-0. The master mechanic of the Lehigh and Mahanoy, Alexander Mitchell, faced a bottleneck on his road: a steep grade that his heaviest engines could surmount only with short and light trains.[40] He wanted to improve operating efficiency by getting long and heavy anthracite trains over the grade, but his existing 4-6-0s and 2-6-0s had proven inadequate for the job. So in 1865 he designed a new, more powerful type: the 2-8-0. When he submitted the design to Baldwin, the firm preferred to avoid the risk of a new design and tried to interest Mitchell in a standard flexible-beam engine. He persisted in his original proposal, which the firm finally agreed to build, although only on a cost-plus basis and without the normal performance guarantee. As it turned out, the design became a great success. Baldwin quickly incorporated the Consolidation into its line of standard products, and the type became "the most popular and widely built freight locomotive in the United States."[41] Mitchell was unusually venturesome—he also experimented with new 2-10-0 and 4-8-0 types circa 1870—but many other master mechanics also took design initiatives during this period.

The final issue contributing to diversity in locomotive design involved personality. Many master mechanics demonstrated a kind of technical virtuosity, taking a self-conscious pride in their own mechanical expertise.[42] System-focused historians overlook this factor, but men created railway systems, and the mere fact that the master mechanics were employed by railroads did not subsume their egos. Quite the reverse. These men had risen through the ranks, often from apprenticeships, and their careers in railway mechanical engineering armed them with strong personal predilections on technical questions.[43] Even the title *master mechanic* was something of a prideful boast. While necessarily devoting much of their considerable expertise to impersonal system building, the master mechanics' creative impulses could have free reign and tangible expression in locomotives.[44] In the mid-nineteenth century, locomotives constituted the most visible element of railway engineering—in turn, the single most important and influential branch of mechanical engineering generally. Such leading mechanical officers as James Millholland, John Brandt, Alexander Mitchell, and John Wootten achieved their stature by experimentation and innovation, creating new engine designs that improved operating efficiency while securing their own reputations among their contemporaries.

This element of ego related closely to the empiricism of locomotive design development, with the two factors fostering each other in many cases. The main motivation in founding the Master Mechanics Association in 1868 was to curb the growing diversity in design that originated in empiricism and ego. One can sense these issues in the prospectus the embryonic organization sent out in its second year to enlist new members.

> Eminent railway mechanics entertain diverse opinions regarding important details of construction of locomotive engines; regarding the correct proportion of various parts; the strongest mode of building . . . boilers . . . the most efficient form . . . of fireboxes . . . the relative value of iron and steel plate . . . the causes and prevention of boiler explosions and a multitude of kindred topics. . . . These diverse opinions we seek to harmonize by interchange of thought and systematic investigation.[45]

This quotation reveals the systematizing and standardizing goals of the association's founders while also suggesting the extent to which ego and trial and error ruled locomotive design.

Over the next thirty years the association worked exhaustively to determine best practice in locomotive design and operation. By 1900 it had approved seventy-seven recommendations for standard locomotive component designs and operating practices—an impressive attempt at systematization. But the crucial issue was not the promulgation of standards but their adoption. To examine this question, the association created a committee in 1899 to poll the members "on the extent to which the recommendations of the Association have been put into practice." The committee sent out six hundred circulars and received only twenty-five responses—implying minimal interest in the standards and recommendations. The twenty-five carriers that answered indicated unanimous acceptance of only seven of the seventy-seven recommendations. Twenty-seven others received majority support, while forty-three did not garner a consensus either for or against.[46] Whether motivated by ego, empiricism, or economic factors, the survey provides telling evidence of the degree to which the master mechanics resisted standardization in steam locomotives and succeeded in preserving their own autonomy in design questions.

The master mechanics' lack of interest in steam locomotive standards was not an isolated case. While historians tend to emphasize the standardizing and systematizing efforts of Gilded Age railway officers, the actual practice of these men was more complex and perhaps less rational than is generally thought. For example, while much is made of the adoption of a standard track gauge on most U.S. railroads in the 1880s, the rails themselves were hardly standard. Here master mechanics and other officers revealed their full range of discretion. In 1881 a leading engineer and authority on the steel industry, Alexander Holley, found that the Bessemer steel mills of the United States rolled rails in 119 different patterns or cross-sections. He noted some technical reasons for this diversity, but an "intrinsic cause" was "the egotism of certain engineers and officers of railways [who] . . . vary from all standard patterns for no reason whatever except to inflict their own individuality upon some feature of the interest confided to their care." While rail patterns have little to do with locomotive design, Holley's remarks show the influence of mechanical officers' egos and technical adventurism in a clear-cut case.[47]

With their six thousand parts, continuous refinement, and center-stage place on American railways, locomotives provided far more latitude for such personalism in technological innovation. The locomotive builders were more circumspect on this point than Holley, and they made no known criticism of egotism in locomotive design. Insulting the customer was bad business.[48] But the railway trade press did comment on the master mechanics' technical adventurism from time to time. *Machinery* noted in 1896, "The building of locomotives is a rather peculiar business and one that has not yet reached the manufacturing stage. . . . The trouble is that every master mechanic has some pet scheme for saving money or becoming famous and he is bound to have this idea carried out, although it often adds to the cost of the engine and as often is rather a detriment than an improvement."[49] The builders had no choice but to adjust to the master mechanics' power, their own loss of design control, and the consequent flowering of diversity in their own products. Baldwin's response is outlined below and in subsequent chapters. But first we must consider the specific methods by which the master mechanics took the initiative in design between 1850 and 1900.

THE MEANS AND ENDS OF MASTER MECHANICS'
CONTROL OF DESIGN

The first method master mechanics used to exert their influence on design involved issuing detailed specifications such as Latrobe's list of 1847. In that case the builder retained some latitude in determining precisely how to meet the specifications. Where possible, standard components were incorporated into the custom design. The practice of issuing detailed specifications that mandated key component dimensions such as driver diameters, cylinder and boiler sizes, and maximum weight limits continued into the Gilded Age.

By the 1850s, however, some master mechanics were taking a more directive role, drawing up complete engine plans that they then furnished to the builders, often with patterns for major castings. In such cases, known as blueprint jobs, the builder had no discretion at all in component designs but had some latitude regarding the production methods used to make given parts.[50] Only the most technically confident master mechanics took the step of issuing complete blueprints, and most likely this practice had peaked by the 1890s, although such lines as the Pennsylvania continued it down to the 1940s.

In the 1890s the railroads and locomotive builders developed a third means for defining the characteristics of new engines: performance specifications. The *Railroad Gazette* noted the rise of this method in 1898; under it a road would specify the load to be hauled, the speed desired, weight of track, type of terrain, and quality of fuel and then "ask the locomotive builders to build the engine that will do this work and also guarantee that the engine will perform it." [51] For example, Baldwin agreed in 1900 to furnish the Philadelphia and Reading with ten 4-6-0 passenger engines that "shall be at least 10 percent speedier than your present engines of the 1027 class." [52] Clearly, in such cases developmental initiative remained with the railroad, which set the performance specifications, but great design and production latitude was given the builder, who had to develop an appropriate engine and guarantee its performance.

If such were the means of the master mechanics' control, what was their end? The obvious answer, better locomotives, begs the question. A major theme here has been that master mechanics' control over design led to diversity in locomotive construction and types—especially diversity between railroads, but also within a road's fleet. Despite this variety some patterns are evident in the design goals of the master mechanics. Most commonly they sought to assure that their roads' motive power ideally suited the traffic it pulled and the terrain it traversed, as in the case of Mitchell's *Consolidation*. Since locomotives were integral parts of vast, complex, and differing technological systems, it made sense to the leading master mechanics to take the design initiative.

The consequences of ensuring that a carrier's motive power was suited to its traffic and terrain included improved performance, more power, less damage to track, better reliability, and simplified maintenance. These improvements sprang as much from the creative drive of the master mechanics as from the drive for economic efficiency. The master mechanics felt sure of their own knowledge and capabilities, believing that they alone knew what was best for their roads. They probably were correct most of the time. Until scientific investigation was harnessed to locomotive improvement, however, their design choices were largely personal, based on their own experience. This promoted design diversity between various roads. But so proprietary were the master mechanics that when this position changed hands on a given road, so did design imperatives. Hence on the B&O in the 1850s and 1860s the Samuel Hayes 4-6-0s were followed by the Henry Tyson Ten-Wheelers, which in turn gave way to the Thatcher Perkins 4-6-0s.[53] And so it went.

This succession of incremental advances resulted in better locomotive performance, but at a large cost in maintenance expenses and carrying charges for spare parts. A noted mechanical authority, Fred Colvin, discussed this problem in his autobiography. Colvin described a visit to a repair shop on the Cleveland, Cincinnati, Chicago, and St. Louis Railway (known as the Big Four), where he encountered "one of the neatest and most impressive storage or stock piles I had ever seen. Tires, driving wheels, cylinder castings, parallel rods, main frames . . . and hundreds of other locomotive odds and ends were stacked in orderly array but in frightening quantities." Colvin asked the superintendent the reason for this vast display. His response: "There is nothing I can do about it. We have over fifty different types of locomotives running on this road and I have to carry spare parts for each type. Maybe there's only one or two engines of a certain make out of the thousand on the line, but I gotta carry a complete stock of spare parts for them just the same." The superintendent valued the parts at $0.5 million—and the Big Four's situation was shared by many other lines.[54] Circa 1885 the Union Pacific stocked parts for seventy-four locomotive designs, and in 1893 the motive power fleet of the Philadelphia and Reading numbered 795 engines in eighty different classes or designs.[55]

Various lines tried to fight this problem, which raises a third goal behind some master mechanics' technical initiatives: the creation of standard designs. The Pennsylvania was the earliest and most successful American carrier in pursuing standardization. In 1868 its mechanical engineer, John B. Collins, pre-pared designs for eight classes of standard engines: two switching models, four American types for various services, and two Ten-Wheelers. In these designs Collins established a number of standard components, using them across classes to lessen the size of repair-parts inventories. Such standard parts could be made with jigs and templates to secure another advantage: interchangeable parts. This allowed the PRR to procure more engines in each class from any builder or its own shops while retaining the advantages of perfect interchangeability throughout the class.[56]

The Pennsylvania enjoyed notable success in its standardization program. As the largest road in the country and the "Standard Railroad of the World," it had some unique reasons for pursuing the program with such vigor. It was also the leading railroad in the country to build its own motive power. To a lesser extent other roads followed its standardizing example—notably E. H. Harriman's Associated Lines, which took up locomotive standards around 1900 during the Progressive-era drive for engineering efficiency.[57] This acknowledged work in standardizing power seems to contradict the earlier emphasis here on diversity. But variety soon reappeared on the Standard Railroad and all other carriers. An inescapable problem with standard designs was that they froze performance at one level while traffic volume steadily increased over time. Even a most determined standardizer like the Pennsylvania had to meet this increased load with new, more powerful designs, multiplying the standard types in use. Over their twenty-five-year service life the original eight standard designs of 1868 were supplemented by thirteen new classes; two of the original classes were modified, and the road acquired a miscellany of nonstandard types, either new power direct from builders or secondhand engines acquired as neighboring lines were absorbed into the system.[58]

Such growth in motive power diversity to meet heavier traffic volume faced all carriers that attempted to maintain standards, and it invariably negated most of the original goals of that effort. Most large lines compromised between their desires for advances in motive power and for economy in repairs by establishing road standards that mandated common detail parts but left the path clear for the creation of largely custom engines. Not surprisingly, however, these motive power standards varied from road to road, increasing the challenge of diversity for the locomotive builders.

The long-term growth of freight and passenger traffic was an unyielding refrain of American railroading in the nineteenth and early twentieth centuries. Meeting this secular trend was the fourth goal of those master mechanics who controlled the initiative in locomotive design. Recollect that their first purpose listed here was ensuring that engines were well adapted to their systems at a given moment. Over time these systems evolved as traffic volume grew, new lines were built or bought, heavier track was laid down, clearances were enlarged, freight and passenger cars grew in size and capacity, and new services such as fast freights and suburban commuter lines were instituted. Each development provided incentives to design new and varied types of power.

One result was annual growth in the total number of locomotives in service from 1830 to 1924.[59] On the Santa Fe, for example, the largest heavy freight locomotives grew from a 58-ton Consolidation (2-8-0) in 1878 to a 144-ton 2-10-2 or Santa Fe type in 1903.[60] Maximum boiler pressures increased from 120 pounds per square inch in 1871 to 200 psi around 1915. With the growth in boiler size and pressure, larger cylinders, and more driving wheels, the tractive force of typical road locomotives multiplied fivefold between 1870 and 1915, to about 50,000 pounds.[61] Such were the results of the incremental developmental advances that the nation's leading master mechanics accomplished in concert with the locomotive builders.

BALDWIN'S RESPONSE TO THE DYNAMIC OF INNOVATION, 1870–1915

Turning from the master mechanics to the builders, how did Baldwin transcend an apparent constraint—its lack of primary control over the design evolution of its chief products—to become the largest locomotive builder in the world? The challenge should not be underestimated. Locomotive sales fluctuated wildly within the secular trend of long-term growth. Yet the general trend in locomotive innovation was the master mechanics' preference for larger, more powerful engines. Thus the locomotive builders had to weather frequent sales depressions and then emerge from them prepared to build more and larger products than ever before, despite their inability to predict or control much of the technological content of these products. These challenges accounted for Joseph Bryan's plea, recounted at the opening of this chapter, for some stability in locomotive design. After 1865 Baldwin adapted its operations to accommodate the leading master mechanics' influence over technical change in locomotives while also preserving and even enhancing its design discretion whenever possible. Not all its customers were technological pathfinders, even among large mainline railroads. The company deferred when necessary but led where it could.

In 1847 Matthias Baldwin had only grudgingly followed the specifications of the Baltimore and Ohio. The necessity of building to market preferences was reinforced in the 1850s, when most customers turned their backs on Baldwin's flexible-beam engine and complicated valve design. The company responded by redesigning its American-type 4-4-0s, updating them to contemporary preferences, and sales improved (its revised design appears on page 60). But junior partner Matthew Baird's initial rejection of Mitchell's Consolidation design in 1865 indicates that the company still disliked following the flowering design initiatives of master mechanics.

The company's attitude changed in the late 1860s, just as that flower took root and hybrid designs multiplied. Within a year of Matthias Baldwin's death

TABLE 3.1 Decennial Breakdown of Baldwin Production by Market Segment, 1850–1900

Year	Total Output	U.S. Railroad Market (units)	(%)	U.S. Industrial Market (units)	(%)	Export Market (units)	(%)	Variety of Classes Built
1850	37	37	100	—	—	—	—	12
1860	80	77	96	—	—	3	4	18
1870	280	253	90	20	7	7	3	23
1880	515	422	82	45	9	48	9	54
1890	939	691	74	76	8	172	18	91
1900	1,147	706	61	111	10	330	29	118

Source. Table constructed from actual counts of engines listed in the Baldwin Order Books and Registers of Engines, BLW-NMAH.

Notes: Data for 1850–80 are by fiscal year (ending June 1), to correspond with the U.S. Manufacturing Census entries on Baldwin. Data for 1890 and 1900 are calendar-year totals. All totals are for engines built, rather than merely ordered, in the year. Note that at each decennial Baldwin's total output was either rising or peaking within the contemporary trade cycle for locomotives. There are some discrepancies in the annual outputs given here compared with those provided by other sources (listed in appendix A), but the margin of error is relatively small.

Baldwin assigned a class designation to different types of engines in production, and the table gives this measure of the growing diversity in its product line. But these designations accounted only for cylinder size, total number of wheels, and number of drivers. Therefore two engines of the same class could vary widely in hundreds of other details. For example, in 1890 Baldwin turned out engines in ninety-one classes, but the total number of different designs was 316.

in 1866, Baird had taken on two new partners, and three more members joined the firm in 1870. Three of the six owners—Edward Longstreth, William Henszey, and Charles Parry—had substantial experience in locomotive design and in updating Baldwin's practice to market preferences. By 1871 the enlarged partnership had developed the capability to create products for four discrete sectors of the locomotive market, and the firm pursued each sector vigorously. These market divisions were custom engines for technically venturesome master mechanics; standard road and switching locomotives for the majority of mainline railroads that believed that pioneering didn't pay; motive power for industrial, agricultural, and other nontraditional customers; and a potentially vast export market.

Table 3.1 outlines the size of three of Baldwin's market segments—U.S. mainline carriers, industrial buyers, and export customers—and their growth over time. American mainline railroads accounted for most of Baldwin's sales throughout the period, but during the Gilded Age the company spurred its growth by meeting the particular motive power needs of industrial and export customers. While a breakdown of custom versus standard production is impossible, custom engines accounted for much of the U.S. railroad and export segments tabulated here.[62] Inferential evidence also suggests that the percentage of total output made to custom designs increased substantially over the years shown in the table.

While Alexander Mitchell had to persuade Baldwin to build his *Consolidation* in 1866, soon thereafter the firm was soliciting such custom work. In its 1871 catalog Baldwin noted that its standard engines "admit of modifications, to suit the preferences of railroad managers, and where machines of peculiar construction for special service are required, we are prepared to make and submit designs, or build to specifications furnished."[63] By this time the firm had already taken blueprint jobs for Pennsylvania Railroad engines, and it had developed new narrow-gauge types to meet the specifications of the Denver and Rio Grande.[64] In the 1870s the firm built to railroad-standard designs for the Pennsylvania and to detailed specifications for the Philadelphia and Reading and other lines. It turned out one-off custom types to master mechanics' specifications, such as a new light tank engine of the 2-4-4 type for the Camden and Atlantic, and it also made freak engines to specifications from single-rail lines in the late 1870s.[65] At this point Baldwin would build whatever the customer desired, be it a design advance or a bizarre offshoot.

The buyers of freak engines paid cash and then disappeared into mechanical obscurity. On the other hand, when the leading master mechanics developed locomotive advances, Baldwin closely followed the performance of their new engines. Once the *Consolidation* proved itself on the Lehigh, the design became a standard product for Baldwin, listed in its catalog and available to all customers. In this way one road's innovation often became incorporated into general railway practice. This happened repeatedly for many different types and sizes of engines. A particular type could even be further refined as a custom product and then reintroduced as a standard one.[66] For example, in 1889 the Philadelphia and Reading's motive power superintendent, G. W. Cushing, designed a new, heavy Consolidation type. Baldwin built the custom order and then sold thirteen more engines of Cushing's original design to the Northern Pacific.[67] Here a western line, completed only six years earlier, took advantage of the technical expertise of the fifty-year-old Philadelphia and Reading.

This incorporation of the master mechanics' initiatives helped Baldwin's standard product line to grow from fifteen models in 1846 to forty in 1871 and to seventy-two only ten years later.[68] Most mainline American railroads were quite content to forgo extensive design experimentation in locomotives and buy the builders' standard products. Benefiting from the leading master mechanics' empiricism, their motive power fleets could advance in step with their traffic

The diminutive character of narrow-gauge locomotives becomes evident when juxtaposed with standard-gauge equipment. Here the Rio Grande Railroad's first engine, fresh from the Baldwin erecting shop in 1872, rests on the standard-gauge flatcar that took it west. A 2-4-2T type—the T signifying the lack of a separate tender—this wood-burning locomotive hauled light passenger trains on 3-foot-6-inch-gauge track. It sold for $6,450. RAILROAD MUSEUM OF PENNSYLVANIA (PMHC)

The *Col. A. J. Wilcox* of 1878 was an unusual engine for an unusual concept. The Bradford and Foster Brook was a monorail line, an idea promoted for its savings on construction costs compared with normal railway practice. So the engine had two in-line driving wheels, with balancing guide rollers mounted below. The concept quickly faded into the obscurity it deserved. SMITHSONIAN INSTITUTION

The heavy Consolidation design developed by G. W. Cushing of the Reading, which Baldwin then sold to the Northern Pacific as well. Bearing Baldwin construction number 10,000, it sold for $12,500 and included such innovations as the Westinghouse air brake, the Eames vacuum brake, and the Westinghouse train signal. The Northern Pacific placed this order on April 13, 1889, and Baldwin shipped it out on June 14—eight weeks being its normal production schedule. RAILROAD MUSEUM OF PENNSYLVANIA (PMHC)

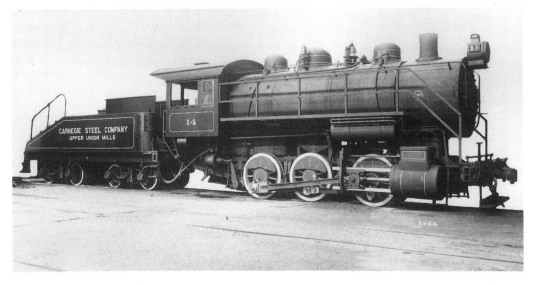

During the Gilded Age high-volume manufacturers like Carnegie Steel learned how to boost productivity by using locomotives to move raw materials and integrate stages of production. Here is a largely standard Baldwin heavy 0-6-0 switcher in 1914 for Carnegie's Upper Union Mills. RAILROAD MUSEUM OF PENNSYLVANIA (PMHC)

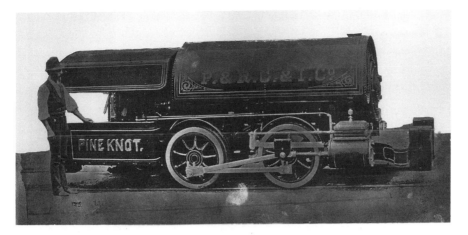

The *Pine Knot* ran inside the mines of the Philadelphia and Reading Coal and Iron Company, hence its small stature. The first model built by Baldwin to this unique design, this little steamer was finished in September 1873. It sold for $5,000, ran on 3-foot-8-inch-gauge track, weighed 17,000 pounds, and could haul 340 long tons on level track. Since it burned anthracite, the *Pine Knot* could not stay long inside the mine workings. Such service better suited the compressed-air engines Baldwin also developed during this period. RAILROAD MUSEUM OF PENNSYLVANIA (PMHC)

loads and in an orderly and relatively economical fashion. Over time most of the carriers that bought builders' standard products created their own road standards for locomotive details and fittings, promoting further economies. They profited by this combination in three ways: the developmental risks were borne by others, the major components of their engines were Baldwin standards, and the minor parts in need of frequent repair or replacement were road standards. While no breakdown in total output or comparative costing between custom and standard engines is feasible, the evidence suggests that standard mainline locomotives provided the bulk of Baldwin's business and profits. The learning curve for managers and workers producing these engines was well advanced, and Baldwin-standard components maximized production economies.

The second market segment Baldwin sought after the Civil War was a host of new nonrailroad customers who discovered the value of locomotives in industrial, agricultural, and mass-transit applications. As America's industrial transformation accelerated after 1865, steel mills, lumber companies, mining outfits, and many other industrial firms boosted throughput and productivity by adopting locomotives. Many of these customers bought switching engines from Baldwin's standard line; in other cases the builder designed new types specially adapted to their needs. These included narrow-gauge models of standard switchers and 5-foot-tall steam and compressed-air locomotives for use inside coal and gold mines. Baldwin sold industrial engines to such companies as Carnegie Steel, Plymouth Cordage, American Dredging, Cannelton Coal, and Pacific Guano. Baldwin's sales to industrial and agricultural customers increased from zero units in 1860 to twenty in 1870, and thereafter this segment accounted for roughly 9 percent of annual output.

Baldwin also developed a market among the streetcar companies and commuter rail lines that arose out of America's urban growth after the Civil War. As a speculative venture, Baldwin built its first steam streetcar in 1876. Within two years the company offered seven different streetcar models, and its steam cars were running on the streets of Cambridge, Philadelphia, Baltimore, Richmond, Memphis, Dubuque, and Havana.[69] The firm created other special locomotive types for the new commuter lines reaching out to developing suburban areas. By 1881 Baldwin offered twenty-eight engine models specially adapted for short-haul passenger services like elevated railways and suburban branch lines.[70]

In total, the market segment of industrial and agricultural companies and short-haul passenger lines represented an important opportunity for Baldwin to expand its product line and sales volume. While these small engines produced lower unit profits than mainline road engines, their share was important. For most of this segment of demand, Baldwin could follow its own design preferences without extensive customer direction. Thus the company economized by

For many decades Baldwin enjoyed regular sales to agricultural producers like the sugar plantations of Cuba and Hawaii. Here is the *Hanapepe* of 1892, built for the Hawaiian Sugar Company. As was common in export sales, a commission merchant house arranged this transaction and received 5 percent of the $4,550 sales price. The 2-foot-6-inch-gauge engine had outside frames (frames outside the wheels), often a necessity on narrow-gauge models with their short axles. This photograph was taken inside the erecting shop, with the photographer using a white curtain to block out the background. RAILROAD MUSEUM OF PENNSYLVANIA (PMHC)

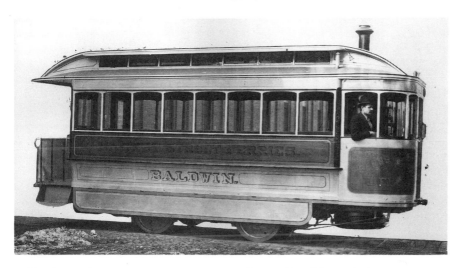

Baldwin's 1876 demonstrator steam streetcar poses here on Broad Street. Such cars were known as "Dummies," since their exhaust sound was muted and their machinery hidden to avoid frightening horses as they passed. Such a car carried passengers and could also haul an unpowered trailer car. Until the rise of electric trolleys in the 1890s, Baldwin sold a number of Dummies to urban operators around the world. When the new technology came into favor, the locomotive builder made electric trolley trucks. RAILROAD MUSEUM OF PENNSYLVANIA (PMHC)

using standard components across a number of sizes and types. These Baldwin-developed designs for nontraditional customers were a successful initiative by the firm to broaden its market base and utilize its productive capacity in labor and machinery more intensively.

Exports constituted Baldwin's third distinct market sector. Overseas demand was vital to the firm's growth after 1865 and proved particularly helpful in tempering the effects of U.S. business depressions. From the standpoint of design, however, the export market was particularly challenging. On the one hand, the overall design characteristics of U.S. locomotives compared with British engines made them attractive to customers in developing nations, such as South American countries, whose terrain mirrored American conditions. But most foreign customers also required modifications from U.S. practice for the fundamental reason that "the locomotives of nearly every country possess national characteristics."[71] As a result, Baldwin designers had to create hybrid engines that generally had the basic attributes of American design traditions with graftings from the technology of the recipient country. The latter often originated in the predilections of the British railway engineers who had laid out most of the lines in South America, Africa, and Asia; hence a third national design tradition impinged on export work.

The melding of these three influences was evident in design negotiations

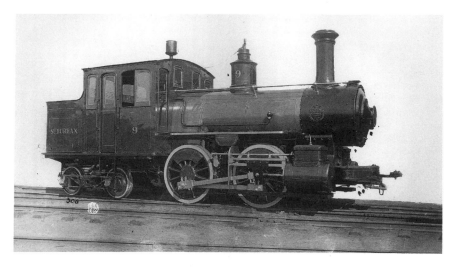

Suburban Rapid Transit number 9 was an 0-4-4T, a Forney type (named for its inventor) suited to short-haul commuter service, in this case out of New York City. Part of an order of eight, it cost $4,700 in 1886 and bore Baldwin construction number 8,167. The very next engine to be finished, tested, and numbered in the erecting shop was the large 2-10-0 freight locomotive for the Northern Pacific shown on page 38. Baldwin's production methods and skilled workers were as adaptable as its design capacities. RAILROAD MUSEUM OF PENNSYLVANIA (PMHC)

between Baldwin partner William Austin and Miguel Tedim, an officer of the Provincial Railways of Buenos Aires, in 1888. The road ordered eight of the common U.S. 4-6-0 types, but with a number of modifications. The Ten-Wheelers were built to a 5-foot-6-inch gauge, with copper fireboxes, brass flues, English-style reversing gear, and English buffers.[72] Such detailed specifications, differing greatly from American practice, predominated in the export trade. Designing to foreign specifications was demanding, giving Baldwin only slightly more latitude on technical issues than it had with custom blueprint jobs for leading U.S. railways. On the question of track gauges alone, practices around the world varied so widely that in a single year, 1890, Baldwin built engines to fifteen different gauges, from 2 feet to 5 feet 6 inches.[73] Despite such complexities, the export trade proved increasingly important to the firm, as table 3.1 shows.

In total, the three segments in Baldwin's late-nineteenth-century business show a deft touch at diversity for such a massively stolid business as locomotive building. Custom work for the leading U.S. railway locomotive innovators and standard motive power for the majority of the American mainline market; industrial and rapid-transit engines built to largely standard designs; and the esoteric specifications of the foreign trade—these diverse endeavors demanded an adaptability in design and production that was characteristic of many other capital goods builders and unknown in mass-production manufacturing.

Although technically venturesome American master mechanics took the initiative in locomotive design development from the builders, this did not relieve the locomotive industry of design functions or the capacity to innovate. While Baldwin did not control the overall rate or direction of technical change in locomotives, the company responded to mandated specifications, it took design initiatives in special markets, and it constantly updated its standard product line in response to market demands and opportunities. As a result, the firm's product diversity exploded. In 1860 the company made a total of 80 engines to 18 different designs; in 1890 it constructed 946 locomotives in 316 different varieties.[74]

THE FULL GROWTH IN LOCOMOTIVE DIVERSITY, 1890–1920

In the ten years following Matthias Baldwin's death in 1866, the firm's partners made the decision to expand into all available or potential locomotive markets. Creating a diversified design capability was essential to that decision. Matthias Baldwin had been "lukewarm . . . to the necessity of a Draughting Room," and he employed only three draftsmen in 1860. By 1878 that number had grown to 16 men, and thirty years later the company employed over 125 designers and draftsmen.[75]

Once it achieved the design and productive capability to diversify its product line, Baldwin forever spurned the idea of national standards for locomotives. When the Master Mechanics Association was founded in 1868 to promote standardization, it was Baldwin's partner in charge of sales, Edward Williams, not the men charged with design or production, who attended its annual conventions. In later years more Baldwin partners joined, but none ever spoke out for the adoption of national design standards. When American railroads came under the control of a federal agency, the United States Railway Administration during World War I, the USRA revived the idea of national standards in motive power. By this time the virtues of manufacturing and of technical standards had become seemingly self-evident truths, given the success of such American volume manufacturers as the Ford Motor Company. Nonetheless, Baldwin initially opposed the concept of standard steam locomotives, although in the end it cooperated in developing twelve designs for use on all mainline roads.[76] Baldwin had opposed standards because it believed that competition among railroad master mechanics to design faster, more powerful locomotives brought more work to the firm. Such competition also drove progress in the railroad industry as leading carriers vied to increase speeds and services. Like Baldwin, the master mechanics disliked the World War I standards. While more than eighteen hundred engines were built to the USRA designs between 1918 and 1920, once the roads returned to private ownership in 1920, "the whole structure of standardization began to fall apart and in a short time it was no more than a memory."[77]

Adopting standard locomotives or even standardizing their major components would have greatly curtailed the design initiatives of master mechanics as well as the diversity in locomotive types. Although the mechanical superintendents' continuous innovations demanded constant reworking of its products and production methods, Baldwin remained largely content to follow their lead through the end of the steam locomotive era. In two cases in the 1890s, however, the builder tried to seize the initiative with two novel types of engines. The firm launched a new model, the Columbia or 2-4-2 type, at the 1893 World's Columbian Exposition. While aspects of the design were generally adopted, as a whole the master mechanics soon rejected it because of its insufficient traction.[78] This episode demonstrated the safety of following the design preferences of mainline roads rather than attempting to anticipate their needs.

In a notable exception to this reactive stance, Baldwin did develop a design initiative in mainline motive power that enjoyed ten years of widespread

Baldwin's drawing-room staff pose here sometime in the 1880s. Thirty years earlier, draftsmen's jobs often led to positions as mechanical engineers. While such advancement was still possible in the 1880s, college training in engineering was on the rise. As draftsmen's ranks grew during the Gilded Age, the jobs increasingly went to machinists who had acquired the considerable skill of reading and interpreting mechanical drawings. The group shown here included many former machinists now pursuing this respected white-collar occupation. AUTHOR'S COLLECTION

This Mikado, or 2-8-2, was one of the twelve standard engine designs developed by a committee of builders and designers under orders from the United States Railway Administration during World War I. Beyond mandating that all mainline American railroads accept these designs, the USRA also established a number of standard components for common use among the twelve engine types. This Mikado was the first USRA-standard engine built. Baldwin rushed production of the 292,000-pound locomotive, building it in only twenty days in 1918. Today the engine is preserved at the B&O Railroad Museum in Baltimore. RAILROAD MUSEUM OF PENNSYLVANIA (PMHC)

popularity: compound locomotives. In 1889 the firm's general superintendent, Samuel Vauclain, patented a unique locomotive design with four cylinders instead of the normal two. In the Vauclain compound, steam from the boiler was first used in two high-pressure cylinders and then reused in two larger low-pressure cylinders before being exhausted up the stack. By using steam twice, this design cut fuel and water consumption by about 15 percent while lessening the problems of forcing and consequent boiler repairs.[79]

Vauclain had not originated the compounding principle, and other builders soon offered competing designs, but his was the most successful of the era. Of the 1,896 compound engines built in the United States through April 1900, 70 percent were to Vauclain's design, an impressive success for a builder-sponsored innovation in mainline locomotives.[80] After this point the principle fell into disfavor because of maintenance difficulties; soon thereafter the carriers adopted simpler means of economizing fuel such as feedwater heaters and superheaters. Although the master mechanics ultimately decided against Vauclain's design, its decade-long run in the market was no small accomplishment, given the dynamic of continuous refinement in locomotives. Before leaving this topic of Baldwin-derived innovations, note that the firm frequently took developmental initiatives in component designs while mostly leaving novel mainline engine types to the master mechanics. Between 1877 and 1900 the builder took out fifty-three patents for innovations in engine components, some of which were

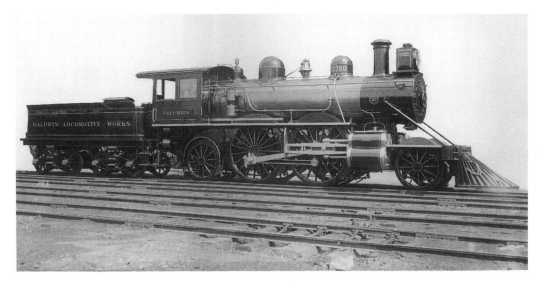

Baldwin occasionally tried out new engine designs in speculative demonstrator models, which also served to market the innovation through in-service tests on a variety of roads. This 1893 demonstrator incorporated two major innovations: the 2-4-2 wheel arrangement advanced by Baldwin's chief designer, William Henszey, and the Vauclain compounding principle. Named *Columbia,* the engine was designed for fast passenger service, having 84-inch-diameter drivers. But American railroads found the design too light to pull useful trains. Ultimately the builder donated its innovative but unsalable prototype to Columbia University for engineering instruction. Certain design features, however, including its two-wheel trailing truck and piston valves, did become common in American railway practice. RAILROAD MUSEUM OF PENNSYLVANIA (PMHC)

incorporated into general locomotive practice.[81] While its customers directed the general direction of technical change in its product line, Baldwin could and did affect aspects of innovation.

Circa 1900 the company regained much of its power and initiative in design issues thanks to two developments that originated in the mid-1880s and grew in importance after the turn of the century. This period saw many railroads adopt performance specifications as a new method of mandating innovation in their orders for engines. These specifications set overall performance targets such as desired speed, power, and maximum weight, but they left it to the builders to derive the specific designs needed to achieve these goals. The second development was the concurrent rise of science-based techniques to test locomotive innovations and materials and assure optimal performance. By 1885 Baldwin had created a test department to analyze scientifically the materials used in locomotive construction to assure their quality. This department expanded in 1891 to doing on-the-road performance tests of locomotives.[82] Scientific testing of locomotive performance under controlled conditions in a laboratory began in 1892, when Purdue University appointed W. F. M. Goss as its professor of locomotive engineering. Goss ran the first U.S. laboratory test plant for locomotives; it performed analyses of fuel and water economy, optimal draft conditions, and precise tractive effort measurements.[83] By 1905 Cornell University and the Pennsylvania Railroad had similar locomotive laboratories.[84]

Controlled performance testing of locomotives quickly replaced the master mechanics' ad hoc experimentation with detailed scientific analysis. The data uncovered were "particularly valuable in designing engines . . . interpreting the performance of engines in service, and . . . form[ing] an accurate basis for reasoning about new designs."[85] Such incontrovertible analysis had the potential of establishing optimal designs for given services, thereby curbing the design diversity that empiricism had fostered.

As it happened, however, scientific testing became coupled to performance specifications, and the combination promoted further diversity in design. These developments also returned much of the initiative in innovation to the builders. In 1900 the trade periodical *Railway and Locomotive Engineering* commented on this trend in an article titled "The Growing Diversity in Locomotive Designs."[86] The Baldwin company was a leader in this development, which occurred in the following manner. As railroads took up the practice of ordering new power through performance specifications, they sent Baldwin the relevant data on the character of their roads—the weights of rail in use, curves, grades, and the like—and on the proposed service, power, and speed anticipated for the new engines. Samples of coal and water were also often sent in to the firm for scientific analysis.[87] The data provided were as extensive as possible.

Armed with this information, Baldwin used or adapted an existing design, or it created an entirely new model to fulfill the mandated service and particulars. With novel designs the drawing room created tentative plans, known as tracings, to provide a basis for negotiations with purchasers. The use of scientific data coupled with performance specifications allowed Baldwin to exert more control over design while giving the railroads new motive power tailor-made to meeting the needs of different divisions and even different services within a division.[88] As a result, the design diversity in American carriers' locomotive fleets expanded as never before. Baldwin's product line grew so large that for a number of years around 1900, the firm supplemented its catalogs of standard models by issuing a series of *Records of Recent Construction* to outline its extensive custom work. While the chief mechanical officers of the major railways still held the general initiative in innovation, their use of performance specifications greatly increased builders' influence on design.[89] In turn, the need for close technical collaboration eased price competition in the industry somewhat while cementing long-term relations between particular builders and the major carriers. Circa 1900 Baldwin's drafting department was creating 125 completely new locomotive models every year.[90] Many of these custom designs then entered Baldwin's standard product line. Given this capacity, no wonder the company saw little production or marketing advantage in the USRA standard engines proposed in 1918.

HISTORICAL LESSONS FROM LOCOMOTIVE INNOVATION

Nineteenth-century buyers of heavy capital goods typically exerted a strong influence over the design of such machines, regardless of their applications. Why leading railway master mechanics took over the responsibility and initiative in promoting locomotive development is therefore an issue with ramifications extending well beyond the railway sector. But two more immediate reasons account for the central emphasis accorded this question: it was important to the railways, and it vitally affected operations at Baldwin. First the railroads' perspective. Locomotives were such an essential part of railway operations that the carriers' mechanical officers almost inevitably sought to adapt them to the particular needs and growth conditions of their companies. Such an effort should come as no surprise to technological historians, who have learned that technologies cannot be studied disembodied; machinery designs incorporate the context and values of designers and users. A similar dynamic of user initiative in innovation occurred in steam locomotive development throughout the world. Indeed, during the nineteenth century such technical flux was common in heavy capital equipment like machine tools, ships, steam boilers, and mill machinery.[91]

Economics offers a partial explanation for this ceaseless innovation. Capital goods buyers believed that improved operating efficiencies more than offset the costs of design change.[92] But the diversity in the master mechanics' designs also flowed logically from the varied impact of geographical, business, and social contexts on the makeup of a technology. Roads varied physically and in the character of their traffic, while the predilections of prideful master mechanics were often quite diverse. Unfortunately this is generally overlooked by historians who have focused on railways as standardized systems and managerial bureaucracies. While their work offers important findings, it also ignores the personal aspect of human endeavor so essential to understanding both technology and history. The USRA's Progressive administrators may have been correct in arguing that the great variety of locomotives on American railways was inefficient and costly. If so, that is a question for historians of railway operations. But if we are to understand the challenges of the systematizers and standardizers, and their frustrations, we do them a disservice when we foreshorten the past by our own imposition of order on historical complexity.[93] Disorder and diversity, empiricism and ego, took the 25-ton 4-4-0 of 1855 with its 7,000 pounds of tractive force and transformed it into the workhorse 4-6-2 of 1915, weighing 150 tons and capable of 50,000 pounds of tractive effort.

The master mechanics' diverse initiatives in locomotive design were also of fundamental importance to the locomotive builders. Their power was the defining aspect of Baldwin's relationship to its leading mainline customers. The mechanical officers' control determined how, where, and when Baldwin could advance its own product design initiatives. Building to customers' specifications was common for most nineteenth-century capital goods firms. This has been overlooked by technological and business historians who have centered on mass-production manufacturers facing very different design and production issues. For these reasons, Baldwin's record of product diversity is described here in detail. In the 1860s and 1870s the firm's acceptance of master mechanics' technical leadership and its concurrent initiatives in new locomotive markets were fundamental strategic decisions that led to the company's growth. The managers who made these decisions are our next focus. Baldwin's success depended upon their creation of managerial controls suited to such a diverse product line.

4

MANAGEMENT AT BALDWIN

1850–1909

IN 1850 MATTHIAS BALDWIN'S FACTORY WAS VALUED AT $250,000, AND HIS four hundred workers built thirty-seven locomotives in twelve different classes during the year.[1] Notwithstanding its size, most of the company's success to this point derived from its early entrance into the locomotive market and from the mechanical talents of its founder. Matthias Baldwin's character and interests still dominated the firm, and he was a mechanician, not a manager. Powerful semiautonomous foremen largely conducted the company's day-to-day operations. This decentralized system did not last long, however. As railroading matured in the 1850s, two new trends challenged Baldwin's early advantages and caused fundamental changes in the structure of the industry and in Baldwin's operations. The first was heightened competition as new entrants crowded into locomotive building. At the same time, railway master mechanics began exerting primary influence over the pace and character of technical innovation in locomotives. These two developments largely negated the advantages Baldwin had enjoyed in early market leadership and the mechanical ability of its founder. As a result, Baldwin's market share fell from 21 percent in the 1840s to 16 percent in the 1850s.[2]

Bouts of intense price competition, highly volatile demand cycles, and customers' influence over design became common challenges for most nineteenth-century capital equipment firms, not just locomotive builders. These market-imposed demands, which arose in the 1850s, added to Baldwin's internal concerns of managing a large, skilled workforce in labor-intensive production of complex and unwieldy products. Such challenges marked another fundamental divergence between this sector and American System consumer product manufacturers. Between the 1850s and the 1870s Baldwin's partners adopted new managerial systems and tactics in response to these market forces and internal production challenges. They sought to improve coordination and control while maintaining the managerial and productive flexibility required to cope with burgeoning product variety and highly fluctuating demand.

While the evolution of Baldwin's management practice between 1854 and 1872 illuminates how the firm grew to great size and dominance, those changes also have a more general significance. Business historians date the origins of industrial management to the late 1870s and 1880s.[3] Most believe that until that period firms had little need or incentive to create managerial controls because industrial companies were relatively small, their markets were restricted geographically, and their owners were preoccupied with mastering the technology of volume manufacturing. For example, Alfred Chandler and others place the origins of systematic management practice among mass producers following the depression of the 1870s.[4] In skeletal form, this standard account argues that the demands of mass production impelled American System and other manufacturers to create novel oversight systems.

Unfortunately for this analysis, Baldwin and other capital equipment firms had created new organizational controls ten to thirty years before their American System consumer product cousins took up such concerns. Precisely because they could *not* mass-produce, machinery builders sought to improve productive efficiency by developing systems of oversight and coordination. Their record of innovation in this field is so extensive that historians may need to reformulate their basic understanding of the systematic management movement, placing its origins in the capital equipment sector of the 1850s and 1860s rather than among the high-volume manufacturers of the 1880s.[5]

Although much of Baldwin's eventual success in its industry resulted from astute managerial development, the company never became one of the "center firms" described by Chandler.[6] As a major supplier to the railroad industry, Baldwin was closely acquainted with the divisional structures and managerial hierarchies first developed in railroading and then adopted by center firms in high-volume manufacturing. But the locomotive builder largely rejected the "visible hand" of managerial control. Wherever possible, Baldwin avoided creating a corps of salaried middle-level managers, concentrating instead on the development of oversight and control systems. Its partners knew that the locomotive market frequently collapsed in sales droughts that a managerial bureaucracy and marketing tactics were powerless to prevent. The growth of custom designs after 1850 blocked exponential growth in production volume or throughput. Baldwin's essential strength lay in reacting quickly to market desires, adapting to shifts in demand or changing technical needs, rather than seeking control through managers. Although never becoming a center firm analogous to U.S. Steel or Singer Sewing Machine, Baldwin learned to adapt to the structures of its markets and the requirements they imposed on production.[7] The fact that it grew to become the largest capital equipment firm in America

This group portrait includes seven of Baldwin's fourteen owner-managers of the last half of the nineteenth century. Above Matthias Baldwin (*middle*) is George Burnham. Proceeding clockwise from Burnham are Charles Parry, Edward Williams, John Converse, Edward Longstreth, and William Henszey. HAGLEY MUSEUM AND LIBRARY

suggests that Baldwin's management structures admirably suited the particular conditions of the capital equipment sector.[8]

Cultural and social factors also hindered the creation of a managerial class at Baldwin. First among these was the company's 1830s heritage. Jacksonian Americans held "a widespread belief that a firm managed by anyone other than its owner could not but fail."[9] This distaste for salaried personnel lacking any direct stake in a company continued for decades, lasting at Baldwin into the twentieth century. Managers in our modern sense remained so rare in the mid-nineteenth century that the term itself had little useful descriptive purpose. Until late in the century American trade unionists spoke of producers versus accumulators, not managers and workers. Some "managers" existed before the Civil War, and at various points labor and management did consciously identify their interests as separate and opposed, as in Baldwin's 1860 strike. But business managers in the middle of the nineteenth century had prerogatives and sensibilities very different from those of their counterparts in 1900 or today.

In their early form, managers were tolerated as a necessary outgrowth of industry—a very small head on the wealth-creating body of work and workers. Company owners themselves generally believed this, and they tried to run their firms with as few overseers as possible,[10] Where managers existed in the mid-nineteenth century, at Baldwin and other industrial companies, frequently they had risen from the ranks of labor.[11] In passing from labor to management, many of these men came to espouse a producer ethos, which combined regard for the work ethic and the manual skills and labor of their blue-collar days with an individualism that their own promotions and rising fortunes seemed to justify.[12]

In an age without an ideology of professional management or a class of salaried managers, the producer ethos became a tenet of management practice at Baldwin and the foundation of its company culture. Most of the firm's partners worked their way up through the ranks, starting out as teenage apprentices, draftsmen, and clerks. These men valued work and workers, not overseers. They saw the company's business as wealth creation, commonly measured in that era by value-added—the difference between the cost of raw materials in a product and its final selling price. The profit derived from transforming raw materials into finished, merchantable goods chiefly resulted from labor's efforts. Decades passed before industrialists perceived that nonproducing managers and other personnel could make a sufficient contribution to value-added to earn their keep, let alone improve profits.[13]

Throughout the nineteenth century this producer ethos, or culture, kept Baldwin's managerial ranks quite small and very active. In 1879 a visitor to the factory wrote, "Owing to the magnitude of the establishment, a stranger would naturally suppose that it would require almost a regiment of Bosses and any amount of 'bossing,' but such is not the case by any means. There are very few lookers-on to be seen in the shops. . . . The proprietors, superintendents, gang, and track bosses all work themselves so there are very few, if any, drones in the hive."[14] As late as 1910 Baldwin's president, William L. Austin, spent over half his normal workday checking locomotive drawings in the drafting room where he had begun his career forty years earlier.[15] Although Austin headed a fourteen-thousand-man company, under Baldwin's producer ethos the president and all other managers defined their own work in furtherance of the company's single goal: building locomotives. Arising from the partners' careers and beliefs, this culture informed all of their management practice. The company led in developing systematized managerial controls between the 1850s and the 1870s, but it was no birthplace for a white-collar class.

THE BALDWIN PARTNERSHIPS, 1854–1909

Between 1831 and 1909 the Baldwin Locomotive Works was owned outright
by a succession of fourteen proprietorships or partnerships, listed in table 4.1.
Three of these reorganizations occurred shortly after the 1837 panic. That
depression caught the young company at a vulnerable time with large debts
outstanding. By 1846 the firm had weathered these problems, and it reverted
to the sole ownership of Matthias Baldwin. Although the company saw nine
more reorganizations before finally incorporating in 1909, not one of these
restructurings resulted from depressions, debt, or a need for capital.

In using the partnership form, Baldwin typified many large industrial
companies of the nineteenth century. Historians have said little about this
device of ownership and management, beyond commonly emphasizing its
impermanence, since partnerships had to be reconstituted or abandoned when
any member of the firm withdrew. This is contrasted with the unlimited life
of corporations.[16] But contrasting these legal forms is in fact not very useful or
enlightening. In the first place, it pits a nineteenth-century institution against
a twentieth-century phenomenon.[17] More important, the emphasis on form
masks substance. While Baldwin's owners restructured the firm every 5.4 years
on average, they directed a company with a continuous record of production
and expansion from 1831 to 1909. During those seventy-eight years employment
grew from thirty men to over eighteen thousand, and when the firm eventually
did incorporate in 1909, its asset value exceeded $18 million. The impermanence
of the partnership form evidently created no great obstacle to Baldwin's growth
and success in the business environment of the nineteenth century.[18]

Baldwin's owners believed in the partnership form precisely because of the
managerial continuity that private ownership allowed. As member of the firm
Alba Johnson noted circa 1907, this type of organization "gives a permanency
in the conduct of the business."[19] To achieve that goal the Baldwin partners
adapted the partnership to provide needed managerial skills, promote the
liquidity of their ownership stakes, and maintain continuity in management.
To secure men qualified to run the company after his death, Matthias Baldwin
looked to his own employees. For over fifty years his successors followed this
practice of securing continuity in ownership and management by grooming
a number of able, experienced employees for elevation to partnership rank.
Chart 4.1 shows the extent of this de facto partnership training system.
Excluding Baldwin himself, the men listed had an average of thirteen years of
experience in the company before becoming partners and seventeen years as
owners. Five of the firm's Gilded Age partners—Austin, Burnham, Johnson,
Longstreth, and Parry—began working at the company as teenagers, three as
apprentices.[20]

TABLE 4.1 The Baldwin Partnerships, 1831–1909

Year	Company Name	Partners
1831	Matthias W. Baldwin	
1839	Baldwin, Vail & Hufty	M. W. Baldwin (33⅓%), George Vail (33⅓%), George W. Hufty (33⅓%)
1841	Baldwin & Vail	Baldwin, Vail
1842	Baldwin & Whitney	Baldwin, Asa Whitney
1846	M. W. Baldwin	
1854	M. W. Baldwin & Co.	Baldwin (66⅔%), Matthew Baird (33⅓%)
1867	M. Baird & Co.	Baird, George Burnham, Charles T. Parry
1870	M. Baird & Co.	Baird (33⅓%), Burnham, Parry, Edward H. Williams (12½%), William P. Henszey (12½%), Edward Longstreth (12½%)
1873	Burnham, Parry, Williams & Co.	Burnham, Parry (16⅔%), Williams, Henszey, Longstreth, John H. Converse
1886	Burnham, Parry, Williams & Co.	Burnham, Parry, Williams, Henszey, Converse, William C. Stroud, William H. Morrow, William L. Austin
1891	Burnham, Williams & Co.	Burnham, Williams, Henszey, Converse, Stroud (18%), Austin
1896	Burnham, Williams & Co.	Burnham, Williams (19%), Henszey, Converse, Austin, Alba B. Johnson, Samuel M. Vauclain, George Burnham Jr.
1901	Burnham, Williams & Co.	Burnham, Henszey, Converse, Austin, Johnson, Vauclain, Burnham Jr.
1907	Burnham, Williams & Co.	Burnham (14%), Henszey (20%), Converse (20%), Austin (16%), Johnson (15%), Vauclain (15%)
1909	Baldwin Locomotive Works Inc.	(Privately held)
1911	The Baldwin Locomotive Works Inc.	(Publicly held)

Source: BLW, *History of the Baldwin Locomotive Works*, p. 4.

Notes: Percentage figures following partners' names indicate their ownership interest in the firm. During the life of the 1854 Baldwin/Baird partnership its ownership was restructured to 50-50.

CHART 4.1 The Baldwin Partners, 1831–1909

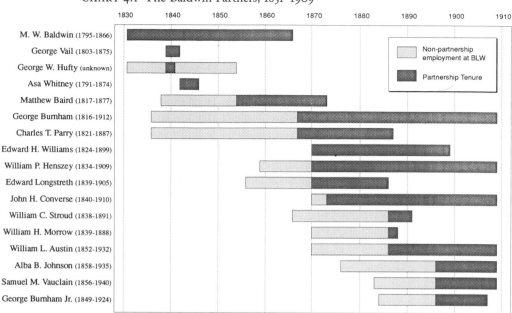

In elevating such men to partnership rank, the senior partners also secured stability in policy and control while promoting the liquidity of their own ownership stakes. In a type of Gilded Age leveraged buyout, the younger partners used the ongoing profits of the firm to buy out the interests of the older men as they withdrew. The chief threat to managerial stability came in the aftermath of a partner's death, since surviving partners had to pay off his interest to his heirs. Here too Baldwin's owners devised methods, perhaps common in the nineteenth century, to promote continuity. When a partner died the firm's partnership articles stipulated that his "interest . . . be drawn out so easily and slowly that there is no hardship on the survivors [in the firm]. The payment of that estate is no more onerous than the payment of dividends would be in a corporation."[21] In order to assure that compensation to heirs would not cripple the company, the partners frequently named one another as executors of their estates.[22] Thanks to the adaptability of its legal form, the Baldwin company grew and prospered for almost eighty years.

The actual concerns within Baldwin's nineteenth-century partnerships centered on three issues the partners saw as closely related: assuring a managerial succession, securing credit for working capital, and protecting assets in the absence of limited liability.[23] These issues united in a focus on character, on the personal integrity and sound business judgment of each member of the firm. The partners believed that integrity and judgment were the sine qua non

of success in the commercial world. While many businessmen fell short of
the mark, perhaps including the Baldwin partners at times, the importance of
stainless character to nineteenth-century business success cannot be discounted.
Recollect J. P. Morgan's assertion in 1912 that the true basis of commercial credit
was "character . . . before money or anything else."[24]

Character or integrity coupled with good business sense had to be the
foundation of nineteenth-century commerce; it literally could not have been
otherwise. In the absence of bank loans for working capital, accounting
standards for assets, or even public reporting of a partnership's assets, loans
were secured largely by good reputation. What was true of face-to-face dealings
became even more important when conducting contractual obligations at
a distance; hence the importance to the partners of their description in the
R. G. Dun & Co. credit reports as "men of excellent character and business
capacity."[25]

The issues of character and judgment also figured prominently in the
selection of new partners. After its 1842 reorganization, the firm had sufficient
cash reserves and access to short-term credits that it never again had to admit
a partner to acquire an infusion of working capital. Matthias Baldwin sought
his next associate in 1853 so that he could be relieved of some managerial duties
and occupy a seat in the Pennsylvania State Legislature. He chose his longtime
foreman of the boiler shop, Matthew Baird, indicating that he valued expertise
over new capital. For a one-third share in a $300,000 company, Baird initially
paid only $30,000 to the firm.[26] After Baird, twelve other men joined the
Baldwin firm during the nineteenth century, and the evidence indicates all
were admitted without any initial cash payment.[27] Because partnerships have
been little studied, it is unclear if Baldwin was typical in this regard, although
another nineteenth-century firm with extensive need for capital, J. P. Morgan
& Co., also consistently took on new partners without expecting or requiring
any initial contribution of capital to the firm.[28]

Like Morgan, the Baldwin principals followed this course because "integrity
was more important than any other quality," and the firm valued able,
experienced, and dependable partners over wealthy ones.[29] Integrity in each
was vital protection to all partners in the absence of corporate limited liability.
As one of the Baldwin principals noted, a partner having the ability to "sign
for the firm" in contracts had obligated the company and all of its partners
individually. He could "make an improvident contract . . . which would plunge
his partners into debt. [But] we have a way of getting over that, and that is by
knowing thoroughly the men before they become partners. We have grown up
together, we have labored together for many years."[30] Of the thirteen partners
admitted after 1853, only one, Edward Williams, had not previously worked in a
subordinate capacity in the firm.[31]

Matthew Baird's life was a classic American success story. While a child, he emigrated from Ireland with his family. He followed his father's trade as a coppersmith, apprenticed in locomotive work at the New Castle Manufacturing Company, and became Baldwin's boiler shop foreman in 1838. Matthias Baldwin took him on as a partner in 1854, and Baird became senior partner after Baldwin's death in 1866. By then he was a member of a wealthy group of investors known as the "Philadelphia interests." On his death in 1877 he left an estate worth $3 million, including the value of his steam yacht. WILLIAM P. SCOTT

The partnership was similar to an exclusive club in that members preferred to select men of similar backgrounds. Most had middle-class origins, but they ranged from Edward Williams's comfortable childhood in a Vermont political family to emigrant Matthew Baird's upbringing in the household of a skilled coppersmith.[32] Following Matthias Baldwin, many were active Presbyterians, but the firm also included Swedenborgians, Quakers, and an evident atheist, Baird. Those joining the firm generally had two other traits in common: they were very hard workers, and like Horatio Alger's heroes, they had a mentor. Baird served that role for the young Charles Parry, who in turn nurtured Samuel Vauclain's career. William Henszey guided William Austin up through the ranks.[33] On the job, each partner had charge of a given department—drafting, finance, sales, or production—which he largely controlled. Although they had weekly partners' meetings to discuss general problems, they were the masters of their own specialties.

This introduction to Baldwin's partnership operations suggests the importance of perspective in understanding the partners' needs and beliefs. A twentieth-century viewpoint emphasizes the impermanence and risks of this organizational form, but in their own time Baldwin's owners believed that the partnership form met their commercial needs while mirroring their own values. The recruitment pattern for new partners was deeply rooted in the firm's producer ethos, first established by Matthias Baldwin. The rights, profits, and risks of management and ownership properly accrued to those whose qualities, training, and efforts made them deserving. The ethos was grounded in an intensely capitalistic society, and certainly not all of high merit within the company received the reward of ownership. But in the decades after the Civil War the partners tried to replicate this conception of individual merit and advancement for middle managers and workers. The producer culture that justified their own careers guided their managerial decisions as they sought new methods to control the company's operations and improve its performance.

MANAGEMENT OF THE COMPANY, 1850–1866

For many years Matthias Baldwin had as little of management and managers as possible. Once he had worked clear of his financial troubles after the 1837 panic, he ran the company for eight years (1846–54) with no partners. In central management he was assisted by George Hufty, who had briefly been a partner (1839–41) but now served as general superintendent. In addition Baldwin relied heavily on George Burnham, who had begun working for the locomotive builder as a teenager in 1835. In 1850 Burnham oversaw the company's finances, largely because the mechanician Baldwin had no interest in or aptitude for such matters. Assisting the founder in design was William Pettit, whose thirty-five-year career at the firm began even before Baldwin first entered the locomotive

Baldwin's partners' room circa 1895. Note the builders' photos on the walls and partner Edward Williams at the far end. Since the early 1880s the office had had electric lighting and telephone service. In this relatively small office the partners conducted the firm's business and met with clients. The room reflected the partners' collective power, in contrast to the managerial hierarchies at incorporated firms. H. L. BROADBELT

market. Beyond Matthias Baldwin, these three men constituted the central management of this four-hundred-man company in 1850.[34] They were expected to do their work with few if any assistants, since their boss "did not approve of employing either clerks or draftsmen and regarded them as drags on the business."[35] With so few at the top, the foremen exerted great authority in the decentralized structure described in chapter 1.

This system of limited management met the company's needs until the railway boom of the early 1850s. Early in that decade workers' output failed to keep up with mounting demand for Baldwin engines. To spur productivity, the company switched from hourly to piecework wages.[36] By the 1860s piece rates were applied to most tasks throughout the plant. Piecework required a thorough subdivision of work tasks, with separate rates set for each job. Although rates might need subsequent adjustment, once they were set, pieceworkers did "not need much looking after," since they had a monetary incentive to work hard.[37] Without any increase in managers, piecework achieved Baldwin's desired improvement in productivity. Of itself this decision to use piece rates was hardly unique to capital equipment builders. But with so many varied components in locomotives, standardizing work was more difficult to accomplish in this sector, just as its labor-intensity made workers' productivity a pressing concern.

Although piecework enabled the company to meet its goal of building an engine a week in the early 1850s, demand for locomotives continued to increase, creating further pressure to improve productivity. When Matthew Baird became Baldwin's co-partner in 1854, his rise marked a transition from the entrepreneurial company centered on its founder to a new emphasis on internal management controls to achieve growth and improve efficiency. Because the new junior partner had to devote much of his time to sales negotiations as the industry's competitive pace mounted, Baird engineered the appointment in 1854 of a new general superintendent, Charles Parry, to promote internal change. Parry had first entered the factory in 1836 as a fifteen-year-old pattern-maker's apprentice, and he was working as a draftsman when Baird selected him for promotion.[38] As superintendent, one of Parry's first initiatives was to order the redesigning of all locomotives in the company's standard product line. This accomplished two goals. Baldwin engines were simplified and updated to suit contemporary market preferences—changes that also streamlined construction and improved the firm's productivity.

With this effort under way, Parry also turned to the managerial problems created by the master mechanics' mounting insistence on engines built to custom specifications. Until he took over as general superintendent, "locomotive building had not been reduced to a system. Each engine was constructed without much reference to those which were built before or those which

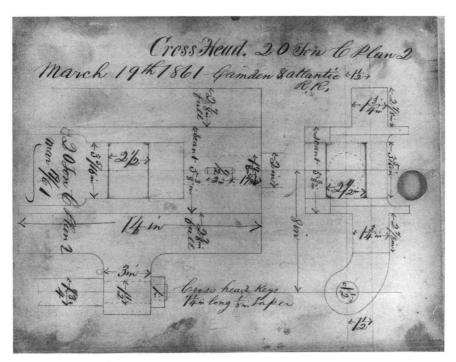

Card drawing from 1861 for a crosshead, part of the mechanism transmitting power from the piston rod to the driving rod. The card was identified by the part name and the Baldwin designation for a particular class of engine, the 20-ton C class to plan 2. Soon after this era, cards were simply numbered. Such drawings established standard designs for parts, speeding drawing room work and improving shop workers' productivity. H. L. BROADBELT

would come after it. Complete drawings were almost unknown."[39] This was "building" technique at its most basic level. Such unstructured production had been manageable in the days of limited output and standard designs, but as product variety mounted in the mid-1850s, Parry found this lack of system intolerable. He had come to a somewhat paradoxical insight: the optimal way to produce custom engines to customer specifications required the use of as many standard components as possible. Standard parts became a corollary effort to the standardization of tasks under piecework. To establish and maintain these component design standards, Parry instituted another change with great managerial implications: he required that workers follow drawings in making all parts. This directive placed the controlling power over production in the drawing room. Such consolidation of coordination required more draftsmen, to Matthias Baldwin's displeasure, but it also curtailed both the need and the ability of foremen and workers to act independently.[40]

In themselves these three steps—piecework, product redesign, and standard parts designs—were relatively minor changes in the management practice of the company. They resulted, however, in enhanced power for the superintendent

CHART 4.2 Organizational Chart of the Baldwin Locomotive Works Management, December 1859

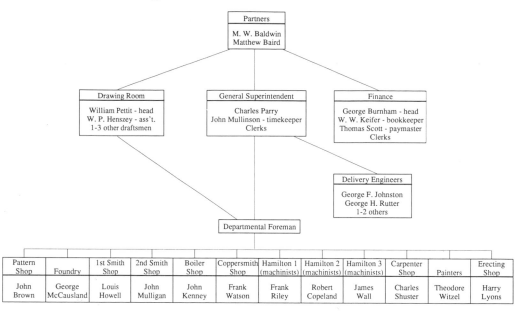

Baldwin's department heads posed for a group portrait in December 1859. The key personnel sat in the front row: Louis Howell (number 2), smith shop foreman and thirty-year Baldwin veteran; Charles Parry (3), then superintendent and later a partner; Matthew Baird (4), partner; Matthias Baldwin (5); William Pettit (6), chief draftsman and another long-term veteran; and George Burnham (7), who oversaw the firm's finances. HAGLEY MUSEUM AND LIBRARY

and the drawing room, and they were fundamental in giving Parry, Baird, and Baldwin improved control over operations. Control was particularly important because the company did grow during the decade of the 1850s, despite its loss in overall market share. By 1860 Baldwin had moved back up in the market, largely because of the changes described here. In that year its 675 workers made eighty-three engines, placing it second in overall sales. To oversee this enlarged force, new managerial positions were created during the decade, as seen in chart 4.2.

The chart shows an extensive managerial hierarchy for an industrial company of the late 1850s, with roughly thirty men in management.[41] Despite Matthias Baldwin's desire to limit the ranks of nonproducing personnel, it appears that Baird created a managerial system capable of running such a large company and securing further growth. The system arose from conditions particular to the capital equipment sector. In 1860 Baldwin purchased over $360,000 of raw materials and made products worth $750,000—a huge sum. Locomotives of the period had roughly five thousand parts each, and Baldwin made seventy-nine engines in seventeen different classes during the year. Their construction required labor-intensive production in eight skilled trades. To ensure adequate supervision of workers, Baldwin created two smith shops and three machining departments. Generally the firm sent out its own engineers to deliver, set up, and run new locomotives for a trial period before receiving final payment from the customer. These were the demands that gave birth to this large managerial structure.[42]

Having established this hierarchy, the partners' attentions returned to improving control systems. During the early 1860s Baldwin built upon the foundation established by piecework and standard parts designs by instituting standards of interchangeable-parts production. Interchangeability was sought largely for its advantages in marketing and production, but it also had a relation to management practice worth noting here. Until the achievement of interchange standards, "the work was done by what may be called a 'rule of thumb' method [and] a good deal had to be thrown away because it 'didn't come right.' "[43] In addition to these problems of waste and duplication of effort, the cut-and-try method of building engines placed strains on interdepartmental coordination. For instance, if the foundry made a cylinder casting that was too large, it required extra machining work, or the machine shop foreman had to clear some outsized dimensions with his counterpart in the erecting shop. Multiply this example by the thousands of parts in each engine, and it is not surprising that "when it came to the erection of the work, filing and fitting and making over, and perhaps conferences of the various fallible parties to apportion the blame for mistakes, were not unusual."[44]

All of this changed with the adoption of interchangeable parts. Once the Armory system of gauges, jigs, and templates was installed in the mid-1860s, every department knew the allowed tolerances for its work. Parts either were made correctly in each shop and sent on to the erecting shop, or they were rejected and scrapped.[45] Workers and foremen could focus on the tasks within their departments, knowing that Armory practice ensured a measure of coordination between shops. In locomotive building, these new production techniques were essential to promoting and "preserving administrative conditions of order and simplicity."[46]

Interchangeable-parts production, piecework, and standard parts designs improved productivity and work coordination without increasing the ranks of middle managers.[47] This accorded with the prevailing industrialist's ethic, which saw little value—and much expense—in paying managers who supervised work but created no wealth themselves. The partners also saw piecework pay in particular as a logical extension of the company's producer culture: effort and output, not mere hours worked, would determine compensation. In establishing standards of work, design, and production, these three policies also represented first stages in the systematic management movement that most historians date to the 1880s.[48] While admittedly small steps in that direction, they were notable nonetheless, especially in a firm whose principals had begun their careers in the preindustrial setting of craft apprenticeships.

GROWTH AT THE TOP AND THE EXTENSION OF SYSTEMS, 1866–1872

Matthias Baldwin died in 1866. Shortly thereafter Matthew Baird took on long-serving lieutenants George Burnham and Charles Parry as partners in a new firm known as M. Baird & Co. Little is known of the mechanics of this partnership transfer or of the managerial practices instituted by the new firm.[49] One known innovation of this period was the installation, circa 1867, of an internal telegraph system that linked the shops with the drawing room and superintendent's office.[50] Providing instant communication between managerial staff, this network particularly aided interdepartmental coordination.

In 1870 Baird, Burnham, and Parry restructured the firm again by taking on three new partners: William Henszey, Edward Longstreth, and Edward H. Williams. The evidence suggests that two related factors motivated this reorganization. First, the partners perceived the long-term growth potential of the locomotive market, and they sought to increase top management's functional specialization to meet that demand. The new partnership certainly had admirable breadth and depth of experience. Baird had been connected with the company since 1838. Burnham had largely overseen its finances from the 1840s on. Production was Parry's concern, to which he brought experience as an apprentice, patternmaker, draftsman, and general superintendent. The drawing room came under Henszey; he had worked there since 1859. Longstreth was the firm's mechanical prodigy; before completing his apprenticeship in 1862, he had already become a foreman. While in that position he oversaw the installation of Armory practice. Finally, Williams, a new arrival to the company, brought extensive contacts and experience in the railroad industry, so quite naturally he was charged with sales in the new firm.

These broad qualifications in the enlarged partnership also support a second rationale for the reorganization. Evidently Baird initiated this managerial transition with an eye to his own retirement. The new partners ensured that the company would survive him and be able to pay off his interest—as they did when Baird withdrew from the firm three years later. The sequence of events demonstrates how the partnership form could enhance both managerial continuity and the liquidity of an ownership interest.

One aspect of the 1870 partnership appears somewhat paradoxical. By admitting three new partners in that year and thereby doubling the number of principals, the earlier partners reduced their ownership shares. In Baird's case his 50 percent interest in the co-partnership with M. W. Baldwin decreased after the 1870 reorganization to 33⅓ percent.[51] This would seem to represent a real loss for Baird, since he was not immediately compensated by the new partners. Henszey, Longstreth, and Williams came into the firm under "contribution shares," which gave them each a 12½ percent stake in the company, to be paid from their portions of future profits.[52] But Baird and the other partners did benefit by this seeming disadvantage. Although doubling the partnership caused the annual returns to each member to decline in the short run, the enlarged management was expected to create growth for the firm. With a larger pie, each piece would shrink only relatively, not absolutely. According to the R. G. Dun credit reports on the Baldwin company, exactly this took place. The firm of M. Baird & Co. grew in total value from $1 million in 1868 to $3.5 million in 1873.[53] The new partners brought growth to the firm through a number of new ventures. They also directly fueled some of this impressive increase in the company's capital. Their payments on their contribution shares were in essence retained earnings for the firm, with the funds available for reinvestment to achieve further growth. The admission of new partners and Baird's declining share prepared the firm not only to survive his retirement but to prosper.

The two years following the 1870 enlargement of M. Baird & Co. saw the most extensive changes to date in Baldwin's management systems. These policies had a number of roots: the infusion of new talent into the partnership and its newfound ability to specialize, the boom conditions prevailing in railroading, and the rising design diversity in locomotive types. After 1870 the Baldwin company vigorously entered new industrial and mass-transit markets for locomotives, expanded its export efforts, and solicited custom work from technically venturesome master mechanics. Although the enlarged partnership facilitated these efforts, they arose in part from the need to broaden markets and improve profits to provide for Baird's retirement. Any partnership sought to plan in advance for the exit of its largest shareholder.

The managerial systems instituted at Baldwin between the late 1860s and 1872 focused on both intradepartmental productivity and interdepartmental coordination to improve the company's capacity for growth. While changes

Part of the Specification Book entry for a 2-8-0 narrow-gauge engine originally ordered by the Denver and Rio Grande but diverted to the East Broad Top Railroad. The locomotive was Baldwin's first to this particular design. The specification filled a folio-size page and included dimensions and details for scores of parts. This basic record guided the draftsmen making plans for new construction. The entry also provided a record when replacement parts were ordered (see the notation about a new boiler ordered in 1891). DEGOLYER LIBRARY

The 3-foot-gauge *Mosca* was the final product resulting from the specification entry shown in the previous illustration. Baldwin received this order from the Denver and Rio Grande on March 13, 1873, but a crush of business slowed its completion until October 20. By that time the 1873 panic had struck, and the engine went to the East Broad Top instead. Baldwin initially priced the *Mosca* at $11,750 in D&RG bonds; in the end the EBT paid $10,500 in cash, a difference reflecting both the relative desirability of cash sales and the panic's effect on profit margins. RAILROAD MUSEUM OF PENNSYLVANIA (PMHC)

were substantial in both areas, they also reveal a continuity with the firm's producer culture and its bias against paid managers. Wherever possible, the new practices sought closer control and oversight without increasing the number of middle managers. The first battery of reforms aimed at improving coordination and cutting production costs by a thorough restructuring of practices and procedures in the drawing room.

As noted earlier, Baldwin lacked extensive shop drawings and systems of production until the mid-1850s. With comparatively small levels of output before 1850 and no need to ensure parts interchangeability, many components were crafted without any drawings at all. The various departmental foremen ensured that parts had serviceable dimensions, and as engines came together in the erecting shop, skilled erectors filed, fitted, and adjusted components as necessary. The engineering data that survive from the pre-1850 era, notably books of engine specifications and ledgers of engine weights, mostly provided a historical record for the firm rather than a guide to ongoing or future work.[54]

As Baldwin's product line broadened during the 1850s, systematic production became more important. The drawing room created new ledgers to record dimensions of some standard components for use across a number of sizes and types of engines.[55] To manage the rising variety of custom designs, in 1854 the firm started a new records series known as Specification Books. These books became the foundation for reordering drawing room practice and coordinating work among departments.[56] The battery of procedures to control production

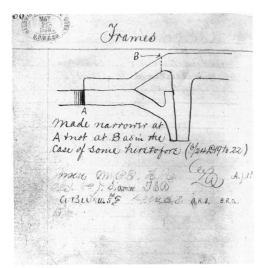

This law mandating the cross-section of frame rails was instituted on May 21, 1879. The change strengthened engine frames at a critical point where two forged pieces joined. In this case the drawing room was probably responding to customer reports of frame breakage. Sketches typically illustrated the laws, since draftsmen thought and worked in a world of three dimensions. When laws were established, all draftsmen had to initial them, showing they were aware of the change in practice. The bold W.L.A. stood for William Austin, then an assistant to William Henszey in the drawing room and later a partner.
DEGOLYER LIBRARY

through the drawing room was developed under Parry, Henszey, and Longstreth in the late 1850s and the 1860s, with the entire system in place by 1872. The new system was based on two elements that reconciled the design variety in engines required by customers with a set of Baldwin-standard components and production procedures. Here is how it worked.

Once the business office received a firm order, all correspondence regarding its engineering and design details was sent upstairs to the drawing room. "Here the work of construction proper commences. First the order, in all its minutiae, is entered in the Specification Book." These entries reduced each engine order to a terse verbal description that included notations for all conceivable varieties of components, materials, and dimensions. Using this detailed list, a draftsman made the preliminary elevation drawing of the engine, in 2-inch scale.[57] This drawing included all special details and parts ordered by the customer.

To prepare the elevation drawing, the draftsman turned to two reference guides: a set of volumes known as Card Books and another series called the Law Books. The Card Books were collected sets of copies of all drawings made to date for every component in a locomotive.[58] They included drawings of various sizes and types of drivers, connecting rods, valves, smokestacks, sandboxes, steam domes, throttles, reversing gears, brakes, and the like. Using the Specification Book entry as his guide, the draftsman found the appropriate sizes and types of component drawings in the Card Books, drew them on his elevation drawing, and recorded the card numbers on a separate sheet. If a specification required an entirely new size or type of part, the draftsman made a card drawing for it, placed it in the appropriate Card Book, and then followed the same procedure. In this way draftsmen created entirely new engine designs largely out of preexisting component drawings. This Card System speeded drawing room work and caused the standardization of much work on individual parts on the factory floor.

Beginning around 1872 the draftsman also followed instructions mandated in the Law Books. These volumes gave general guidelines for drawing new cards, instructions on altering old cards to update older components, and lists of standard detail parts. They included rules for designing new components within the framework of existing design or production standards. The Law Books also allowed the factory floor to alert the drawing room to the capacities of production tools, and laws showed draftsmen the sizes of templates and gauges in use, ensuring that new component drawings would conform to established interchange standards. Examples of Law Book entries are shown at left and below.[59]

The Specification Book entries delineated the customer's exact requirements, the Card Books allowed them to be met wherever possible by standard drawings, and the Law Books ensured that all card drawings—and therefore all work—conformed to evolving production practices. These production controls required that all work in the factory be done to drawings, a systematizing reform in and of itself. By mandating the use of drawings, management established that production workers would follow its standards. The partners' goal was relatively benign; they sought to cut wastage and boost productivity by curbing variations in parts. But the universal use of drawings curtailed workers'

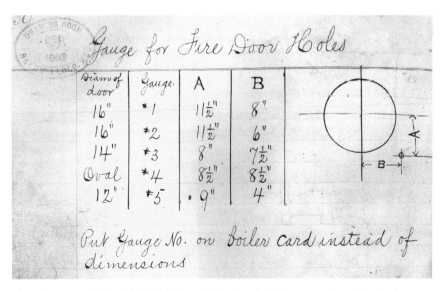

Shop floor practice lay behind this law, which alerted draftsmen to the boiler shop's gauges for flanging fire door holes, ensuring that new boilers conformed to the standards. Draftsmen wrote the gauge number on their cards because the actual dimensions often became irrelevant to shop workers once gauges were in use. DEGOLYER LIBRARY

BALDWIN LOCOMOTIVE WORKS										
201.										
Engine.	Date Finished.	Card.	Card.	Card.	Card.	Card.	Card.	Card.	Card.	Card.
8-22 D 71										
8-22 D 72										
8-22 D 73										
8-22 D 74										
10-34 E 270										
10-34 E 271										
6-8½ C 13										
6-11 D 6										
6-11 D 7										
8-28 C 491										
8-28 C 492										
8-28 C 493										
6-16½ C 14										
6-16½ C 15										
10-32 D 73										
10-32 D 74										
6-16 D 6										
6-16 D 8										
10-34 E 300										
10-34 E 301										

A blank Shop List from the fall of 1882. The superintendent's office filled in these lists with the name of the shop across the top and the card drawing numbers it required to complete the work. The engines listed by their Baldwin class designations ranged from 0-4-2 types to 2-8-0s. Note, however, that batch orders were grouped together when possible to improve productivity in each shop. SMITHSONIAN INSTITUTION

discretion and control over production, and it increased the draftsmen's ranks from three in 1860 to sixteen in 1878. Although the draftsmen were not managers in a modern sense, their work was essentially managerial. Standard parts and practices mandated in or through the drawing room guided all work throughout the factory.

Once the draftsman completed the erecting or elevation drawing with its list of card drawings for each component, Baldwin's purchasing agent ordered from outside suppliers the raw materials and semifinished components required to build the order.[60] Concurrently, a clerk made out a list of the cards required by each shop to complete its work on that engine. These lists formed the basis for another system, this one establishing a standard production schedule for all orders. Run by the general superintendent and known as the List System, it had been fully instituted by 1872.

Under this system, each week the general superintendent's office sent out "Shop Lists" to every foreman in the factory. These lists described each shop's work for the following two weeks and listed the card drawings foremen needed to requisition. A surviving list from 1872 for the brass shop allotted all brass foundry work required for eight engines for the week ending June 15 and also outlined the work forecast for the following week on eight more locomotives.[61] On the basis of this list—which enumerated each engine by its Baldwin class designation—brass shop foreman Thomas Billingsfelt estimated the size of the castings that would be required during the next two weeks. With this information he ensured that his shop would have sufficient labor and raw materials to get the work out on time. Referring to the list, Billingsfelt then ordered up copies of the necessary card drawings from the drawing room, his brass shop workers cast the required parts, and they sent them on to the machine shop by week's end. In the following week machining work on those castings appeared on the Shop List sent to the brass machine shop. Other Shop Lists concurrently directed other departments to proceed with the casting and machining of iron parts for the listed engines, forging work, boilermaking, and so on. In subsequent weeks the same engines appeared on the lists for the erecting shop. This system allowed relatively rapid construction of both standard and custom locomotives. While the elapsed time often varied, the List System established an eight-week schedule for the drafting and production work required to build each order.[62]

Beyond promoting speed and order in production, the List System also heightened central control in the plant. While foremen of the 1850s had told the superintendent how and when to schedule work, by 1872 the lists had reversed these roles. The List System bore similarities to the Shop Order system of production management, described in an 1885 paper by its chief publicist, Henry Metcalfe, and credited by historians as an important first step in systematic management practice.[63] But a Baldwin rival, the Norris Locomotive Works, had employed a variant of the List System as early as 1855.[64] Regardless of who originated such systems, it is not surprising that builders of complex, customized capital equipment required production management controls earlier than mass-production manufacturers, which despite their large output derived some order and regularity from their standard designs.

The Card System, Law Books, and List System built upon Baldwin's first efforts in the 1850s to systematize production through piecework, standard parts designs, and interchange standards. The locomotive builder had adopted all of these policies and controls by 1872, a date and place that challenge tenets of business management historiography. For instance, Alfred Chandler places the impetus for the systematic management movement in the depression following the 1873 panic, when high-volume manufacturers with excess capacity "began

FORM 31.

Baldwin Locomotive Works.
SMITH SHOP.

Work not finished to date of List, week ending.................................. 188

KIND OF WORK.	NO. OF ENGINES.	KIND OF WORK.	NO. OF ENGINES.
Axles, Driving,		Reverse Levers,	
" Truck,		" Lever Rods,	
" Tender,		" " Catches,	
" Collars,		" Shafts,	
Brake Work,		" Shaft and Lever Bolts,	
Back Bumpers,		Radius Bars,	
" Bumper Braces,		" Bar Clamps,	
Connecting Rods,		Running Board Edges,	
Centre Pin Links,		Rock Shafts,	
Crossties,		" Shaft Pins,	
Crossheads,		Rocking Grate Work,	
Crosshead Pins,		Stub Straps,	
Cylinder Cock Arms,		" Keys,	
Cab Lagging Angle Iron,		Smoke Box Braces,	
Drop Bars and Cranks,		" " Brace to Bmpr.	
Driving Spring Links,		Swing Beam Crossties,	
" " Staples,		Swing Truck Work,	
Draw Bars, Front,		Sand Box Arms,	
" " Back,		Truck Equal. Beams,	
Damper Work,		" Channel Iron,	
Equal. Beams,		" Spring Staples,	
" Beam Fulcrums,		Tender Truck Frame Work,	
" " Links,		" " Frames,	
Engine Pedestals,		" Frame Channel Iron,	
" Truck Frames,		' Strap Bolts,	
" " Work,		Throttle Levers,	
Eccentric Rods,		Valve Rods,	
Furnace Bearers,		" Yokes,	
" Bearer Clamps,		" Rod Pins,	
Feed Cock Hangers,		Water Space Frames,	
" Pipe "		Waist Bearer T Iron,	
" Water Work,		" Braces,	
Frames,		Whistle Work,	
Frame Braces, Middle,		Wrist Pins,	
" " Back,		Wrenches,	
" Front Rails,			
Front Bumpers,			
Guides,			
Guide Blocks,			
" Bearers,			
" Bearer T Iron,			
" " Knees,			
Links,			
Link Blocks,			
" Lifters,			
" Pin Plates,			
Mud Drum Work,			
Pilots,			
Piston Keys,			
Pump Plungers,			
Rail Clamps,		EXTRA WORK.	

By the 1880s Baldwin had developed another set of forms for use when a shop fell behind on its listed work. They allowed a foreman to alert the superintendent's office of a bottleneck, so that the same work would appear on the following week's Shop List for that department and more men could be sent into that shop to clear up the backlog. This form for the smith shop shows the range of forged parts common to most locomotives. SMITHSONIAN INSTITUTION

to turn their attention from technology to organization." By that time Baldwin had been focusing on the organization of production for over twenty years, during which it had erected a comprehensive structure of managerial controls.[65]

Baldwin partner Charles Parry was the chief architect of the List System, while drawing room practice came under another partner, William Henszey. The policies created by these men promoted central control and allowed further growth for the company. By 1873 Baldwin employed twenty-eight hundred men; only a few textile mills and iron-making plants were larger. The Card and List Systems enabled Baldwin to grow without creating an extensive layer of middle management, since these systems required only some extra draftsmen and clerks to keep them running. Though the company's workforce grew fourfold between 1860 and 1873, the period saw no increase in foremen's ranks. Because of their demonstrated effectiveness, the Card and List Systems and the Law Books continued at Baldwin with only slight modifications until the 1930s.

Although their numbers did not grow, foremen's managerial duties evolved considerably between the 1850s and the 1870s. The List System ended their authority over scheduling, while the universal use of drawings curtailed their influence on design. Given the swelling number of workers under their direction, the foremen's main duty became personnel administration. They remained responsible for hiring new men, setting their piece or day pay rates, and assigning all work.[66] Hiring and firing were particularly time-consuming tasks, given the great volatility of Baldwin's employment levels, which rose and fell according to the volume of orders for new engines. Apart from their responsibilities for personnel, foremen had other administrative functions such as inspecting finished work and ordering shop supplies and tools. Fortunately, piece-rate pay predominated for most tasks, reducing the need for direct oversight of workers. By the early 1870s foremen were largely shop personnel administrators with little time to oversee work. Parry perceived this problem and created another structure to increase production management oversight: the inside-contracting system.

MANAGEMENT FOR PRODUCTIVITY:
THE INSIDE-CONTRACT SYSTEM, 1872–1910

Many nineteenth-century American industrial companies relied upon inside contracting as an important managerial tool.[67] Under the system, a company's top managers took bids from its own leading workmen to produce a given number of parts, subassemblies, or complete products for a certain price per unit. The firm provided the winning bidder with factory space and machinery, power, labor, and raw materials but left him responsible for wage rates and supervision. The contractor sought to organize his workers and improve his production methods to bring in the job at a labor cost below his bid price.

If he succeeded, the difference was his profit. Contractors occupied a some-what contradictory position—neither purely managers nor workers. This indeterminate status accorded well with the producer ethos of the Baldwin partners.

Charles Parry decided to establish contracting at Baldwin in 1872. In June of that year labor unrest swept northeastern manufacturing cities as thousands of workers in scores of trades took up the Eight-Hour Movement. Although Baldwin's men did not strike, they did seek to reduce their workday from ten to eight hours with no cut in wages.[68] The partners refused. Instead they adopted the contract system, in a further effort to link the remuneration of labor to the amount of work accomplished. The company had evidently experimented with contracting as early as 1854, and its adoption in 1872 was a logical outgrowth of piecework pay.[69] At times contracting was called the "gang piecework system." In establishing contracting, it seems that Parry ultimately sought improved intradepartmental productivity, while the nearly simultaneous adoption of the List System enhanced interdepartmental coordination.

After 1873 Baldwin's labor and production management systems combined various practices. Foremen remained in control of their departments, con-tractors oversaw many tasks, and the labor force operated under a mix of pay systems: day, piece rate, and contracting. Because very few payroll records or inside contracts survive, determining what kinds of work fell into each category is something of a puzzle.[70] Contracted work seems to have centered on subassemblies such as tanks, trucks, tender brakes, and erecting shop work, and on complex machining and forging jobs including springs, gears, cylinders, frames, and boiler braces. As Baldwin's output soared, more tasks appear to have been switched over from piece rates to contracts.[71] By 1914 approximately 75 percent of the labor force were on piecework or worked under contractors.[72]

Although Baldwin's contract system evolved over time, a surviving contract shows how it began.[73] On November 1, 1872, Charles B. Allen agreed to supply all the forged springs required by the firm, receiving a rate of 4¾ cents per pound of finished springs. To complete the contract, Allen had control of Baldwin's fully equipped spring shop, he received all required materials, and he could employ "such workmen as shall not be objectionable" to the company. Allen had to pay for all coal, oil, and borax used in forging springs and for wastage of bar steel and iron above a given allowance—provisions assuring economical operations. He selected his own workers but had to report their names and wages each week. This significant provision allowed Baldwin to determine what profit Allen made on the contract, since the firm, rather than the contractor, paid his workers.[74] This contract continued indefinitely, and there is no way to know if Allen made a profit or remained a contractor.

Originally Baldwin let contracts out among its workers for bids, and a number of skilled men competed for the work.[75] After some years veteran contractors gained an advantage from their knowledge and managerial experience, and the company was soon drawing its contractors from a relatively small pool.[76] By 1900 bidding had been discontinued, and the firm alone established the rates for contract work, although probably after discussions with contractors. Until roughly 1907 all hiring was done by foremen, with contractors requesting men for a given job. After that point Baldwin created an employment department to which contractors applied for men.[77] Although the contract system evolved over time, it served Baldwin's needs sufficiently well to continue in use until 1942.[78]

As a management system, inside contracting provided a range of advantages, and it proved particularly useful in the production of capital equipment. At the time Parry instituted the system, Baldwin was broadening its product line, soliciting industrial, export, and custom work. The number of different engine classes grew from twenty-two built in 1870 to fifty-four in 1880.[79] With their experience in production and skill in labor management, contractors provided the sort of flexible specialization needed to cope with this variety in design.

Because Baldwin built engines to order and early delivery dates provided an important competitive advantage for the firm, the firm also looked to contractors to turn out work quickly. Contracting encouraged fast production because the system amounted to a kind of managerial piecework. As a company officer noted in 1903, "The contractor is a piece-worker on a larger scale. As he is paid by the job, he has an incentive to turn out his work as quickly as possible and to get as much work as possible out of the men under him."[80] Unlike American System consumer products like typewriters, locomotives were composed of a number of very large, technically complex, and unwieldy subassemblies. Building engine tanks, boilers, and trucks and then erecting the complete product required the synchronized work of large numbers of skilled, semiskilled, and laboring men. Contractors had the experience and the incentive to ensure that this work was done accurately and quickly.

The fact that Baldwin instituted contracting only in 1872, rather than at some earlier point in its history, provides further insights into the benefits Parry sought. As a response to potential labor unrest, contracting had the advantage of splitting up the labor force and transforming some leading skilled craftsmen into quasi managers. It is said of contracting elsewhere that owners adopted it as an expedient because they "generally had little technical knowledge," but this was clearly not true at Baldwin.[81] Contractors there were productivity managers rather than technological pathfinders for the company.[82] By 1910 they often had two hundred to five hundred men under them. As a contemporary observer noted, "The contractor usually has an administrative force of his own—one or two right hand men, known as his percentage men. They are virtually his personal representatives, or sub-bosses, and they are paid to get the best results from the labor employed."[83] The superintendent's office oversaw technical changes in the production process, departmental foremen and the List System ensured interdepartmental coordination, and the contractors oversaw complex operations within departments.

Finally, there is the question of rate setting and, by extension, the character of relations between contractors and the company. Although other historians have found that the bidding process created antagonism between management and contractors, Baldwin seems to have avoided such ill will. Elsewhere this problem arose from contractors fearing rate cuts while top management "had no accurate way of knowing what the new jobbing rates should be."[84] Baldwin avoided such uncertainty by paying contractors' men directly, which let it determine contractors' profits on any job.[85] Thus the superintendent had accurate data to use in adjusting rates either up or down. Furthermore, by 1878 the company had started recording all labor charges *per job,* allowing accounts of labor costs and profits for every engine built.[86] Because the firm knew its contractors' profits and its unit labor costs and profits, the company had a sound basis for fixing equitable contract prices.

With all this cost data, Baldwin learned to set its own contract rates rather than rely on a potentially adversarial bidding contest. Since the same components could vary greatly in size and details from one design to the next, setting separate rates for every different part also became impractical with the growth of custom production. So by 1900 the company had established schedules or scales of prices for each different class of contracted work. These were formulas "for working out the cost of a given part . . . [taking] into consideration lengths, widths, weights, holes, areas, etc."[87]

An example will illustrate the system. If the company needed twenty tender water tanks of a certain size, the schedule would give a price per unit—say $500, for a total contract value of $10,000. The contractor and his workmen then set about making the twenty tanks, working to a deadline under the List System. Workers all received an hourly or piece rate set by the contractor and approved by the company. The contractor sought to complete the work before his wage bill on the job exceeded $10,000. If the contract's wage account with the company came out at $9,000, the profit was $1,000. Theoretically the contractor could take all of this sum, but if he did, his workers would avoid him in the future. So he shared the profits with some or all of the men, according to predetermined portions set by the contractor, approved by a foreman, and paid directly by the company.[88]

Rather than suspicion, antagonism, and inexorable rate cuts, Baldwin's managers sought to maintain a sense of cooperation and partnership with the contractors. The growing size, weight, and intricacy of locomotives exerted upward pressure on many schedule rates. Furthermore, to meet the cyclical requirements of the locomotive market, Baldwin sought more than simply a reduction in labor costs from its contractors. In booming markets the company charged more for locomotives, and its competitive posture in the industry centered on early delivery dates rather than prices. At such times the firm could afford higher profits for contractors, providing their incentive to complete more work quickly. During sales lulls Baldwin sought cost-cutting efficiencies. Knowing contractors' profits, the firm could seek economies without entirely killing their financial incentive. No doubt the contractors disliked such cuts, but they knew that strong markets and higher profits would return. Contractors valued their positions and sought them for the long run. John Curry, a veteran erecting shop contractor circa 1900, had begun with Baldwin as an apprentice in 1860.[89]

All this is not to say that inside contracting was faultless. The system was particularly susceptible to favoritism. Contractors had conflicting loyalties— to management and to workers—or they might look out solely for their own interests. Baldwin could do little about this, since essentially it used contracting to derive managerial oversight on the cheap, by a system whose internal contradictions prevented all its participants from seeing contractors as bona fide managers.[90] But for the most part contractors and their managerial superiors were allies, not adversaries. The firm valued their labor management and productivity enhancement, particularly because these men still contributed directly to value-added. By adopting the contract system, Baldwin's partners further developed the producer ethos within these key production managers. The contractors were then to impress it on the rank and file.

This portrait of contracting argues that it offered a number of particular advantages in building capital equipment. The system promoted a type of flexible managerial control that readily adapted to the constantly evolving products and gyrating markets characteristic of capital equipment companies. This argument is confirmed by the widespread use of contracting in this sector. Locomotive builders employing it included Richmond, Schenectady, Brooks, Taunton, Lowell, Hinkley, and other New England shops.[91] Other capital equipment firms using the system included the Whitin Machine Works, Bement, Miles & Co., General Electric, Pressed Steel Car, Pullman Car, Cramps Shipyard, and Hoe Printing Press.[92] Although mass producers began abandoning contracting in the 1880s, many builders retained it well into the twentieth century.[93]

OTHER MANAGERIAL DEVELOPMENTS, 1872–1909

The policies described above—coordination through the drawing room and the List System, and productivity enhancement under contracting—were the most important management policies instituted at Baldwin during the Gilded Age. But other managerial changes also deserve mention. We left off consideration of the partnership in the aftermath of the 1870 reconfiguration of M. Baird & Co., from three to six principals. This expansion, the List System, and the adoption of contracting all aimed at improving the company's profitability, in part to prepare for the withdrawal of Matthew Baird's interest. With the locomotive market booming, Baldwin expanding into new product lines, and its managerial reforms well under way, Baird retired in April 1873. His old partners reconstituted the firm as Burnham, Parry, Williams & Co.[94] Under the terms of Baird's retirement, the new firm had five years to pay off his one-third interest, valued at $1.125 million. This debt was secured by five notes from the company, each for $223,000, due annually through 1878.[95]

Six months after Baird's retirement the Panic of 1873 overwhelmed the nation's finance and commerce, with devastating results for locomotive building. Output at Baldwin fell from 437 engines in 1873 to 162 a year later, and the 1873 volume was not reached again until 1880. In the immediate aftermath of the panic the firm slashed its weekly payroll from $40,000 to $10,000 while dispatching Edward Williams to Russia to secure orders. Despite this contraction in sales and operations, the firm met each of its payments to Baird on schedule, fully retiring his interest in May 1878.[96] This smooth transition in the midst of a deep depression testified to the vitality and expertise of the enlarged partnership, the benefit of expanding into new markets, and the value of improved coordination and productivity through the List System and contracting.

The firm's profitability during the 1870s also stemmed from another internal change. During this period of financial stringency Baldwin improved its control over costs by instituting unit cost accounting. Although no financial records survive from this period, changes in the company's timekeeping and pay regulations clearly indicate that Baldwin had begun aggregating labor costs on a per-unit basis by 1878.[97] This was for the evident purpose of developing unit cost accounts.

In developing such controls by 1878, Baldwin was relatively early compared with mass-producing manufacturers.[98] But that date also seems like a rather late beginning for cost accounting in the capital equipment sector. These companies normally secured contracts through competitive bidding, making knowledge of production costs a vital concern. As the leading historian of accounting notes, "What was more logical than to take the next step; that is after accepting a contract for a certain project, to keep some sort of collective details as to the costs of executing [it] . . . to ascertain the profit or loss thereon, and to provide information for future estimates."[99] John Souther's Globe Locomotive Works in Boston was assembling such unit cost data, including labor, materials, and some overhead charges, as early as 1851.[100]

Until the 1870s depression the Baldwin partners evidently believed that the potential insights from cost accounting were not worth the trouble involved in collecting the data. Such work required clerks—drags on the business, in Matthias Baldwin's view. Instead the company priced engines according to what the market would bear until the market collapsed in the mid-1870s, making knowledge and control of costs more important than ever before.[101] The post–Civil War trend toward custom locomotives also heightened the importance of cost accounting as an aid to bidding. After 1878 Baldwin's cost controls grew increasingly sophisticated and detailed.[102]

By 1879 Baldwin's markets finally recovered from the panic, and its sales tripled between that year and 1890. Notwithstanding such growth, this era saw little change in the firm's existing internal management structures, which proved well suited to expansion. As the firm built new shops, the foremen's ranks increased, growing from thirteen in 1873 to twenty in 1904.[103] Over the same period employment grew from 2,800 men to 10,500, so clearly the company maintained its producer-culture tradition against top-heavy management. That ethos is also evident in another area. Like the pay of shop floor personnel, managers' compensation was linked to the volume of work in the plant. When Baldwin's output fell from 2,655 locomotives in 1907 to 1,024 in 1909, shop managers saw salary cuts averaging 31 percent.[104] When business picked up in the following year, so did their compensation. Given the great volatility of locomotive demand, the Baldwin company needed to exert direct control over personnel costs—for managers as well as workers. But these fluctuations in managerial salaries also arose from the firm's culture. Pieceworkers, contractors, foremen and other managers, even partners knew that their incomes depended on the volume of their own work and that of the company as a whole.

This unvarnished regard for the work ethic raises the question of how conservative or innovative the Baldwin partners were in their management practices. Many assume that partnerships took a conservative course by nature, since they operated without limited-liability protection, and because any capital investments made by the firm represented forgone profits for the partners. But the record of Baldwin's owner-managers shows frequent innovation. The partners adopted the List System and other standardizing reforms well ahead of production management controls by mass-producing manufacturers. In the mid-1880s Baldwin had created a materials-testing laboratory, an uncommon department in American manufacturing companies of the period. Starting in

Baldwin built this little 3-foot-gauge electric engine in 1904. Note the two round builder's plates—one for Baldwin, and one for the Westinghouse Company, which made the electrical equipment. By the middle of the twentieth century Westinghouse had taken a controlling stock interest in Baldwin. The view here was inside the factory, on land formerly occupied by a city street that Baldwin took over in its 1902 expansion. RAILROAD MUSEUM OF PENNSYLVANIA (PMHC)

1890 the company began "the first important application of electric power to drive machinery" in the United States.[105] Concurrently, it launched a major initiative in locomotive design, the Vauclain compound. In 1895 Baldwin entered a joint agreement with the Westinghouse Electric and Manufacturing Company to "unite in developing electric motors [locomotives] suitable for railroad work."[106] Finally, in 1906 the company broke ground on an entirely new satellite plant, located in the Philadelphia suburb of Eddystone. By the early 1900s Baldwin's partners presided over a giant among American industrial companies, which they owned outright. It became and stayed large precisely because of the partners' willingness to innovate in a number of areas. But they would also ascribe much of their success to the firm's longstanding culture of individual initiative and reward.

BALDWIN AND THE MANAGERIAL REVOLUTION

Between 1890 and 1915 American industry underwent a number of basic trans-formations that were so extensive as to constitute a managerial revolution.[107] Many industries saw horizontal combinations to reduce competition in the aftermath of the 1893 depression. Mass-production manufacturers integrated backward to acquire sources of supply and forward to control their distribution networks. Industrial companies incorporated to take advantage of the developing public markets for securities, issuing equities and bonds to finance further expansion. Over time these publicly held companies came to rely on a professional class of salaried managers to integrate their operations and oversee expansion into new ventures. Some of these executives also turned to Scientific Management, welfare work, and personnel management in an effort to solve the "labor problem," as it was known. In combination these developments created the modern form of American corporations and made them dominant in the national economy.

How did these transformations affect the Baldwin Locomotive Works, and how did it respond to them? A resume of events at the firm suggests varied answers. For many years Baldwin ignored the rising market for industrial securities that developed after the late 1880s, promising new sources of investment capital through incorporation. The partnership was reconstituted four times between 1891 and 1907. Notable among the exiting partners were Edward Longstreth, who retired, and Charles Parry and Edward Williams, whose deaths caused reorganizations. The firm gained talented new partners like William Austin, Alba Johnson, and Samuel Vauclain to counterbalance these losses.[108] Despite their close ties to the investment banking community, the partners held back from incorporating for many years. As Johnson said at the time, the partners truly believed that the partnership form gave "a permanency in the conduct of the business."[109] No doubt such beliefs contributed to keeping Baldwin aloof from the American Locomotive Company (Alco) consolidation of 1901.

But in the booming market of 1905, the Broad Street factory became clogged by success. Both the overall volume of work and the growing size of individual road locomotives greatly taxed Baldwin's physical plant, causing productivity problems. The new Eddystone plant promised to alleviate these difficulties, if capital could be raised for its expansion. Other capital needs also loomed, with three of the six partners being over the age of sixty-five in 1905. So the firm finally incorporated in 1909, becoming a publicly held company in 1911. Although the partners had held off for eight years, once Alco gained access to the public capital markets in 1901, Baldwin had to follow sooner or later.

Incorporation did not change Baldwin's top management personnel until 1929. Former partners Converse, Austin, Johnson, and Vauclain filled its presidency for twenty years after 1909. For other executive positions the firm still favored internal promotion of longtime employees over outside hiring. As late as the 1920s four Baldwin vice presidents had risen from positions as office boy, blacksmith's helper, laborer, and clerk.[110] Beginning in the 1890s Baldwin

Baldwin's owner-managers in 1902 on their annual inspection trip to their other factory, the Standard Steel Works in Burnham, Pennsylvania. From right to left are William Austin, John Converse, Samuel Vauclain, George Burnham, unidentified, George Burnham Jr., and Alba Johnson (the three men at the left are unidentified). Standard made specialty iron and steel products like axles, wheels, and locomotive driver tires for Baldwin and other customers.
SMITHSONIAN INSTITUTION

recruited some professional managers from outside, largely railroad men, who generally took up sales positions. Their numbers were small, and they had little effect on the producer culture shared by the majority of Baldwin's midlevel managers who had risen from craftsmen's, drafting, or foremen's jobs.

The company's internal departments also saw little change before World War I in response to the managerial revolution, which offered little of value to Baldwin's operations.[111] As a builder of custom products, Baldwin could gain few economies by divisional restructuring. Beginning in 1902 Baldwin attempted to relieve the press of work at Broad Street by transferring much foundry and forging work to its Standard Steel plant in Burnham, Pennsylvania. While expansion at Standard and subsequently at Eddystone did ease congestion at Broad Street, this dispersal of operations hindered coordination more than it promoted efficiency.[112] The only other divisional innovation before 1915 was the creation of a central personnel office circa 1907. It took over the hiring prerogative from foremen—who kept their power to fire men—and consolidated the functions of the old time and pay offices.

Beyond this departmental consolidation, personnel management at Baldwin in the early twentieth century was largely unaffected by trends elsewhere. The company had never been a paternalistic employer, and it ignored the development of welfare work at other industrial companies. The great employment fluctuations it faced in building capital equipment probably prevented the adoption of such welfare reforms as a cafeteria, medical department, and recreational activities.[113] These benefits simply cost too much for a firm whose total employment could shrink by 75 percent over seven months, as happened after the Panic of 1907.

Baldwin's managers also saw little value in the Scientific Management movement then being trumpeted by its chief publicist, Frederick Taylor. The partners particularly disliked Taylor's personnel policies. Scientific Management of the 1890s and later is generally seen as an outgrowth of the systematic management movement of the 1880s. Baldwin had little interest in the former, partly because it had adopted systematic policies between 1855 and 1872 and found them to work well. In the partners' opinion their existing systems were not broken and therefore needed no functional foremen, time and motion studies, or differential piece rates as fixers. These planks of Scientific Management were also poorly suited to conditions in the capital equipment sector. In particular, the variety of component designs and sizes faced by workers in the daily course of production hindered any effort to establish standard times or one best way of completing tasks.[114] But Baldwin's partners did not just decline to adopt Scientific Management policies; they opposed Taylorism with a measure of dislike, even antagonism. Their opposition arose from the firm's producer ethos. As the partners saw it, Scientific Management was antithetical to that culture.

Many of Taylor's planks sought to increase oversight and control of the workforce. This required an enlarged executive force with route, time, cost, and inspection clerks and other additions to the central managerial staff. And as efficiency reformer Harrington Emerson noted of these men, "It is the business of staff, not to accomplish work, but to set up standards and ideals, so that the line may work more efficiently."[115] The Baldwin partners looked on these ideas with disdain. Consider Alba Johnson's views, given in 1914 to the U.S. Commission on Industrial Relations: "Every premium system and every efficiency system that I know of involves a large number of nonproducers and the question is whether you can so spur the men on to increased production as to not only pay the cost of their maintenance but in addition thereto increase the produce of labor over and above the additional load you have put on his back."[116] Johnson went on to describe efficiency clerks and other "nonproducers" as "dead help" whose "burden becomes intolerable to the workingman and it also becomes intolerable to the employer." When asked about the cost savings and improved productivity claimed at the Watertown Arsenal as a result of Scientific Management, Johnson dryly responded, "Well, I think that that reflects very badly upon the management before it was introduced."

Johnson's views arose from the producer culture that had guided his firm's management since the days of Matthias Baldwin. The firm highly valued such systemic policies as the Card and List Systems and cost-accounting controls.

But these were simply managerial tools to promote individual effort and responsibility among managers and workers. Such individualism and hard work marked the partners' own careers as they advanced up through the company to become its owners. Their producer ethos seemed entirely justified by the triumphs of their own lives and by the business success of the Baldwin company. Now it is time to consider how the men on the factory floor responded to the firm's individualistic producer culture.

5

THE BALDWIN WORKFORCE

1860–1900

THROUGHOUT THE NINETEENTH CENTURY ALL FIRMS IN THE CAPITAL EQUIP-
ment sector relied heavily on skilled labor in the construction of heavy
machinery. Such work required more expertise from many more trades than was
true in mass-production manufacturing. In locomotive building, the massive
size of the product and its continuous technological evolution mostly prevented
the adoption of mechanized production techniques right down to the demise of
the steam locomotive in the 1940s. Baldwin's management could never forget
that its skilled workforce was an essential partner in production. Who were
these men who fashioned engines out of wood, iron, brass, and steel? What
skills and trades did they bring to the work of locomotive building? Other
issues considered here include recruitment of workers, the organization of work,
methods of payment and relative pay scales, overall growth of the workforce,
promotions and layoffs, and the character of workplace hazards. Finally, how
did management view the workforce, and, much more difficult to answer
satisfactorily, what did workingmen think of their employer?[1]

Two great strikes occurred at Baldwin, in 1860 and in 1911, conflicts involving
union organizing, the presentation of workers' grievances, and their rejection
by the company. The strikes largely failed to achieve their goals. These episodes
are considered elsewhere, because of both their importance and the need to
examine them in the context of concurrent developments in the firm and
the industry. During the forty-year period examined here, the company saw
very little overt workplace conflict, and Baldwin operated as a nonunion
factory. During the Progressive era circa 1900, labor-management conflict and
its resolution became a widespread national concern, and magazine articles
frequently cited Baldwin as having "solved the labor problem."[2] The judgment
is ironic, given the events of 1910 and 1911, but this juxtaposition provokes
an important question of perspective. Was conflict the defining experience of

the nineteenth-century workplace? Was a continual divergence of interests between workers and managers inevitable? Put another way, were the years from 1861 to 1909 a time of barely submerged hostility at Baldwin, with labor and management glaring at one another like boxers from opposing corners of the ring?[3] Or were they years of accommodation and compromise? If so, how was this reconciliation reached?[4]

In fact this half century saw a reconciliation of interests between capital and labor at Baldwin. Because of the demands of its industry, Baldwin's managers knew that shop floor workers played an essential role in creating many of the firm's competitive advantages.[5] Since locomotives were made to order, management sought to avoid strikes and work slowdowns if at all possible. Given the custom character of its products, workers could not be replaced by production machinery. These realities argued powerfully for avoiding an open break with the workforce.

If conflict could hurt the company, the Baldwin partners also knew that consensus would help their business. At points of high demand, the firm relied on workers to provide a crucial competitive advantage: achieving early delivery dates. At all times product quality also rested squarely in workers' hands. Finally, labor's goodwill and exertions were required to deal with the long-term trend toward bigger and more complex locomotives. Between 1850 and 1900 the average weight of Baldwin's products nearly tripled. But the company maintained its regular eight-week production schedule thanks to a fourfold increase in labor's productivity over the period. Much of this improved efficiency derived from investments in new facilities and equipment; the remainder came from increased effort by workers. In both cases cooperative labor-management relations were vitally important in achieving the productivity gains.[6]

The character of Baldwin's managerial policies simply required it to foster autonomy and expertise in the workforce. Its determination to limit managers' ranks and overhead costs necessarily gave workers a relatively free hand. Piecework secured efficiency and productivity, but through a financial incentive designed to obviate the need for close oversight. Other controls like inside contracting and the Card System sought to optimize Baldwin's productive capacities and workers' skills. Such internal policies promoted market success by marshaling the firm's chief asset, its human capital. To succeed, these policies again depended upon cooperative relations with workers.

How was such cooperation elicited? The firm adopted as its company culture the producer ethos of individualized productivity, responsibility, and reward that had marked the partners' rise through the ranks. Through the use of piecework, relatively high wages, a long-term trend of increases in real wages, the adoption of the contracting system, the assurance of job security for the most skilled men, and avenues for promotion for a minority, Baldwin's managers created a positive vision of the workplace. It had sufficient validity and appeal to attract at least a minority of the workforce, including most of the labor aristocracy of skilled men within the plant. Even though this vision largely ignored the majority of semiskilled and laboring workers, it defused overt conflict for fifty years.

Managers fostered the producer ethos, and many workers supported it out of a shared belief that management and labor necessarily stood as interdependent partners. As Charles Parry told the workers in 1872, "We are as much your *employees* as you are ours."[7] Examples of this reliance on workforce skill abound in Baldwin's history: its craft apprentice program in the 1850s and 1860s, a similar program created in 1901, and its practice of taking locomotive contracts at a loss during periods of low demand to maintain the skilled nucleus of the organization. Obviously the partnership more resembled a divorce during the two major strikes. But those conflicts were precipitated by specific breakdowns in the company's producer culture as competitive pressures in the industry spilled onto the factory floor.[8] For example, the 1860 dispute arose after the locomotive builder cut its overtime rate in response to pitched competition following the 1857 panic. Over the long run, however, Baldwin could ill afford such enmity, and for a half century it succeeded in preserving peace.

The notion that Baldwin's managers and workers were partners in production, the espousal and acceptance of the producer culture, and the absence of open conflict over this fifty-year period did not mean that the company was a paradise for craft workers or that the Baldwin partners ultimately lacked the power to control the workplace. Throughout the period detailed here Baldwin's production managers continuously redefined the character of work, trying to boost productivity. This goal required the firm to lessen its dependence on craft-trained skills. The shift out of a craft organization of work began in the 1850s with the adoption of piecework. The failure of the 1860 strike and the demise of Baldwin's craft apprenticeships in 1868 were further steps in this transformation from craft to industrial work. Between 1870 and the late 1890s the workforce became industrially organized, with only vestiges of its old craft orientation. Workers' skills and autonomy were narrowed during this period by the extension of piecework, the adoption of standard parts, and the drawing room's influence over work. These policies allowed the company to move into new markets and boost sales while enabling employment to grow from one thousand men in 1866 to thirty-five hundred in 1896. Baldwin's great expansion thereafter to upwards of 18,500 men (1907) marked a third stage for workers. Although many jobs still required skilled hand and machine work, locomotive building had become so carved up into specialized tasks that most workers were essentially industrial operatives, rather than craftsmen.

Until 1910 workers did not unite in opposition to these changes, for a number of reasons. They were too divided among themselves—by ethnicity, trades, skills, income, and other factors. Furthermore, Baldwin's managers acted rather carefully when changing the character of work, for example avoiding the charged labor management reforms of Scientific Management.[9] The firm's long-term expansion also created opportunities for promotion and caused absolute growth in the number of skilled jobs, even as their percentage in the total force declined. Finally, since a substantial number of Baldwin's skilled men subscribed to the firm's producer traditions, this forestalled disaffection in the aristocracy of skilled workers who would have been crucial in leading any organized opposition to management. For these reasons the firm largely avoided open breaks for fifty years, although the failed strikes of 1860 and 1911 imply that the Baldwin partners retained substantial latent powers throughout the intervening years. But to reiterate the broader point: For the most part consensus, not conflict, ruled on Baldwin's factory floor from 1860 to 1910, thanks largely to a workplace culture fostered by management and aligned to the beliefs and aspirations of a large body of workingmen. Most of Baldwin's partners, foremen, and contractors had started out on the factory floor. To paraphrase Bill Haywood, these men were workers who put on a manager's cap, and the producer ethos they created or supported arose from that combination of a laboring background and managerial needs.

LABOR-INTENSITY, TRADES, AND SKILLS IN LOCOMOTIVE BUILDING

A key feature distinguishing builders of capital equipment from high-volume manufacturers of machinery involved the character of skills and work in each sector. Manufacturers of such mechanisms as sewing machines, watches, and firearms responded to the vast consumer demand for their products during the Gilded Age by adopting automatic and semiautomatic production machinery—using capital to replace skilled workers or to degrade their work. But builders like Baldwin could never emulate this capital-intensive and labor-saving approach to production.[10] The custom character of locomotive designs, the huge size of the product, and its narrow, evolving, and highly fluctuating demand all precluded such a course. Skilled men, an army of craftsmen, were irreplaceable. Though this army required such expensive production machinery as lathes, planers, slotters, and power riveters to accomplish its work, these machines still required the skills and adaptability of human hands and minds. This labor-intensity drove builders like Baldwin into the top ranks of American industrial employers for much of the nineteenth century.

Locomotive building also required a broader range of trades and skills than did the mass production of consumer goods. Table 5.1 lists the variety of craft, semiskilled, and unskilled workers employed at Baldwin in 1876. The table

The *Alexander Mitchell* of 1869 gives evidence of the skills and trades required to build
such a sophisticated product. Carpenters made the wooden cab with its curved moldings;
machinists achieved precision and polish in the valve motion and connecting rods;
boilermakers worked up the steel firebox and installed 166 iron flues; sheet-metal workers
formed the highly polished sheet-brass casings for the steam domes and sandbox; and
ornamental painters finished it all off with filigree on the cab, floral designs on the headlight,
and a rainbow of vivid colors throughout. RAILROAD MUSEUM OF PENNSYLVANIA (PMHC)

includes eight distinct trades and various specializations within crafts. While
the divisions by skill level are somewhat arbitrary, they reveal Baldwin's great
reliance on craftsmen, who accounted for almost half of the total force. Also
note the importance of laborers, the largest single category, whose strong backs
and brawny muscles manhandled raw materials and finished components
through the extensive shops.

Many of the occupations found on Baldwin's factory floor long predated
industrialization or the rise of the factory system. By 1876 blacksmiths and
their helpers, molders, sheet-iron- and coppersmiths, carters, and laborers were
already venerable occupations. Other trades arose from industrial development
itself as machinery-building companies trained men in the new skills required
to construct such novel technologies as locomotives. The machinist's craft dated
only to the 1830s and 1840s, representing "a kind of cross between a millwright
and whitesmith, a fitter, finisher, locksmith, etc." [11] Other trades at Baldwin that
were new to the industrial era included draftsmen, whose collective identity
in training and skills dated roughly to the 1850s, and ornamental painters of
machinery, whose embellishments of engines were in vogue from the 1840s to
the 1870s.

As the factory system in machinery building matured during and after the
1850s, the variety of trades and the character of skill in each continued to evolve.
Thus the achievement of interchangeable-parts production at Baldwin in the

TABLE 5.1 Baldwin Workers' Trades and Levels of Skill, 1876

Trade	Number	Trade	Number
Skilled (46%)		*Semiskilled (31%)*	
Machinists (best)	217	Blacksmith helpers	175
Machinists (ordinary)	171	Boiler helpers	106
Blacksmiths	98	Machinists (inferior)	92
Patternmakers and carpenters	41	Riveters	32
Painters	28	Chippers and caulkers	21
Iron molders	24	Tank makers	15
Sheet-iron workers	20	Stationary engineers	9
Turners	20	*Unskilled (23%)*	
Boilermakers	9	Laborers	254
Flangers	9	Flanging helpers	27
Brass founders	8	Holders-on (riveting)	16
Core makers	4	Carters	12
Fitters	2	Casting cleaners	12
		Patternmakers' assistants	10
	Total workforce	1,432	

Source: Pennsylvania Industrial Statistics Report, vol. 4 (1877), p. 546.

Notes: Beyond the white-collar draftsmen, the eight basic production trades in the factory since the 1830s were patternmakers, molders and founders, blacksmiths, sheet-iron- and coppersmiths, boilermakers, carpenters, machinists, and ornamental painters. The rise of specialties within these crafts due to labor specialization and piecework is clearly evident here. The classifications by skill are my own, based on a knowledge of the work in each trade and on the wage rates, also given in the *Industrial Statistics Report.*

1860s brought forth a new class of machinists, skilled tool and gauge makers, who made the precisely fashioned jigs and templates then used by general machinists in turning out interchangeable work. Once these general machinists began working to gauges, the tolerances and precision required in their work became more exacting as well. Finally, when Baldwin began die-forging in the blacksmith shop of the 1880s and in the boiler shop of the 1890s, the highly skilled trade of die makers joined the workforce. The rise of such trades as the machinists, ornamental painters, draftsmen, and tool and die makers shows how factory employers such as Baldwin created demand for lines of skilled work.[12] These examples underscore why Baldwin's managers looked on the labor force as an essential partner in production. While the firm often narrowed skills through specialization and piecework, skilled men themselves could seldom be replaced.

CHARACTERISTICS OF THE WORKFORCE

Beyond the range of skills required to build locomotives, two essential developments affected Baldwin's workforce between 1860 and 1910: substantial long-term growth in size coupled with intense fluctuations in the short run as the company struggled to match employment levels with the fluctuating demand for locomotives. Overall the force increased from 600 men in 1860 to 18,500 in 1907 before falling back to 14,500 in 1910. The employment peaks for each decade within this fifty-year period are shown in table 5.2.

Compared with other industrial employers, Baldwin had been a relatively large firm since the mid-1830s, with three hundred men on the payroll in 1837. With the economic recoveries following each trade cycle after the Civil War, locomotive demand and the number of men employed reached new heights. Growth occurred in relatively small increments until the recovery during the late 1890s from the 1893 depression. After 1897, employment at Baldwin rocketed upward. This period saw strong export sales, a flourishing domestic market, and great growth in the size of individual road locomotives. In combination these factors pushed manpower needs to a remarkable pace of growth. Baldwin's employment levels, before and after the post-1897 takeoff, placed the company in the front ranks of American industrial companies from the 1830s to the 1910s.

Despite this pattern of long-term growth, many of Baldwin's workers faced very uncertain job security over the short term, since their jobs were directly tied to the mercurial character of the locomotive market. Experienced employees knew that every period of booming demand, hard work, and high wages invariably gave way to a severe bust with short hours and large-scale layoffs. For example, August 1873 saw three thousand men on the payroll. After the panic struck in the fall, railroad capital expenditures, the locomotive market, and Baldwin's workforce all fell back sharply. By December fewer than thirteen hundred men were working at the plant, many on half time. Not until 1881 did Baldwin's employment return to the August 1873 level. Similar booms and busts marked the entire period from 1880 to 1897, as indicated by chart 5.1.

The chart raises a number of issues. First, it clearly shows how Baldwin's

TABLE 5.2 Baldwin Employment Peaks by Decades, 1860–1910

	1860s	1870s	1880s	1890s	1900s
Peak year	1869	1873	1889	1899	1907
Employment	1,700	3,000	3,579	6,336	18,499

Source: See appendix A.

CHART 5.1 Baldwin's Annual Employment and Output, 1880–97

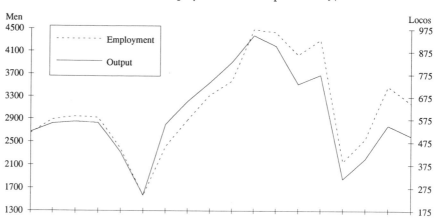

managers exerted strict control over employment levels to meet prevailing
market conditions. The firm met boom conditions by flooding the factory floor
with men, and it slashed the rolls as orders dried up. Both courses are shown
in stark outlines in the chart, which might seem to indicate that Baldwin's
management simply viewed the labor force as another factor of production,
to be expanded or contracted at will. But underneath the aggregate data,
many factors came into play, and not all workers were the same. Highly skilled
machinists, molders, boilermakers, and blacksmiths accounted for almost half
the workforce, and the company's survival rested upon those skills. Although
Baldwin tried to lessen this dependence, it never really succeeded, and skill
levels touched every issue from labor recruitment to pay, promotion, layoffs,
and turnover.

RECRUITING THE WORKFORCE

Labor recruitment hinged on two central issues: how did men get jobs at
Baldwin, and how did the company secure and retain workers with the skills it
needed? The cornerstone of Baldwin's employment policy was high wages. For
a half century the company paid more for each given class of work than any
other Philadelphia employer. This reputation for paying well enabled Baldwin
to draw off the skilled forces of other employers whenever it needed to expand.

Through the 1860s the policy of high wages also had a second rationale:
attracting men with the requisite quality of skills and training, which in practical terms meant men who had apprenticed in their trades. In skilled work such
as machining and smithing, Baldwin found that the need for precise, high-
quality work could not be skimped. In 1860 the company's business manager,

George Burnham, wrote privately, "We find hands that command the highest wages are the most profitable to us"; Baldwin preferred "good machinists who have served a regular apprenticeship to the trade."[13] Although the company had begun a program of labor specialization under piece-rate pay, it still depended on the general training and skills acquired through craft apprenticeships.

Given Baldwin's preference at this point for guild-trained craftsmen in trades like machining, smithing, molding, and boilermaking, the company had notably little difficulty filling its labor needs. In the normal course of affairs Baldwin readily augmented its force from Philadelphia's large pool of skilled craftsmen, without resort to ads or recruitment from afar.[14] Foremen had the responsibility for hiring new men once the partners determined that the volume of incoming orders required increasing the force. Nineteenth-century industrial companies commonly vested foremen with this power, for a number of reasons. Avoiding a central hiring office accorded with the predispositions of the Baldwin partners against adding any managers who did not contribute directly to the work and wealth of the company. Also, foremen could best judge applicants' qualifications for work ranging from skilled handcrafts to laborers' tasks. In cases where laid-off hands sought to be rehired, their old bosses knew their abilities, and foremen could quickly judge new hands by an inspection of their tool chests and a brief trial on lathe, vise, or other work.

Evidently Baldwin relied almost exclusively on word-of-mouth advertising to fulfill its labor needs throughout the nineteenth century.[15] The firm posted hiring notices at the factory gates, and foremen put out the word to men in their departments. Most of Baldwin's workers lived in or near the Bush Hill neighborhood where the plant was located, a district also crowded with other metalworking firms and the residences of their workers. News of job openings traveled quickly in such a tight community bound by common interests and occupations. Thus when Baldwin was hiring new men in the early 1870s, the news was discussed among the residents of Bush Hill boardinghouses like Eliza Preston's, where Baldwin machinist John Skelly lived with eight other metalworkers.[16]

Because most skilled hiring was done in face-to-face encounters between applicants and foremen, these supervisors often considered a number of personal and social factors beyond an applicant's simple possession of skills. A job seeker with friends or relatives already working at Baldwin, or one from the same church congregation, ethnic group, or neighborhood as the foreman, was more likely to get the job than a stranger. Such ties also affected life on the job and one's likelihood of staying on the payroll when others were laid off. No records survive to demonstrate this; indeed, no such records were ever kept, but the strength of such personal and social ties should not be doubted.[17]

BALDWIN'S CRAFT APPRENTICESHIP PROGRAM

While the hiring of journeymen and laborers left little imprint for the historian
to study, Baldwin also utilized another avenue of recruitment and training
through the 1860s: craft apprenticeships. Here sufficient records survive to give
insights into the character of Baldwin's skill needs, the relationship between
personal ties and securing work, and the attributes of at least a minority of the
men who made locomotives. Baldwin-trained apprentices were a relatively small
group compared with the total force, but apprentice program graduates had a
disproportionate influence on the factory floor, thanks to their special training
and status, which heightened their chances for promotion.

Although already an industrial giant by 1835 with its three-hundred-man
workforce and three-story factory, the Baldwin Locomotive Works evolved
directly, if quite quickly, out of Philadelphia's preindustrial heritage as a center
of skilled, artisan craftsmanship. Matthias Baldwin himself had apprenticed as a
jeweler during the War of 1812, and he employed apprentices before entering the
locomotive business.[18] Given the novelty of the trades and work in locomotive
building, continuing craft apprenticeships in the new factory setting was a
logical, even essential, tactic to secure the skills needed. There is no reliable
gauge of the number of apprentices in the plant before the 1850s, but by then
a Baldwin apprenticeship had become something to prize. In 1849 Baldwin
wrote, "So great is the desire for situations that some parents send their sons
to me and [they] will work the 5 years without receiving any compensation."[19]
Most apprentices were in fact paid, but at barely subsistence wages. Rather than
money, the program offered boys expert training in "the art and mystery" of
one of the many trades required in building locomotives, a high-technology
growth industry of the era. The program gave Baldwin a ready supply of labor,
trained in the skills it most needed and able to serve either in the top ranks of
skilled men or in management.

Apprenticeships at Baldwin were as close to their preindustrial form as the
factory setting would permit, as illustrated by the company's usual terms: Boys
first worked a five-week trial period without wages. Apprentices served for
either a five-year term or until reaching age twenty-one. A boy was "not to
be connected either directly or indirectly with any Military or Fire Company
during his term of service." He was "to be put under the care of a friend [adult]
who will be responsible for his conduct out of the shop." Apprentices' pay was
stopped for misbehavior or violations of rules. Finally, Baldwin held back a
portion of each week's wages, which was paid to the boy only upon completion
of his term—the $130 of "freedom money" being a substantial inducement to
finish in good standing.[20]

Baldwin's apprenticeship system closely resembled its eighteenth-century craft shop antecedents. The rule against joining the rowdy and disruptive fire companies and the requirement for adult supervision outside the shop represented efforts to continue the master's role of moral guidance *in loco parentis*. Until the 1850s Matthias Baldwin personally oversaw the apprenticeship program, indicating that the company took its responsibilities quite seriously. Misbehavior by boys frequently resulted in letters from the company founder to their parents.[21] When Matthew Baird became co-partner in 1854, he took on this role from Baldwin. Even before this Baird, who had trained as an apprentice himself, acted to strengthen the program. In 1853 he instituted formal, signed indentures that bound the boys to the company.[22] This adoption of a formalized relationship between the firm and apprentices provides clear evidence of both the value the Baldwin company placed on the institution and its continued vitality in the factories of the capital equipment sector.[23] The system was modified somewhat from its preindustrial form—most obviously in that foremen and experienced journeymen, rather than a master craftsman, trained the boys. And although the company had some rules governing behavior outside of work, its supervisory role in this area was necessarily limited. Since three-quarters of the boys lived at home, however, most came under parental supervision during nonworking hours.[24]

In one key respect the system exactly mirrored its eighteenth-century form: the boys were literally bound for a period of servitude. If an apprentice ran away, the company tried very hard to track him down, reporting his indenture to his new employer and seeking his return.[25] When Baldwin learned that it had inadvertently hired runaways from other firms, such boys were invariably fired. The firm considered its apprenticeship obligations inviolable, even into the 1860s, when a number of preindustrial trades had discontinued indentures. For example, a week after the 1860 strike began, with the company desperate for skilled men to replace the 170 striking blacksmiths and machinists, Baldwin did not hesitate to fire a young man after learning that he had run away from another firm.[26] Such a system of bound labor was certainly exploitative, just as it had been in the eighteenth century. But apprenticeships provided far more than cheap labor. The obligations of the program were mutual ones, and the system frequently worked to heighten the mutuality of interests, the partnership, between management and labor.

With the institution of formal indentures, record keeping on Baldwin's apprentices became far more complete, providing an accurate portrait of the program between 1854 and 1868, the year formal apprenticeships were discontinued.[27] Over this period Baldwin took on approximately 310 apprentices in every skilled trade involved in building locomotives. Who got these jobs, so prized that parents would have placed their boys without pay?

Three-quarters of all apprentices came from Philadelphia or surrounding

This proud young man was a Baldwin apprentice machinist circa 1860, posing with the tools of his trade: file, hammer, chisel, and calipers. He was Edward Longstreth, who successively served as a foreman of a machining department, foreman of the erecting shop, general superintendent, and partner by 1870, only eight years after completing his indenture. FRANKLIN INSTITUTE

counties.[28] Among city residents, boys whose fathers worked for Baldwin had a high priority for admission to the program. Although no payroll records survive, other sources reveal that fifteen apprentices were the sons of Baldwin employees, and the actual number probably exceeded a hundred. That these fathers placed their sons in the program is strong testimony to their positive view of their employer. The second favored group of applicants were the sons of railway master mechanics and other business associates of the company.[29] Giving positions to these boys helped cement their fathers' business ties to Baldwin. The third group receiving a measure of favoritism were the "deserving poor" who could somehow get a personal plea to Matthias Baldwin.[30]

Even more than their varied class backgrounds, the apprentices' ethnicity reveals something of the character of the program within the larger social context of Philadelphia at midcentury. In 1860 immigrants accounted for 30 percent of the city's population as a whole, while only 14 percent of those boys in the program were foreign-born.[31] This would seem to suggest that this point of entry to skilled work was largely preserved for native whites. But the ethnicity of apprentices' fathers qualifies that interpretation, since 60 percent were non-native, with the Irish alone outweighing the native-born.[32] This suggests that Baldwin's apprentice program significantly promoted the assimilation of second-generation Americans into the industrial economy.

Analyzing the occupations of apprentices' fathers gives further evidence of the boys' varied backgrounds.[33] Fathers across a wide social and economic spectrum saw the Baldwin factory as a place of opportunity for their sons. Seventeen percent of the boys came from middle-class families and 60 percent were sons of skilled tradesmen, while 23 percent had unskilled parents. Of the fathers who worked in craft trades, two-thirds were metalworkers, plying trades quite similar to those in which their sons apprenticed.[34] This suggests that

Justus Johnson

Item No. 1. COMMENCED APPRENTICESHIP *Second*
day of *February* 18 *65*

" **2.** To learn the Business of *Machinist*

" **3.** Born *Twenty Eighth* day of *October*
Eighteen hundred and *Forty Eight*

" **4.** From *Philada* County, State of *Pennsylvania*

" **5.** To serve the Full Term of *Five* Years, exclusive
of all lost time, (except in case of severe sickness.)

" **6.** The wages shall be as follows: (except five weeks on trial without pay.)

" **7.** During the First Year, $ *2.25* per week.
" " Second " $ *2.50* "
" " Third " $ *2.50* "
" " Fourth " $ *2.75* "
" " Fifth " $ *3.25* "

" **8.** When absent from the establishment, either from sickness or other causes, such
lost time will not be paid for.

" **9.** All lost time to be made up before commencing the succeeding year.

" **10.** In case of sickness, satisfactory proof must be produced.

" **11.** And it is agreed, nevertheless, that the said M. W. BALDWIN & CO. reserve
to themselves the right (in case of disorderly conduct or violation of any rules
of the shop) of withholding such sum or sums from the above-named wages as
they may think proper, and also, of discharging the said *Justus*
Johnson entirely from their service.

" **12.** If these conditions shall be satisfactorily fulfilled *M W Baldwin & Co*
shall pay to *Justus Johnson*
the sum of *One Hundred & Thirty Dollars.*

We agree to the above conditions, and promise to honorably fulfil them.

Witness:

Wm R Mullison

Jno Fallon
acting guardian
Justus Johnson

Justus Johnson signed this indenture in 1865 as a sixteen-year-old orphan. By 1870 he had completed his training, continuing at Baldwin as a journeyman machinist. By that time he headed a household that included his two sisters and a domestic servant, and he had $1,000 in personal property, according to census records. Johnson went on to become a foreman at Baldwin, where he continued working well into the twentieth century. HISTORICAL SOCIETY OF PENNSYLVANIA

many of these fathers worked at Baldwin and looked to the company to train their sons.

The substantial representation of both middle-class and unskilled laboring families is also noteworthy. Professional and white-collar fathers saw a Baldwin apprenticeship as a point of entrance for their sons into the high-technology industry of the generation, precision metalworking. In this industry, theory and basic principles could be learned only by practice, and yet "to those who could see the handwriting on the wall, the steam engine and the growth of machine culture were destined to become a major intellectual challenge of the nineteenth century."[35] The engineering colleges of their era, apprenticeship programs like Baldwin's promised an entree for middle-class boys into the new professional world of mechanical engineering whose "shop culture" originated at places like Baldwin. Such hopes paid off in many cases, and not just for middle-class families. Of Baldwin apprentices who entered the plant between the 1830s and the 1860s, many entered the top ranks of mechanical engineering, including three partners of the company itself and at least four railway master mechanics.[36]

Whereas middle-class families wanted more for their sons than lives as Baldwin machinists, the 23 percent of apprentices' fathers from unskilled occupations rightly saw such training as a ticket to upward social mobility over two generations. If they completed their indentures, the boys could look forward to lifetimes of notably better incomes and job security than their fathers had known. But signing an indenture was only the first step up the ladder.

The Baldwin company knew that demand for positions always outran openings, but the firm did not take advantage of this by flooding the factory with low-cost apprentice labor. Instead it kept the apprentice corps at roughly 10 percent of the overall size of the workforce.[37] This was a low ratio for the period, and it confirms that Baldwin saw apprenticeship as a training ground rather than a means to acquire cheap labor. Here too the company's policy matched the interest of its skilled men, who feared that large numbers of apprentices would dilute the skills in craft work and the earnings of journeymen.

Most apprentices were age fifteen to seventeen when they began their training. They indentured in every skilled trade involved in locomotive building, although most became machinists. Table 5.3 outlines the distribution of trades. Machinists were in the highest demand, at 63 percent of all apprentices, because of the painstaking, time-consuming nature of their work and the great number of machined or finished parts in locomotives. Although the other trades seemed less popular with the boys, actually the company's labor needs determined which crafts an aspiring apprentice could enter. Baldwin's requirements corresponded to the boys' long-term economic interests, however, since the five trades taking the bulk of apprentices were also the most prevalent

TABLE 5.3 Trades of Apprentices, 1854–68

Trade	Number	%	Trade	Number	%
Machinists	185	63	Patternmakers	13	5
Blacksmiths	31	11	Ornamental painters	11	4
Molders-founders	24	8	Drafting and designing	2	<1
Boilermakers, sheet-iron-			Various	6	2
and coppersmiths	20	7			
	Total	292	100%		

Source: Data compiled from Apprentice Books, BLW-HSP.

Note: The category "various" includes narrow specialties like machine planing, plied by older boys working short terms who arrived at Baldwin with the rudiments of these skills.

in metalworking throughout the country. Boys who left the firm trained in these skills had expertise that was in wide demand.

The apprentices worked hard, ten-hour days—six days a week when the plant was busy—for a wage barely providing subsistence, particularly in the first year. First-year pay was $2.25 a week, and by the fifth year it had risen only to $3.25. Boys nearing the end of their training and earning such nominal pay worked alongside journeymen making $10 to $15 a week, doing much the same quality and quantity of work.[38] While an apprentice could look forward to receiving $130 of freedom money upon completing his term, he had more than earned these back wages.[39] It seems an exploitative system to modern observers, yet indentures were avidly sought by boys and their parents. Perhaps the most striking evidence of the desirability of such positions lies in the fact that nineteen families indentured two or more of their sons at Baldwin between 1856 and 1868. If older boys had felt exploited, surely they would have warned off their younger brothers. Evidently the reverse was true: boys saw the factory as a fine place to learn a good trade.

From Baldwin's point of view, the best measure of the viability of apprenticeship as a training and recruitment tool was the rate at which apprentices completed the program and joined the company's journeyman force. For the young men themselves, a high number completing their terms would indicate that the program adequately served their desire to learn a trade. Unfortunately these issues are somewhat clouded by the four years of Civil War in the middle of this period, a war that drew off a number of apprentices who might otherwise have completed their terms. Despite the complicating factor of wartime service, the surviving data indicate that the apprenticeship program was relatively successful for Baldwin and for the boys.

The Baldwin Indenture Books give notations for most of the boys, indicating if they completed or left the program. Although such notations as "left" or

TABLE 5.4 Disposition of Apprentices, 1854–73

Listed Disposition	Number	%	Listed Disposition	Number	%
Served full term	136	46	Died (during term)	10	3
Left*	46	15	Gone*	4	1
Discharged	40	13	War*	1	<1
Canceled	23	8	Rebel*	1	<1
Runaway*	22	7	Enlisted*	1	<1
Volunteer*	13	5			
			Total 297	100%	

Source: Data from Apprentice Books, BLW-HSP.
Notes: Asterisks indicate voluntary departures from the program. "Gone" and "war" are
    imprecise dispositions but are put in the category of voluntary departures to produce a
    more conservative assessment of the program's success.

"gone" are hard to interpret—they could apply equally to draftees and runaways—overall the dispositions tell us much about the success of the program,
and they are outlined in table 5.4. I believe that the 46 percent completion rate
shown above was a relatively good showing for craft apprenticeship training
in a factory setting with the flux of war intervening. A second gauge of the
program's success can be constructed by dividing the table into two categories:
those who voluntarily left the program prematurely, for whatever reason (these
dispositions are marked with an asterisk), versus those who served a full term
or left involuntarily ("discharged," "died," and "canceled," the last category
applying generally to cases of illness or physical impairment). Under this
division only 30 percent of the apprentices left voluntarily, turning their backs
on their training and freedom money. Although lacking a comparative basis on
which to improve a judgment, this seems a relatively low figure for voluntary
departures.[40] Craft apprenticeships in the 1860s still proved a viable recruitment
and training tool for the boys and the company.

Many boys who completed the apprentice program between 1854 and 1868
rose to important positions in Baldwin's management. The lack of personnel
records prevents a complete accounting, but at least seven program graduates
became Baldwin foremen; one of them, Edward Longstreth, went on to join the
partnership. Others became draftsmen or inside contractors, positions with a
substantial managerial role, if not in management per se. Those remaining in
the trades of their indentures were key leading men, in the eyes of their fellows
and the company.[41] For example, the shop employees elected seven colleagues
to represent them at Matthias Baldwin's funeral, and at least three of these
representatives were former apprentices.[42]

Despite the success of the program, Baldwin ended formal craft apprenticeships after 1868. No surviving evidence explains the program's demise, but some reasonable inferences may be drawn. In 1868 three men owned the firm: Matthew Baird, George Burnham, and Charles Parry. Parry was the superintending partner, directly responsible for production management, and as such he had the most influence in the decision to end formal indentures. In the late 1850s and throughout the 1860s Parry had focused on standardizing production in a variety of ways: instituting piecework, redesigning engines, overseeing the adoption of interchangeable-parts production, and systematizing design and drawing room practice. With the standardization of parts and the narrowing of tasks under piecework, the firm had less need for the general training in all branches of a trade that characterized the craft apprenticeship program. By the late 1860s many of the skilled men at Baldwin were specialized industrial workers, devoting their labor to particular components or processes. In this setting craft apprenticeships became largely irrelevant.[43]

Throughout the Gilded Age many employers moved away from broad training in the metalworking crafts as a result of specialization, a development criticized by both social commentators and labor activists. But craft training retained notable resiliency in the machinery-building sector as a whole. Capital equipment firms with less specialized production methods and product lines than Baldwin, such as Philadelphia's Southwark Foundry and Machine Company, Kensington Engine Works, and Bement, Miles & Co. (machine tools), were still taking craft apprentices in the early 1890s.[44] By this time Baldwin had grown far larger than any of these firms, and its great size impelled it to develop alternatives to craft training.

FROM CRAFT TO INDUSTRIAL WORK: RECRUITING LABOR IN THE GILDED AGE

The standardization of parts and work tasks that accompanied the demise of Baldwin's craft apprenticeship program stemmed primarily from management's desire to boost productivity. But these policies also related closely to the matter of labor recruitment. The substantial short-term fluctuations in Baldwin's force size, coupled with its long-term growth, made labor recruitment a pressing and recurring issue. With work tasks narrowed and divided under piecework, Baldwin found it easier to take on men quickly during market rebounds, since many new workers needed only a specialized competence. During such periods of strong demand, men with metalworking skills from a variety of backgrounds were drawn to Baldwin by the lure of higher wages and put to work at these specialized jobs. For example, in December 1877 Baldwin received a large order from Russia for forty locomotives. The work had to be completed quickly, since the contract required delivery in Russia by May 1878. As a result, Baldwin increased its force from eleven hundred men to twenty-

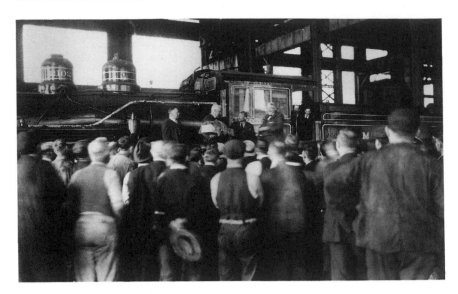

The 1921 ceremony shown here suggests the long tenure common among Baldwin's managers and skilled workers. From the left, the superintendent, John Sykes, looks on as president Vauclain gives a gold watch, box of roses, and purse of gold to William Magee, an erecting shop foreman completing sixty years at Baldwin, while retired delivery engineer George Johnson smiles to the crowd of workers. Like Magee, Sykes and Johnson also started out as workers on the factory floor; respectively they had forty-two and sixty-seven years seniority. The group photo on page 104 shows Johnson (#19) early in his career. RAILROAD MUSEUM OF PENNSYLVANIA (PMHC)

three hundred in less than three weeks.[45] This increase of over twelve hundred men probably exceeded the total force then employed at any other Philadelphia metalworking firm. Even before the contract was announced in a news item in the Philadelphia newspapers, Baldwin's foremen found "hundreds of skillful mechanics at the doors on the lookout for a job."[46] Word-of-mouth advertising and Baldwin's reputation for high wages were sufficient tools for recruiting labor from Philadelphia's large pool, thanks to the subdivision of tasks that had lessened the necessity for broadly trained craft workers. Philadelphia's economy was still somewhat depressed at this time, which no doubt aided recruitment, but men flocked to Baldwin in such numbers that the firm had no need to place help-wanted ads in the major Philadelphia papers.

Workers taken on in early 1878 joined a force that had numbered roughly eleven hundred men throughout the lean four years since the Panic of 1873. This was Baldwin's skilled nucleus: the men whom the firm was loath to lay off and for whom Charles Parry would travel to Russia to drum up business. Although little evidence survives on the question of turnover, it appears that skilled men in particular often stayed with the company for years. An 1883 profile on Baldwin in the *American Machinist* asserted that "comparatively few changes take place among the staff of workmen, especially among those engaged on skilled labor. The same set of men remain from year to year, many of them having homes of their own about the city."[47]

This source supports the notion of low turnover among the skilled force, but layoffs threatened newer hires and less skilled men whenever the locomotive market softened, as it did within a year. Under the company's normal practice in such depressions, "common laborers are first laid off, then helpers and journeymen [if necessary], until only the cream of machinists are retained."[48] During these slow periods the firm tried to limit the extent of layoffs by cutting back on hours to spread more work around. It also cut prices to encourage orders.[49] Nonetheless, between 1883 and 1885 employment dropped by 30 percent, from 2,290 men to 1,563. Men laid off were quick to return when Baldwin began hiring again. Charles Harrah, the president of Midvale Steel, a competitor of Baldwin's in the Philadelphia labor market, commented on this in 1900 when asked about turnover at his own firm:

> I do not think we are as fortunate in that respect as the Baldwin Locomotive Works. The Baldwin Locomotive Works are very peculiar; they must treat their men very kindly, because when work is slack with them we take on a lot of their men; but the moment work picks up again with them the men leave us and go back to Baldwin's, no matter whether they are getting big wages— no matter what happens they go back to Baldwin's.[50]

We turn now to what was so "peculiar" for workingmen at Baldwin.

LIFE ON THE JOB IN THE GILDED AGE

A number of mutual ties bound Baldwin's workers together in various groups and to the company. Most tasks required teamwork among men; foremen and fellow workers trained newly hired employees; and skilled men acted cooperatively in creating a mutual relief fund. The company sought to promote a congruence of interests, particularly with skilled workers, by offering them better wages, job security, and chances for promotion, and by hiring their sons and other relatives for apprenticeships and other positions. Other ties among the men, based in ethnicity and skills, created other groupings. But the chief defining characteristic of life on the job was piecework, and in this crucial area that defined work and compensation, workers stood alone as individuals.

Baldwin's managers instituted and extended piecework—or "payment by results," as it was called—to boost output and foster a sense of personal responsibility, productivity, and reward within the workforce. Piecework served as the partners' chief tool in their effort to instill in the workforce the producer ethos that had guided and justified their own advance up the ladder of promotions to ownership rank. Such a piecework system certainly represented an effort

The most basic sort of piece-rate work. The boy heated steel bar stock in a furnace, then used the machine at left probably to form hex heads on the stock to make bolts. In this case the boy could master the necessary skills in an hour. His foreman set the piece rate, and the boy's take-home pay depended entirely upon his speed and diligence. PENNSYLVANIA MUSEUM AND HISTORICAL COMMISSION

by the partners to hinder the development of a collective consciousness and identity among workingmen, promoting instead an ethic of individualized effort and reward. But the partners did not then corrupt piecework by using it to dominate and exploit workers through incessant rate cutting, as often happened in mass-production manufacturing. The firm generally avoided rate cuts, fearing they would blunt the productivity so essential to gaining a competitive advantage in the industry. Also, the partners genuinely believed in personal effort and reward under piecework. Their belief took tangible form in workers' relatively high take-home pay. This same regard for individualized relations with workers saw Baldwin reject dealings with labor unions on the one hand and employers' associations on the other.[51] Thus the individualism of piecework was a policy of self-interest by management, for improved productivity and profits, but the policy arose from conviction, not just convenience.

The Baldwin company first instituted piecework on a large scale in the 1850s, changing from day-rate pay to improve productivity and curtail workers' soldiering or self-imposed limitations on output. Evidently much of the work in locomotive building had been put on a piecework basis by 1857, including many highly skilled jobs, although probably not erecting work, which was too variable to standardize. The firm deliberately set the rates quite high to encourage and reward the productivity Matthias Baldwin had sought. An account book of a skilled Baldwin machinist, Joseph Shear, details his piecework wages from March 1856 to February 1857. Shear did a few jobs on day rates paying $1.25 daily (or $7.50 a week), but generally he worked by the piece, and in most weeks Shear cleared $12 to $15—almost double the day-rate wage for a week. Shear finished the most precise parts in a locomotive—reversing motions, pumps, and other running gear—largely with hand tools like files, and almost always on a piecework basis.[52] Most tasks in locomotive building were less demanding than Shear's work, easing the adoption of piecework.

During and after the 1860s Baldwin extended piecework to more and more tasks throughout the plant. Beyond productivity gains, it offered a number of other advantages to the firm. Wastage was cut as pieceworkers honed their skills by specializing on a limited number of components. Their heightened dexterity also aided Baldwin's achievement of interchangeable-parts production in the 1860s, as an observer noted shortly thereafter: "The whole system is so ordered that each man has a simple work [task] to do."[53] Although piecework did not wipe out the skills needed in locomotive building, it narrowed the expertise required of many workers.

By no means was Baldwin unique in using piecework. Its competitor, the Grant Locomotive Works, adopted the system sometime before 1871; the Whitin Machine Works, a builder of textile machinery, installed piecework as early as 1860, and in the same industry the Lowell Machine Shop started with it in 1817.[54] By 1870 piecework had become so widespread in the metals trades that the Machinists and Blacksmiths International Union condemned its use, alleging that it resulted in shoddy work and overproduction.[55] Regardless of the merits of that accusation, the large employers in the machinery trades found the system advantageous and had largely adopted it by the 1870s. By the time Frederick Taylor weighed in with his own solutions to "the labor problem," he was largely inveighing against straight piecework—like trade unions, although for very different reasons—in advocating his differentiated rate plan.[56] Even capital equipment builders making machinists' tools—bastions of precise and exacting work such as the William Sellers Company, the Pond Machine Tool Company, and Brown and Sharpe—operated on a piecework basis.[57] While contemporary labor activists cast Taylor as a chief villain in undercutting the machinists' craft organization of work, the large employers in the machine-building sector had undermined the craft long before by adopting piecework.[58]

If piecework fundamentally changed the organization of work, how was it viewed by Baldwin's employees? In the main most men seemed to have

Unlike a boy making bolts, a blacksmith required many years to learn the techniques of iron forging. This smith appears to be holding a smoke-box brace, which he alternately heated in a fire and then worked under the steam hammer into the shape required, according to a dimensional drawing. His helper actually operated the hammer, following the smith's orders on how hard and fast to make the hammer blows. This view dates roughly to 1900. CARL HOOPES

supported it, or at least they were not opposed, largely because it paid well. Piecework was not an issue in the 1860 strike.[59] Even a dedicated craft unionist, Job Barry, a founder of the machinists' union and a leader of the 1860 strike, sought in 1856 to receive piece, rather than day, pay.[60] Fifty years later a leader of the 1911 strike and officer of the blacksmiths' union, John Tobin, who had worked under piece rates at Baldwin for fourteen years, said, "I have no grievances against the Baldwin Locomotive Works. . . . I had a job that paid me $30 a week while I was there, and I quit of my own accord with the promise of a position any time I came back for it." A visitor at Baldwin in 1895 found that piecework also received the approval of workers who were not trade unionists, noting that "in no case was any workman, when asked his views of the system, heard to express any desire for a change. The preference was always given unqualifiedly for the piece-price system."

The fact that piecework paid well and that Baldwin's workers favored this form of compensation challenges two central tenets of nineteenth-century labor historiography. First is the conviction that employers invariably cut piece rates whenever pieceworkers earned more than the prevailing day-wage scale for their lines of work. A corollary of this belief is that workers, knowing the inevitability of cuts if any one of them produced too much, would "deliberately work slowly so as to curtail the output."[61] Such soldiering is described as a calculated collective response to counter inevitable exploitation by employers. Many cases in nineteenth-century industry support these findings, but employers in the capital equipment sector had powerful reasons to avoid this descent into exploitative piece-rate cuts and consequent worker soldiering.

Firms like Baldwin found workers' skills essential to production. Half the men in the Broad Street factory possessed extensive skills not readily replaceable, by other workers or by machines. Management viewed its skilled ranks as partners in production whose goodwill and exertions were essential to

Making coil springs called for strength and skill, which in this instance required two men. While the old veteran controlled the spring machine's chuck and speed, the muscular young helper held the hot bar stock as the machine revolved, forming the coil. Such work required close teamwork, and it allowed young workers to learn the skills of their elders.
PENNSYLVANIA MUSEUM AND HISTORICAL COMMISSION

productivity improvements. Baldwin had instituted and extended piecework precisely to speed production and curb soldiering. The partners knew that if they responded to high pieceworker output and earnings with piece-rate cuts, it would blunt workers' incentive and drive their best men to seek employment elsewhere. The company followed an alternate course, described by a visitor to the plant in 1897: "Piece rates once established are retained . . . some rates have been unchanged for twenty years, and a man's value in that shop is estimated by the money he draws; the higher he makes his pay, the more valuable he is considered." [62]

High take-home pay obviously served labor's interests, and high productivity, motivated by that pay, served management's needs in two distinct areas. First, recollect that during periods of high demand, locomotive buyers often placed a premium on fast delivery of engines rather than on price. Productive workers who did not fear rate cuts as a result of their productivity were essential to building engines quickly during such market booms. Second, the logic of inevitable rate cuts ignores a reality that Baldwin's managers felt keenly: wages were a major production expense, but not their only one by any means. Costs in materials, plant and equipment, and other overhead and fixed charges far exceeded labor costs. Improving workers' productivity resulted in increased output of engines, lower per-unit overhead charges, and larger unit and gross profits.[63] This was not a zero-sum exercise. During boom markets Baldwin could leave piece rates unchanged, workers could take home more pay by increasing output, and the company's profits grew thanks to higher throughput and better distribution of overhead.[64] By astutely giving its workers a larger-than-average portion of the margin between production costs and selling prices, Baldwin derived higher productivity, which drove down unit costs and increased overall profits.[65] This was part of Baldwin's bargain with its workers, a central tenet of the producer ethos it sought to foster. This policy helped draw workers to Baldwin from employers like Midvale Steel, making Baldwin so "peculiar" in the eyes of Midvale's president.

None of this is to say that rates were never cut. They were. But Baldwin did not cut rates simply to curb the high take-home pay that a worker garnered from his own exertions. After deriving its initial productivity gains from switching from day to piece rates, the company had to seek further improvements in methods and tooling. Commonly workers themselves would "suggest improvements which will hasten the work, offering to do the job at a lower piecework rate if the alteration is made."[66] With such improved methods, workers increased their output and earnings while Baldwin lowered its unit costs. When the firm originated a new tool or process, it did not share the productivity gain with workers in the form of higher wages, but neither did the company use such changes to drive wages down, since doing so would have blunted the productive efficacy of the investment in new tooling.

Piecework had other advantages for the company when the recurring recessions hit the locomotive market. As the volume of business shrank, payroll costs also fell proportionately under piece-rate pay. Workers' earnings suffered, often substantially, but the blame for such declines clearly lay with the state of the market.[67] On occasion Baldwin's partners did cut rates across the board during depressions in the locomotive market. For example, in 1893 orders for Baldwin engines fell from 186 in the first quarter of the year to nineteen in the third quarter. Before the year was over the Northern Pacific, the Union Pacific, and the Santa Fe had entered receivership. With important customers on the fiscal ropes, Baldwin announced a 10 percent cut in piece rates in early August. The news touched off a twenty-four-hour strike by 250 boilermakers. This soon collapsed amid massive layoffs, also caused by the market decline. Between July 1 and November 4, 1893, superintendent Samuel Vauclain cut the workforce from 5,052 men to 2,364 while also shortening the workweek. Over the same period the average weekly wage fell from $13.06 to $7.12.[68] Now the men and the company had survival on their minds.

Fortunately for Baldwin and its workers, depressions and the attendant piece-rate cuts were relatively rare. When earnings were high, experienced men banked some of their wages against the rainy day they knew would come. For the most part the firm did not tamper with piece rates, and Baldwin paid its men "the highest wage given anywhere in the world to men in similar lines of employment," according to Horace L. Arnold, a respected writer specializing in industrial management issues.[69]

Although Arnold's statement sounds hyperbolic, other evidence bears it out. A U.S. Census Department survey, the Weeks Report, allows comparisons of Baldwin's wage payments with those of other locomotive builders and metalworking employers. In 1880 Baldwin's machinists received an average of $2.46 a day, while machinists working across the street at the William Sellers Company made $2.00 a day. Baldwin's blacksmiths earned $2.88; Sellers' took home $2.20. Even laborers at Baldwin made 10 percent more than those working at the machine tool firm. In almost every class of work Baldwin's men also earned more than workers at such smaller locomotive-building competitors as Portland, Manchester, Mason, and Pittsburgh. Nor was 1880 an aberrant case. During boom and bust years throughout the 1870s—the only decade for which comparative data are available—Baldwin's workers consistently earned more than the others in the survey.[70] They also enjoyed a long-term trend of increasing real wages. Between 1863 and 1893 the average worker saw the purchasing power of his wages grow by 60 percent.[71]

The men received high pay because they worked hard—very hard. Horace Arnold could find no stints at Baldwin: "The Baldwin workmen do drive themselves, or allow themselves to be driven. . . . Purely hand operations involving the use of purely muscular power are urged and driven and rushed to the limit of the workman's strength, precisely as the tools are driven to the limit of their powers."[72] The pace was driven by both personal motivation, for higher piece-rate earnings, and by company sanction. Men whose speed, effort, and earnings fell below the standard were pressured by foremen to improve their output or quit—or they were in the first ranks of layoffs when sales next slumped. It was a hard system, hard on men and tools alike, but it paid. As Arnold noted, "Notwithstanding this high wage [expense], the work is done very cheaply indeed, and in many cases at incredibly low prices."[73]

WORKER SAFETY AND HEALTH

The fast pace of work brought good wages to the men and low unit costs to the Baldwin company, but these benefits came at a price in terms of workers' safety and health. The Weeks Report noted that locomotive building was "not unhealthful or unusually dangerous," and no doubt this was true by the standards of 1880. Compared with workers in the mining or steel industries, the men in locomotive building led charmed lives.[74] But to modern eyes the toll in deaths and injuries at Baldwin, often caused by the pressure to work quickly, is notable nonetheless. The most reliable statistics on workplace deaths and injuries are from the company itself and date from 1903 to 1913. Over this eleven-year period seventy-four men died in industrial accidents. Men were crushed by boiler shells falling from cranes, struck by falling shafting, or killed by benzene explosions in the paint shop.[75] Injury rates were far higher. For example, in 1911—a relatively safe year when the Baldwin workforce reached upward of 15,500 men—over 1,800 men were injured, roughly 12 percent of the total force.[76]

The dangers arose largely from workers and managers skirting safe practices in their rush to produce engines.[77] In 1900 boilermaker Dominick O'Dowd lost three toes after a chain broke while he was lifting a boiler throat sheet with a crane. He had complained to his foreman about the defective chain but was told, "No; let her go until [it can be fixed on] Monday, go on with your work." And in 1913 machinist John Campbell lost an eye to a steel chip from a defective cutter bar on his boring machine. Campbell showed the defective tool to his foreman, but he was told to "make the best use he could of the old cutter" until new ones were made.[78] Since at least the 1860s men injured on the job at Baldwin had received treatment from a doctor selected and paid for by the company.[79] But this provided scant consolation for men with serious injuries. Disability could threaten their incomes for life, and even if they fully recuperated, how would their families eat while they were laid low?

Baldwin's skilled men addressed this threat early on by establishing the Baldwin Mutual Relief Fund Association in 1866. Its charter stated the association's mission: to provide "mutual assistance to the members and their families in case of accident, sickness or death, out of a fund to be paid by general contribution of the members in season of health and prosperity."[80] Evidently Baldwin's managers had no role in the association, whose leaders were skilled workers with long seniority.[81] Management probably viewed the fund with wary approval; though mutual aid by employees was certainly laudatory, such collective effort could easily evolve into an organized union. The association maintained its original purpose, however, until dissolving sometime after 1888. In exchange for an admittance fee and monthly dues of 50 cents, members who could not work because of sickness or accident received $5 a week for the first six months of incapacity, with lesser payments thereafter. After a member's death his spouse or heir received $60.[82] Such benefits offered only a meager safety net, but aid from fellow workers offered more assured protection than the kindness of strangers.

The bylaws of the association had one provision that underscored the special status of skilled workers at Baldwin: membership was limited to 250 men. When leading skilled hands began the fund in 1866, Baldwin's force totaled over eight hundred workers. The membership limit suggests that skilled men

The inside-contract system suited tasks like riveting up a boiler, since such work required the management of large groups of men working on unwieldy subassemblies. In this view from the 1880s the boy (*left*) removed glowing-hot rivets from a furnace and passed them to the "holder-on" inside the boiler. He placed the rivet in a hole and backed it up while the large fellow (*right*) hammered or peened over the rivet head. SMITHSONIAN INSTITUTION

would associate with each other for their own benefit but felt an unbridgeable gulf of interests between themselves and the semiskilled and laboring men also working in the plant.[83] For years the company had sought to create bonds of mutual interests with its skilled craftsmen; here they acted to reinforce their own group ties and distinct status.

THE INSIDE-CONTRACTING SYSTEM

The Baldwin men who sought protection from accident or sickness in the Mutual Relief Association utilized similar cooperative effort throughout their work. While piecework aimed to foster individual productivity, most tasks in locomotive building required teamwork. Almost all skilled workmen depended on laborers or crane operators for help in setting up their work and moving materials. Tradesmen in boilermaking, smithing, forging, molding, and erecting all worked in groups or gangs throughout the day. In cases of standard tasks, many such gangs were no doubt receiving piece rates by the early 1870s, but group pressures probably blunted the individual incentive for high output Baldwin sought to foster. Other group work such as assembling boilers or

Inside contracting quickly came to dominate in erecting work. In this view from the 1880s workers from a range of trades erect an engine in Baldwin's 1869 erecting shop on Broad Street.
SMITHSONIAN INSTITUTION

erecting engines proved difficult to organize on a piecework basis, yet the firm also needed productivity improvements in these areas to avoid production bottlenecks.

To deal with these and other problems, the Baldwin company instituted the inside-contracting system in 1872. One of this era's few instances of labor unrest in the plant impelled its adoption. During the summer of 1872 the Eight-Hour Movement swept through eastern manufacturing centers, particularly New York City. Workers at Singer Sewing Machine, an American System manufacturer, struck when that firm refused to shorten its hours of work.[84] In June the workforces of the three Paterson (N.J.) locomotive builders—Cooke, Grant, and Rogers—also walked off the job. In Philadelphia two thousand cabinetmakers held a mass meeting on June 11 to lobby for the shorter workday, and shortly thereafter they went on strike.[85]

Baldwin's workforce also became caught up in the movement, but before the men went so far as to strike for the eight-hour day, the company's superintending partner, Charles Parry, wrote them an open letter. It was addressed to "that higher grade of our workmen who unite judgment with intelligence," specifically the skilled men essential to production.[86] Parry argued against the notion that the eight-hour system would result in higher wages by decreasing the quantity of labor in the economy at large. He also noted that such an enforced change would cut the company's return on fixed capital by 20 percent. But he said, "Your right to work eight hours is not disputed by us," noting, "If you want it you can have it, but if you take it, you must bear the loss incident to its adoption equally with your employers." Parry enjoyed great respect among workmen, having first entered Baldwin as a fifteen-year-old apprentice in 1836, and his letter cooled the ardor of the eight-hour advocates. Baldwin's employees stayed on the job.

Parry knew that such a letter might fail to prevent a strike in the future, and so he instituted the inside-contract system.[87] Contracting offered a number of advantages to the company. If the rank and file were broken into productive subunits, general labor movements such as the eight-hour drive would be less

The great variety of work in progress inside Baldwin's 1890 erecting shop suggests why contractors' oversight and adaptability were so important in building heavy capital equipment. In this room dozens of designs stood circa 1900 in varying stages of completion; finishing each required a complex series of sequential operations. In such a shop, foremen found it difficult enough merely to see workers, let alone guide their work, but contractors had the experience and the incentive to get the product out the door. PENNSYLVANIA MUSEUM AND HISTORICAL COMMISSION

likely to enlist the entire force. Contracting also put day-labor tasks under what amounted to group piecework, since the contractor would bid by the piece or job. This would improve productivity, particularly in assembly work. Finally, contracting allowed increased managerial oversight, at a time when the workforce numbered over twenty-five hundred men, without adding to the ranks of central management or increasing overhead costs.

It has been said that companies adopted contracting in "an attempt to control the craft system from the inside" by enlisting skilled men in a managerial role.[88] The Baldwin case suggests another motive, however. By 1872 labor specialization under piecework had already mortally wounded the craft orientation of most work. Evidently Baldwin instituted contracting mainly to extend the productivity and cost-cutting benefits derived from piecework to tasks not easily organized on an individual piece-rate basis. Such assembly work as tanks, cabs, tender frames, brakes, and trucks were difficult to put under piecework, since designs varied so greatly from one order to the next. Assembly and erecting workers could also face delays awaiting parts held up in other shops, delays that again blocked piece rates. By putting these assembly jobs under contracting, Baldwin secured close supervision, a faster pace of production, and a piece-rate price for the work, leaving the contractor to manage any problems, variations, or uncertainties.

As a company executive noted in 1903, "The contractor is a piece-worker on a larger scale. As he is paid by the job, he has the incentive to turn out his work as quickly as possible and to get as much as possible out of the men under him."[89] Contractors were generally older men with long seniority who had worked their way up to the position. On small-scale contracts with only a few employees, contractors worked alongside their men, but in other cases these quasi managers oversaw hundreds of workers and acted solely as supervisors. In either case contractors were leaders in the workforce, princelings in the aristocracy of labor.

Their high income provided one measure of contractors' special status. Circa 1910—a year of strong locomotive sales and much work in the plant— their average annual wages ranged from $1,600 to $3,000.[90] Such totals matched or exceeded foremen's incomes, while the average annual wage of all workers, skilled and unskilled combined, stood at $723 in 1910.[91] Since the company knew its contractors' profits precisely, it evidently paid such high wages to ensure close supervision and high productivity.

Before bidding on a job, contractors selected their workers and negotiated pay rates with them—with their foreman's approval.[92] Contractors could not seek advantage by depressing wage rates very far, since the company prohibited clear exploitation, and skilled men in the force avoided such contractors. As a visitor at American Locomotive noted, the contractor "has to be sharp and use the best methods for getting a good output, or the men will not work for him."[93] This resulted in another set of mutualist ties in the plant, based on the producer culture, as productive contractors developed long-term relationships with hardworking men in the skilled force. Contractors had to entice these men with prospects of good earnings and mutual benefit, rather than chisel them for personal gain. Contractors paid a mix of hourly and piece-rate wages, depending on the task at hand and their workers' receptivity to either form of payment.[94] The contracting system itself encouraged use of piece rates, since a contractor constantly sought per-unit efficiencies while knowing that the size of the contract would automatically limit his pieceworkers' total earnings.[95]

Contractors' anomalous position as both managers and workers and their ties to their men became apparent following the 1893 cut in piece rates. This 10 percent reduction was first applied to the workforce of two joint contractors, Thomas Roberts and Samuel Mettler, making boiler staybolts.[96] When their 250 workers struck in protest, Roberts and Mettler pleaded with superintendent Vauclain to roll back the cut to 5 percent. The cut threatened the incomes of both contractors and their workers, since all had to share the loss. And with their men on strike, Roberts and Mettler earned nothing at all for the duration. As it happened, the strike collapsed within a day after Baldwin announced the first of two thousand layoffs.[97] This brief incident well illustrates how contractors could represent the interests of their men, particularly in less dire situations.

Contractors' bonds with employees represented another set of ties that cemented various groupings among workers. Beneath the individualized producer culture fostered by piecework, ties of trades, skills, family, ethnicity, religion, and neighborhood stretched across the workforce. These mutual relations regularly played their part in labor recruitment, job assignments, pay rates, and layoffs. Piecework set the overall pace of work, but these mutual ties ameliorated the driven individualism the partners sought to foster. Such relationships often served the company's interests as well. Men who had foremen to thank for hiring their sons, contractors who showed loyalty to their workforces and received it in return, and blacksmiths who trained their helpers in the trade were creating bonds that aided the company's long-term strength and stability.

ADVANCEMENT AND PROMOTIONS

Opportunities for promotion were another aspect of working life at Baldwin during the Gilded Age that heightened cooperation between managers and workers and relieved potential discontents among skilled workers. Piecework, the specialization of tasks, and the incessant drive for productivity were pressing realities for workers at any given moment. Over the passage of time, however, Baldwin's great expansion affected men's working lives the most. In every decade from 1870 to 1900 the company's workforce essentially doubled in size, growing from an average of 1,455 men in 1870 to 8,208 in 1900. The firm came to rely upon internal promotions of experienced workers to cope with this helter-skelter growth.

Growth enabled managers to narrow work tasks further, the division of labor expanding as output increased, but it also created new opportunities for advancement. This occurred in a variety of ways. When men with true craft training were drawn to Baldwin by the high pay, they worked in tasks requiring only a narrow portion of their training and skill. That expertise remained, however. As the force expanded, these men moved into higher-paying work that utilized more of their skills. Thus a craft-trained molder could move up from simple castings to pouring locomotive cylinders—still specialized work, but requiring great skill nonetheless. The group orientation of many tasks opened another avenue for promotions through on-the-job training. As the company grew, blacksmiths' helpers became smiths themselves, and rivet holders became riveters.[98] As a visitor noted in 1895, "Every mechanic who shows powers beyond the sphere of his employment is advanced until his limit of usefulness is reached."[99] Throughout the Gilded Age the firm narrowed the tasks and skills required in production, but growth also allowed a minority to move up in skills and pay.

Some blue-collar workers also rose into Baldwin's managerial ranks. The firm's policy was "to make promotions from within, foremen, bosses or superintendents are not imported."[100] Seven of Baldwin's craft apprentices became departmental foremen during the Gilded Age. Others moved into quasi-managerial positions as draftsmen, contractors, and delivery engineers. Draftsmen in turn took up sales positions and jobs in production management. Similar patterns held true for the highest ranks of the company. Five of the firm's Gilded Age partners had entered the company as teenagers, working their way up from the ranks of clerks, draftsmen, and apprentices.

While only a select group of workers moved into top management, substantial numbers rose up limited ladders of promotion. Hard work and loyalty to the firm were required for advancement; these men had decided to cast their lot with that of the company. But even workers who did not identify with the firm knew that most of their managers could partly identify with them, having once been workers themselves. Managers and most skilled men shared a common ethnic background, further binding these two groups, while ethnicity also separated skilled from unskilled workers.[101] These factors helped to heighten the cooperative relations prevailing between management and labor at Baldwin.

It is difficult to summarize the character of life on the job at Baldwin between 1860 and 1900 because so much was in transition, and different employees and trades had very different experiences and opportunities within this context of change. In general terms, however, the workforce passed through three broad stages between 1860 and 1900. In 1860 skilled work retained a crafting character, although artisan traditions had never been firmly rooted in many of the new trades required in machinery building. From the 1860s to the 1890s Baldwin's employees were transformed from craft to industrial workers.

Thereafter narrowly skilled industrial operatives accounted for the majority of the force. Throughout the entire period, management clearly exerted control over the broad character of work, using its power vigorously to recast operations along industrial lines. Its chief instrument was piecework, which subdivided skilled crafts into job specialties, curtailed soldiering, and boosted productivity.

But workers had power too, thanks to their skills and the influence they exerted over the pace of production. Baldwin's severe limits on the number of management personnel also promoted workers' autonomy. The key success of Baldwin's producer culture was its ability to foster the strengths and powers encompassed in management and in labor, so that for a half century both contributed to production without clashing. Each side offered enough value to the other, and each respected the other sufficiently, that both could focus on their common goal: building locomotives. Baldwin's workers derived considerable pride from their work. Rather than calling themselves machinists or boilermakers, many preferred the title "locomotive builder."[102]

BATTLEGROUND OR MEETING GROUND?

Many late-nineteenth-century workplaces, particularly in metalworking, are commonly seen as battlegrounds between capital and labor. According to this portrait, managers sought an increasingly directive and assertive role over the content and pace of work—an effort that skilled craft workers fought bitterly from their power base as the practitioners and guardians of the craft knowledge essential to production.[103] By the 1920s craft labor had succumbed to management's Taylorist assaults. Although a tidy drama, it scarcely applies to Baldwin, the largest metalworking employer in America during this period. Why?

The answer is in three parts. First there is the matter of management and its policies. The Baldwin case suggests that Frederick Taylor came in the middle of a long lineage of managerial systematizers, not at the beginning. Between 1850 and 1872 Baldwin installed piecework, standardized parts, regularized production processes in achieving interchangeability, and created design and production controls over work. In sum, the company took a highly directive role over production twenty to thirty years before the supposed origins of the systematic management movement. Baldwin's Gilded Age growth saw it extend piecework and adopt contracting. By the 1870s the company's workers were becoming industrial operatives, possessing only vestiges of a crafting past. They remained skilled, but theirs was a narrow expertise, defined by the company's needs rather than by craft training.

This raises the second factor: workers lacked sufficient power to resist these managerial encroachments. Piecework and contracting splintered the workforce, hindering group consciousness and action. Boilermakers, for example, could not readily take on management when they were themselves divided

Finished cast-iron cylinders and steam pipes bolted to forged and machined engine frames. Making this relatively small number of parts required the work of patternmakers, coremakers, molders, blacksmiths, machinists, and laborers. HAGLEY MUSEUM AND LIBRARY

Two finished wheel sets stood on this wagon, the larger diameter for fast passenger service and the smaller for a slow freight engine. Notwithstanding their size, such wheel sets called for precise machining. Steel tires were turned and then heat-shrunk onto the cast-iron wheel centers. The crankpins and axles were pressed into holes, measured to the thousandth of an inch, bored in the wheel. Then the whole wheel set, weighing upwards of 5,000 pounds, was turned on a wheel lathe to ensure that all drivers had the same diameter. HAGLEY MUSEUM AND LIBRARY

Above and opposite: In 1887 Philadelphians held a great parade to honor the centennial of the U.S. Constitution. Baldwin mounted an impressive spectacle for the procession. On a dozen horse-drawn wagons the firm displayed examples of all the finished components that entered into a locomotive. They underscored the range of trades and skills found at Baldwin—even after three decades of piecework and labor specialization. The parade also demonstrated how by the 1880s the work of locomotive building had become oriented along industrial, rather than craft, lines. In the machine makers' procession of 1844 described in chapter 1 Baldwin's men walked with other Philadelphia craftsmen in a parade organized by trades. In 1887 the workers marched in groups organized by the shops they worked in, rather than by their varied trades. Although they remained skilled, here was physical evidence of their transformation from craft to industrial workers.

The work of the patternmakers and carpenters. Patternmakers made wooden mockups for the molds used to cast parts in iron or brass, including the cylinder, steam pipe, and driving-wheel patterns shown here. Steam pipe molds in turn needed cores made by other skilled workers to create hollow castings. Carpenters built the engine cab in their own shop, sending the complete cab to the erecting shop. HAGLEY MUSEUM AND LIBRARY

Locomotives required hundreds of machined detail parts like crankpins, finials, nuts, and bolts. Making these parts to precise and interchangeable standards was an idea of self-evident value, although it required new production controls and changed the character of machinists' work. HAGLEY MUSEUM AND LIBRARY

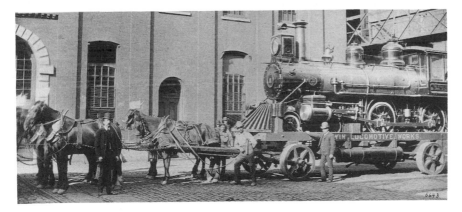

Pulled by a team of thirty-two horses, this locomotive must have been a show-stopper in the parade, with steam up and the drivers slowly revolving on rollers. Built to the meter gauge for the Dom Pedro Segundo Railway of Brazil, this 4-4-0 bore Baldwin construction number 8,780. Part of batch order of two, it sold for $8,200 and came with a guaranteed performance rating to haul 95 long tons on a 2 percent grade. HAGLEY MUSEUM AND LIBRARY

into leading men, helpers, flangers, flanging helpers, chippers and caulkers, riveters, and rivet holders. Employment fluctuations further divided the men into temporary, semipermanent, and long-term workers. These tenure rankings roughly matched skill levels, a further point of division. Management possessed unity, but labor found itself divided, making protest difficult.

Workers did not even try to resist management's mounting influence over the character of work, raising the third point. Baldwin secured their consent through cooperative labor-management relations, high pay, and its producer culture. The partners created these policies to serve their conception of the company's long-term interest. Workers, particularly the skilled men, generally believed that those interests sufficiently accorded with their own needs to be worth supporting. Baldwin's workplace culture required individual initiative and high productivity as its first criteria. In exchange for the piecework and specialization needed to achieve that productivity, workers received high pay, and over time they garnered substantial increases in real wages.

In dealing with workers Baldwin's partners truly believed that "we are as much your *employees* as you are ours," a view echoed in 1903 by another executive who wrote that "the policy of the firm is to make the interests of the men identical with its own." [104] It would be easy to dismiss such words as pandering patter, but they truly summarized the firm's cooperative policies. These policies offered sufficient value to enough workers to defuse conflict and promote accommodation at Baldwin for fifty years. When discord erupted in 1860 and in 1910–11, it followed a breakdown in Baldwin's cooperative relations with workers. In both cases competitive pressures in the locomotive industry caused the firm to curtail its high-wage policy. As a consequence, workers walked off the job.

During the intervening half century of labor peace, Baldwin derived substantial competitive advantages from its cooperative relations with workers and its producer culture. Over that period the company's share of the growing market for locomotives rose from 24 percent to 39 percent, while workers' productivity increased fourfold between 1850 and 1900. Here Baldwin reaped the benefit of its workplace policies. [105] Labor received high wages, and in return Baldwin took firm control over production. How it used that power in building locomotives is our next subject.

6

BUILDING LOCOMOTIVES

1850–1900

THROUGHOUT THE NINETEENTH CENTURY BALDWIN'S MANAGERS MOSTLY FO-
cused their attention on the factory floor, believing that production provided
the essential key to the firm's success. Thus they exerted firm direction over the
evolving character of skills and labor in locomotive building. Issues in finance,
marketing, and product innovation certainly occupied the partners' attention
as well, but circumscribing their autonomy in these areas were design, demand,
cost, and customer-relations issues inherent in making complex, expensive
capital equipment. Production had fewer constraints, imposed largely by the
technology of locomotives. Here the partners and other managers had the
latitude to effect change. And because demand for locomotives increased in
every decade from the 1830s to 1905, the firm had pressing incentive to focus its
efforts on building engines.

Our focus here is on changes in Baldwin's methods of constructing locomotives between 1850 and 1900. In this half century the essential character
of production, common to all machinery builders, remained unaltered. The
company never attempted any production revolution analogous to integration
in steel mills or moving assembly lines in automobile manufacture. Throughout
its history Baldwin employed skilled workers to make large, complex, and costly
products to customer order. Although it initiated no notably revolutionary
changes, the firm continuously modified and updated its production processes
between 1850 and 1900. Pressures impelling change included long-run growth
in demand and in the size of individual road engines, increased variety in the
locomotive product line, new materials, and improvements in reliability and
precision. Technological changes originating outside of locomotive building also
had their effect on Baldwin's factory floor, including the use of interchangeable
parts, new production technologies offered by the developing machine tool
industry, and the adoption of electric light and power in factory production.

Organizational developments also played an important role, particularly in labor management.

While a variety of influences came to bear on the complex and lengthy process of building engines, the rate of innovation on the factory floor was steady neither across time nor among the factors of production.[1] Baldwin's partners first focused their efforts to increase productivity and profits on those particular areas that were easiest or cheapest to change. Past successes guided future efforts. For example, superintendent Charles Parry improved labor productivity by initiating low-cost organizational reforms in the 1850s. These policies laid the basis for further gains through technological changes in production methods in the 1860s. When such reforms exhausted their potential for further benefits or exposed bottlenecks elsewhere, the focus of innovation shifted. This chronological pattern in production innovations is the main concern here. The pattern reveals the partners' perceived latitude for choice in innovation, it indicates how Baldwin grew to dominate its industry, and it is suggestive of stages in industrial development—stages posed by opportunities and constraints many other companies faced in creating the immense production capacity of late-nineteenth-century industrial America.

In their production innovations Baldwin's managers had two essential goals: to promote efficiency while preserving and enhancing productive flexibility. Growth depended on turning out more locomotives with every passing decade, and increasing output required improved efficiency. But Baldwin also sought expansion by broadening its product line and markets, a goal achieved through optimal use of plant, equipment, and workers' skills. This combination of efficient and flexible or optimal production marked another fundamental divergence between capital equipment builders and American System manufacturers of standard products.

This discussion of the evolving process of locomotive building also speaks to a number of other issues in industrial history. For example, Bruce Laurie and Mark Schmitz have found that there were diseconomies of scale among the firms in Philadelphia's mid-nineteenth-century metalworking sector, which raises the question of why Baldwin grew so large at precisely that place and time. Nathan Rosenberg has emphasized the importance of the machine tool industry to the rise of American System volume manufacturing; was there an analogous relationship between machine tool companies and locomotive firms like Baldwin, all builders of capital equipment? Baldwin claimed it was producing locomotives with interchangeable parts by 1865; how does this relate to the accounts of interchangeable production in American System manufacturing advanced by Merritt Roe Smith, David Hounshell, and Donald Hoke? Finally, how does Baldwin's history speak to the debate among historians regarding whether technological change in nineteenth-century industry was primarily motivated by desires to improve productivity versus efforts to displace labor, particularly by decreasing reliance on skilled workers?[2]

A STATISTICAL OVERVIEW OF CHANGE, 1850–1900

Before detailing events on the factory floor in the half century to 1900, it will be helpful to provide some background and give a statistical outline of production during that period. By 1850 the Baldwin company had passed through the difficult stages of startup and early growth, and it was as secure as any company could be in this feast-or-famine industry. But at midcentury the firm hardly dominated the industry as it had in the 1830s and would again after 1865. In 1850 Baldwin ranked fourth among the majors in invested capital, third in employment, and third in number of units produced. Although well established, the company still made a variety of non-locomotive products, such as stationary engines and boilers. Finally, the firm's management had just begun its transition from an informal and entrepreneurial style reflecting Matthias Baldwin's personality to a structured hierarchy with delegated responsibility.

In combination these factors make 1850 an appropriate baseline for the statistical profile given in table 6.1 of Baldwin's production during the remainder of the nineteenth century. The table seeks merely to illustrate the magnitude of growth and expansion in Baldwin's production over the last half of the nineteenth century.[3] Overall it reveals almost continuous growth in all indices of production as the company expanded to meet the demands of American railroading and international markets. The single noteworthy point of contraction, the average weight of engines in 1880 versus 1870, actually connotes expansion as well, since the 1880 figure was pulled down by small engines built for such newly tapped markets as industrial, agricultural, and mass-transit customers. As this example indicates, Baldwin's growth in output and market share partly arose from developing the capacity to build engines to any size or specification desired. Thus the variety in its annual output jumped from 12 classes in 1850 to 118 in 1900. Because locomotives varied so much in size and design details, table 6.1 also renders Baldwin's production at each decennial in a measure of total pounds of output. The increase here, from 1.7 million to 157.6 million pounds, represents a growth rate of ninety-five times over a half century. This is perhaps the most direct measure of Baldwin's production success. To see how its managers and workers accomplished this burgeoning record, we turn to the factory floor and the production processes in use in the 1850s.

BUILDING LOCOMOTIVES IN THE 1850S

Although the bulk of Baldwin's expansion still lay in the future, in 1850 the company was already a very large concern compared with most other industrial firms in America. Production operations centered in the imposing brick factory

TABLE 6.1 Measures of Baldwin's Production, 1850–1900

	1850	1860	1870	1880	1890	1900
Fixed capital (constant 1850 $)	250,000	318,471	1,529,052	2,843,602	5,700,000	12,323,177
Men employed	400	675	1,455	2,648	4,493	8,208
Plant horsepower per man	NA	0.17	0.23	0.17	0.56	0.73
Locomotives built	37	79	273	517	946	1,217
Variety of loco. classes built	12	17	22	54	91	118
Avg. weight of each loco. (lbs.)	44,661	51,812	64,513	64,176	92,636	129,521
Total weight of output (lbs.)	1,652,460	4,093,137	17,611,963	33,178,992	87,633,656	157,627,057
Market share (by decade)	16%	24%	32%	31%	39%	

Source: Manuscript schedules of the U.S. Manufacturing Census. For a full discussion of this source and its liabilities (and other sources used in the table), see appendix B.

Notes: "Men employed" is the average number of production workers employed in the year.

"Plant horsepower per man" measures the horsepower generated to drive production machinery, showing the mechanization of production (this information is not available for 1850).

"Locomotives built" is self-explanatory, but note that Baldwin made non-locomotive products through 1854, during the 1890s it built electric trucks for subways and interurbans, and throughout its history it repaired engines and made replacement parts. Such work cannot be measured and is not reflected here.

"Variety of loco. classes built" gives the number of Baldwin engine classes made in the year, a partial measure of product variety.

"Total weight of output" is the weight of all engines built in the year (engine tenders not included). Because Baldwin built engines in a variety of sizes and classes throughout this period, its output is rendered in pounds to allow comparisons of productive efficiency over time. Although there are some difficulties with this method of homogenizing output (see appendix B), it is a valid and useful index.

built in 1835–36 by Matthias Baldwin, fronting on Broad and Hamilton Streets. Although the historical record of the plant's configuration does not become fully detailed until 1860, evidently Baldwin occupied an entire city block by the late 1840s. The machine shops filled the three floors of the original building; separate structures housed the smith shop, foundry, and boiler shop.[4]

Structures filling a city block and all their tooling cost a substantial sum, and Baldwin valued its fixed capital at $250,000 in 1850. While other locomotive builders were larger still, in the Philadelphia industrial scene this figure made Baldwin a giant. The mean capitalization for all industrial producers in the city was merely $7,078. Beyond its large asset base, Baldwin also had an exceptionally large labor force. While the average Philadelphia producer employed 12.9 workers in 1850, Baldwin had 400 men on the payroll.[5] Such a large force was common in the industry, however. In 1850 the Rogers firm in Paterson, New Jersey, had over 600 men, and the Norris Locomotive Works employed 350 in a factory three blocks from Baldwin's. The size, weight, and sheer variety of locomotive components and the intractability of machining, rolling, and forging iron to build engines called forth these battalions of skilled workers.

Despite Baldwin's large factory and sizable capital base, its production was not markedly capital-intensive at midcentury. While the capital-labor ratio for Philadelphia industry as a whole stood at $549, each Baldwin worker of 1850 was backed up by $625 of fixed capital, a modest difference.[6] In locomotive building, the factory system chiefly aimed to marshal labor effectively, rather than to house machine-intensive production.[7] Baldwin utilized powered machinery only for tasks impossible to accomplish with human labor and skills: tires for locomotive drivers were machined on boring mills, wheels were turned on lathes, and connecting rods and axles were forged under triphammers in the blacksmith shop. The firm bought many of these tools, but it also made its own, such as planers, slide rests and lathes, key seat cutters, and wheel presses.[8]

Baldwin's ability to make such precise and complex machine tools testifies to the skill of its workers while also suggesting the nascent state of the machine tool industry. It also indicates the paucity of capital available in the 1840s for investment in the plant. As late as 1850 the company still practiced the type of capital conservation in its relations with suppliers described in chapter 1. Baldwin relieved itself of substantial burdens in fixed and working capital by buying many finished components from vendors, using ninety-day drafts for payment, and employing a rudimentary just-in-time inventory system. Given its lack of working capital, many needs for investment in plant and equipment must have gone unmet.

Most of Baldwin's production at midcentury depended on the hand skills and muscle power of its workers. While machinists had lathes for turning, they mostly used chisels and files to create flat surfaces in iron parts. This work was so extensive that Baldwin ordered over seventeen hundred files from American and English makers in just three months during the summer of 1850.[9] Such hand skills prevailed in every aspect of locomotive building from patternmaking to painting. Foundry work, blacksmithing, and boilermaking entailed particularly high labor requirements, since skilled men required assistance from helpers and laborers to finish large components. Machinists worked alone, but because of their reliance on hand tools like files, even simple tasks required many hours of labor. This labor-intensity meant that in the late 1840s Baldwin required roughly sixty days to build an engine.[10] The company completed engines only one at a time, probably because so many parts needed to be hand-filed and fitted together during assembly.[11] Baldwin lacked a full erecting shop in 1850, so it could not turn out a number of engines simultaneously. This was the most basic form of machine-building technique.

By 1850 Matthias Baldwin was chafing at the slow pace of production as output failed to keep pace with demand, despite increases in employment. Given the competitive state of the industry, profits were inadequate to provide much capital for reinvestment to boost productivity. Instead the company divided up work tasks and adopted piecework. This boosted labor's productivity while also improving the output of Baldwin's plant and equipment, since the men now had a direct financial incentive to work harder and meet higher production goals. The company moved to increase capital productivity further by installing gas lighting in the fall of 1850.[12] Now overtime work could continue into the winter months, increasing output from the same base of plant and equipment. Thanks to the introduction of piecework—which, as George Burnham put it, "costs no more by night than day"—the capital efficiencies of overtime work were not offset by higher pay for overtime.[13] With these changes, the company was able to meet Matthias's goal of producing an engine a week in the fall of 1850. The sequence of events described here illustrates how an industrialist-entrepreneur facing a need to boost output first chose low-cost managerial reforms in the organization of work.[14] Piecework improved productivity through more intensive use of both capital and labor.[15]

During the decade of the 1850s Baldwin adopted piecework for a variety of skilled tasks, and output increased. But Matthias Baldwin still felt problems crowding around him. In an 1851 letter to an old friend he complained that "the engines have increased in weight and workmanship and multiplied in parts to such an extent [that] they are not worth making."[16] Such increasing complexity arose largely from the master mechanics' mounting insistence on custom engines. The coming railway boom of the 1850s would cause demand for motive power to surge ahead for most of the decade, which promised to improve Baldwin's profits—if demand could be met with increased efficiency in production.

Building upon the foundation of piecework, Baldwin in the 1850s reordered production in two distinct areas. It mainly sought to regularize and standardize operations through a series of low-cost managerial reforms, discussed in detail below. These changes improved profits sufficiently to allow some limited reinvestment. Increasing capitalization was Baldwin's second reforming goal of the decade. The growth in plant and equipment was modest; measured in constant (1850) dollars, Baldwin's fixed capital grew from $250,000 to $318,000. Since workforce growth outpaced this rate of reinvestment, the capital-labor ratio actually declined over the decade. Nonetheless, Baldwin made notable capital improvements, such as a new erecting shop built in the early 1850s.[17] Now the lengthy process of assembly could proceed on a number of engines simultaneously, rather than one at a time. To exploit this expansion in erecting facilities, the firm enlarged its other shops in 1853, increasing plant capacity from fifty-five to seventy-five engines a year. The company expanded again late in the decade, and by 1860 Baldwin's factory buildings occupied almost two entire city blocks facing Broad Street.[18]

A growing inventory of power tooling filled these shops. Although Baldwin continued to build many of its own tools during this decade, orders in the company's letterbooks show an increasing reliance on outside suppliers. Such firms as Sellers and Bancroft, Bement and Dougherty, and others sold line shafting, lathes, planers, and slotters to the locomotive builder.[19] Baldwin also provided indirect support to these new firms. Both Sellers and Bement specialized in large machine tools particularly adapted to railroad needs. Baldwin recommended these tools for its customers and often served as a sales intermediary in their tool purchases. By direct orders and such helpful intercession, Baldwin "greatly influenced" the growth of Sellers and Bement.[20]

In turn, the massively constructed tooling made by these firms became identified as a distinctive "Philadelphia style" in the machine tool industry. By the 1860s that industry had two main branches: firms in the Connecticut River valley and the Philadelphia builders. While the New England makers focused on the relatively light special-purpose precision tools required in American System manufacturing, firms like Sellers and Bement built heavy general-purpose precision machinery for railways, locomotive firms, and other capital equipment builders.[21] Although historians preoccupied with American System manufacturing have largely focused on the New England tool builders, in fact Philadelphia was the leading center of the American machine tool industry, with 30 percent of total U.S. output in 1860 and 1870.[22]

By 1860 Baldwin ranked among the most heavily capitalized firms in the locomotive industry. Rogers outranked it in units produced that year, but Baldwin's investments in plant and equipment laid the basis for future growth.[23] Thanks to a June 1860 newspaper profile of the company, we can gain some insight into the character and extent of powered tooling in use at this time. For the most part this inventory consisted of tools for operations impossible or impractical to accomplish using hand tools alone. These included wheel borers and lathes, flue sheet drillers, cylinder borers and planers, and other lathes, borers, and planers.[24] All of these tools required substantial skill from their operators to achieve acceptable work. Without them locomotive production was nearly impossible, but tools alone did not sufficiently ease the challenges of coordinating production or improving productivity. For these problems Baldwin turned to organizational changes.

CUSTOM DESIGNS, STANDARD PARTS, AND INTERCHANGEABLE PRODUCTION, 1850–1880

Beyond improvements in plant and equipment, Baldwin's second major concern in production during the 1850s lay in creating systems of coordination and control. Through the early 1850s "each engine was constructed without much reference to those which were built before or those which would come after it. Complete drawings were almost unknown."[25] With mounting output in the early part of the decade, such chaotic production became troublesome.

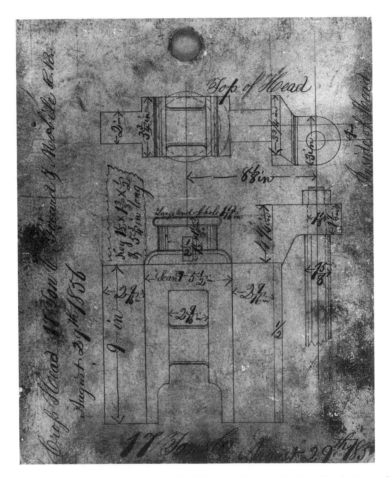

This 1856 card drawing for a crosshead dates to the period when Charles Parry decided to require drawings for all work in the plant. The drawing established a standard design for this part, but such mandates did not automatically lead to production of rigorously standard parts. Note the measurement in the middle of the drawing marked "Scant." Such imprecision in language and drawings betrayed imprecision in production techniques and parts. A machinist using this drawing hung it over his bench, hence the hole at the top.
H. L. BROADBELT

The need for systems grew even more pressing at mid-decade as the demand for customized engines accelerated. As earlier chapters have described, new managers gained power in the company after 1854, and they created new design and drawing room controls to order the chaos. Two other policies accompanied the Card System and other reforms discussed previously: all parts made in the plant would be finished to drawings, and as many parts as possible would be standardized. These policies were closely linked: one could not assure that parts would be made to standard dimensions without mandating these sizes in drawings, and it would have been impractical to provide drawings for all parts of all engines without using standard components.[26] Matthias Baldwin's new partner, Matthew Baird, and the new superintendent, Charles Parry, realized that the optimal way to build custom products involved using as many standard components as possible.

The process of reducing the variety of components to a set of standards, described in lists and profiled in drawings, took roughly ten years to accomplish, starting circa 1855.[27] By the 1870s Baldwin had carried this standardizing principle to great lengths, even standardizing locomotive paint schemes.[28] Documentation from 1889 shows that the principle extended to entire standard boilers, with twenty-five sizes available for standard or semicustom engines.[29] As engines grew in size and variety throughout the Gilded Age, the lists swelled with new sizes of standard parts. But such additions did not overtax the system, as an observer writing in the London journal *Engineering* noted in 1876:

> Of course a very large number of types of engines have [been] and are being built; but in very few cases does it happen that anything entirely novel has to be constructed or any engine the chief parts of which have not been made over and over again. . . . [While] every year certain changes occur in design . . . these modifications in accumulating increase the number of fresh combinations. . . . For this reason very few drawings are made in proportion to the amount of work done.[30]

Baldwin's adoption of piecework in the 1850s eased and complemented the production of such standard components. Specialized pieceworkers quickly gained the skill required to boost their output of standard parts. And as Baldwin's overall output and market share surged ahead, in part as a result of efficiencies from piecework, a further division of labor became possible, as Adam Smith foretold.

Standard parts secured productive efficiencies for Baldwin in building a variety of locomotive designs. Superintendent Parry believed they could also provide efficiency gains when constructing a number of engines to the same design. But he soon ran into a problem, evident in the figure at left. Although the drawings established standard designs, parts made from a given drawing frequently varied quite substantially after being machined, filed, forged, or cut to finished dimensions. Although machinists attempted to follow the drawings,

> they were not absolutely depended upon, for the simple reason that there was no standard of measurement which could be relied upon for accuracy. The unit of measurement was the inch—but what kind of inch? According to the boxwood rules which every man carried in his pocket, it was $\frac{1}{32}$ over or under a standard inch; hence, it became absolutely necessary to leave details larger than the sizes marked on the drawings and fit one piece to another all the way through.[31]

Parry soon realized that securing further benefits from standard parts designs required the installation of production standards to hold variations in finishing to certain minimums or tolerances. His goal was interchangeable parts.

The achievement of interchangeable-parts production was among the most notable feats of nineteenth-century American industry.[32] This basic change in production methods originated early in the century at the United States Armories, which made guns for the army. The new methods, known as Armory practice, encompassed a range of precision metalworking techniques, such as using hardened-steel gauges, templates, and filing jigs to mandate precise dimensions and guide machinists in their work. Later in the century manufacturers of a whole range of consumer products and light machinery adopted variations of Armory practice. These firms soon flooded national and international markets with mass-produced guns, sewing machines, bicycles, and eventually automobiles. Their techniques of volume production were uniquely American and eventually became known as the American System of manufactures.

All accounts to date of interchangeability focus on American System consumer product manufacturers, but capital equipment builders like Baldwin were also concerned with this issue by the 1850s. The manufacturers generally sought interchangeability in production to boost overall output. Baldwin shared that goal, although it also had other motives. In the 1850s Charles Parry took up interchange standards so that work could proceed simultaneously on a number of parts for the same engine. Until this, work on one component often had to await the creation (i.e., the determination) of the final mating surfaces of other parts in an assembly. Interchangeable parts promised to speed erecting work, heighten productivity, and improve the volume of work or throughput in all Baldwin's shops. Parry also saw such parts as a valuable marketing feature, allowing improved customer service and technical support. In particular, interchangeable parts promised to speed engine repairs after breakdowns— an important advantage for Baldwin's railroad customers.[33] In all, the first locomotive builder to achieve these internal and external benefits through

Only a portion of the vast assortment of gauges used circa 1915 on the Baltimore and Ohio to keep interchange standards for its motive power fleet. No similar illustration survives of Baldwin's gauge shop, but this view suggests how extensive such a system of standards had to be. SMITHSONIAN INSTITUTION

interchangeability could well derive substantial competitive advantages in the industry.

While these were potentially important benefits, Baldwin quickly encountered difficulties in achieving interchangeable production—as did American System machinery manufacturers. The fundamental problem centered on establishing and maintaining precise standards of measurement at a time when calibrated measuring tools seldom achieved an accuracy beyond one-sixty-fourth of an inch. Baldwin found its burden of precision somewhat less demanding than did most manufacturers, since it had relatively limited output and because of technical characteristics inherent in steam locomotives. But Baldwin faced an additional complication, unknown to manufacturers, arising from the vast quantity of parts in a locomotive. While most of those six thousand components did not call for great precision, ensuring their interchangeability still required extensive production controls and a major administrative structure of records.

Baldwin began its first work in interchangeability to meet customers' demands for certain replacement parts. Such efforts started circa 1839 and continued sporadically until the mid-1850s.[34] After Parry became superintendent in 1854, Baldwin rationalized its production with new locomotive designs, piecework, and standard parts. Building on these changes, efforts in interchangeability shifted to a focus on new construction—which would also ease the repair-parts problem. By 1865 Baldwin was producing engines with a number of interchangeable parts, and the company's 1871 catalog asserted that all "principal" engine components were "accurately fitted to gauges and thoroughly interchangeable."[35] Over time the locomotive builder further extended the system to include more parts and achieve higher tolerances. Like manufacturers, Baldwin achieved interchangeability in part by adopting the production controls and techniques of precision metalworking known collectively as Armory practice. But Baldwin seems to have developed its own variant of these techniques without any direct transmission from the Armories or other American System manufacturers.

Baldwin first sought interchangeability to improve its technical support for customers ordering tires for locomotive driving wheels. Most early engines had cast-iron driving wheel centers fitted with a separate tread or tire made of expensive but highly durable English wrought iron. Such tires required regular replacement because of normal wear in service. To meet the need for replacement tires, Baldwin in 1848 began to carry a stock of imported English Lowmoor or Bowling iron, which it would hoop and weld to the sizes required for new construction and replacement parts.[36] To establish the correct diameters for these tires, Baldwin used a fundamental tool of Armory practice: gauges. These tire gauges were quite simple: straight bars of iron that machinists used to measure the tires' inside diameter while boring them on a lathe. The first recorded application of such gauges was in 1846.[37] By the mid-1850s Baldwin guaranteed owners of its engines that replacement tires would fit their driver centers without further boring.[38] The sizes of driver centers and the corresponding tire gauges were kept for reference when supplying these parts.

These tire gauges were far better than the poorly calibrated rules of the day and yet much simpler than the complex gauges then used in the U.S. Armories. But they were an opening wedge of Armory-style practice in locomotive building. There is no evidence that Baldwin took this idea from Armory mechanicians; the gauges' actual origin is undeterminable. Soon after it adopted tire gauges, Baldwin began using gauges to assure interchangeability in the diameter of iron boiler flues.[39] Here the firm sought to achieve interchange standards from its flue supplier, which made hundreds of finished flues for use in new construction.

Flue and tire gauges were basic tools to maintain precise interchange dimensions of simple components. Baldwin also met demands for more complex replacement parts before 1860, although here the locomotive builder took advantage of the inherent characteristics of engines. These operational requirements of steam locomotives largely endured throughout their history. To cope with rough and jarring track and thermodynamic stresses, strength and flexibility were higher desiderata for many components, even moving ones, than was precision. Selected parts of the running gear had relatively close fits to precise tolerances, but such fits in wearing surfaces were generally adjustable. Adjustment was a regular and integral part of locomotive operation, one that lessened the builders' burden of precision in producing interchangeable repair parts. Thus Baldwin assured the Clinton and Port Hudson Railroad in 1846 that wheels and axles, eccentrics, and eccentric hooks could be supplied "at any time without drawings."[40] The customer would fit these parts or their mating surfaces as required.

Simple gauges and adjustable parts greatly facilitated Baldwin's ability to supply repair parts throughout the 1850s. But during that decade the firm felt increasing need to extend its interchange standards, thanks to two contradictory pressures from customers. Many railroads began insisting on custom or semi-custom engines, and Baldwin's product variety went from twelve to seventeen classes over the decade. Concurrently, more lines began ordering motive power in batches, generally of two to five engines, rather than simply buying one locomotive at a time.[41] So while Baldwin's product variety mounted, its separate customers increasingly ordered duplicate engines. Parry had already relieved some of the internal pressures in production that arose from the custom design trend by utilizing drawings of standard parts. Now new production controls to make these parts interchangeable became more viable economically as the quantity of batch orders increased. Customers themselves increasingly sought interchangeability within batch orders—again to ease and speed engine repairs. Thus in 1856 the Cleveland and Pittsburgh Railroad ordered two engines from Baldwin with a stipulation that their axle journals be "turned to same gauge as on last engines furnished of this class."[42] This era saw other similar orders.

Despite the slow increase in interchangeable components and the use of gauges in the late 1850s, Baldwin was moving quite hesitantly in adopting Armory practice. In the case of interchangeable replacement parts, Baldwin responded to pressures from customers who increasingly saw benefits to standardization, at least within a group of locomotives of one design. For example, the company had agreed in 1854 to build eight locomotives for the North Pennsylvania Railroad with this contract stipulation: "Each and every engine to be alike in plan and detail, so that the main parts of one engine may be used on another."[43] This is the first known instance of such a general stipulation for interchangeability, an idea initiated here by the customer. It is unknown whether the engines' main parts did in fact interchange, but the builder did not attempt such a contract again for another nine years. Over this period Baldwin learned that haphazard steps in working to interchange standards were little better than no steps at all. The fabrication of duplicate, finished parts in metal required a systematic approach to Armory practice. This involved the creation of complete sets of gauges, jigs, and templates to establish the precise dimensions required. Machinists had to follow new production methods when working to gauges, and inspections were necessary to ensure that they made the parts with sufficient precision to interchange. Like the pioneering U.S. Armories, Baldwin found that interchangeable parts required a thorough reordering of production methods.

Between 1854 and 1864 superintendent Parry oversaw this revolution in production—a revolution requiring changes in locomotive design, in the organization of work, and in production methods. Baldwin introduced standard parts designs and adopted piecework, honing workmen's specialized skills in finishing components. Parry visited other locomotive builders in 1854 to learn techniques that would lower the costs of Baldwin's production and products. He reported that "we who used to make the simplest engine in creation now make the most complicated engine to the eye that I have seen on my whole rout[e]."[44] Between 1855 and 1860 Parry and draftsman William Henszey redesigned all the locomotives in the company's standard product line. In these new Baldwin designs,

> the forms and parts were, as far as possible, laid down in straight lines [which] . . . made them easy to fit and more susceptible of being reduced to standards than the old patterns were. . . . The truck wheels were spread far enough apart to permit the cylinders to be brought down into a horizontal position. Both sides of the frames were made with plane surfaces, so they could be planed from one end to the other. . . . All the details of the locomotive were redesigned.[45]

This effort had three related goals: to update the product line, to stem Baldwin's decline in market share, and to redesign components for lower production costs and easier adoption of standard and interchangeable parts.

With these efforts under way, Parry and a young foreman of machinists, Edward Longstreth, turned to establishing a rational gauging system to standardize production methods. By 1865 they had developed such a system, incorporating sophisticated adaptations of Armory practice to achieve and maintain interchangeable work in locomotive building. The company's standard-gauge shop provided the cornerstone of the system. Here a set of master gauges was kept. The shop's machinist-toolmakers used these masters only to verify the accuracy of the working gauges that guided actual production.

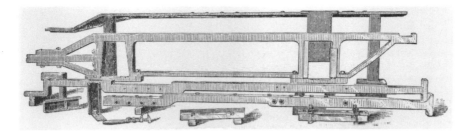

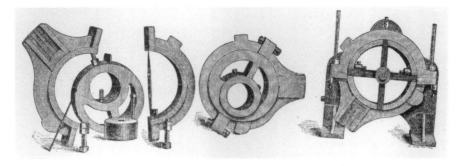

The Grant Locomotive Works used these jigs and gauges in production during the 1860s. The top set includes a spacing jig for the pedestal jaws in a locomotive frame and a frame drilling template with hardened bushings. The set below includes collars and rigid calipers to measure reversing-gear eccentrics. Baldwin used very similar tooling. HAGLEY MUSEUM AND LIBRARY

In making the masters and the various working gauges, the machinists used as measurement references a set of United States standard-measure gauge blocks manufactured by the Rhode Island toolmaker Brown and Sharpe. Out in the factory, each department requiring them had a closet "containing hundreds of all kinds of standard gauges and templates for boring, turning, and planing the work."[46] Every evening, toolmakers checked these working gauges against the masters for wear. Baldwin used a range of different types of gauges and other measuring tools, from simple measuring bars for boring tires to precision-ground limit gauges for turning tapered bolts. The choice of measuring technique depended upon the degree of precision that was technically required and economically justified. In the early years only the "principal parts" of engines were interchangeable. By 1880 almost all parts with machined surfaces were made to gauges, and Baldwin's 1881 catalog asserted that "all work is accurately fitted to gauges."[47]

In operation the gauging system worked much like the shop card system. As with the shop cards, the drawing room became a centralized point of power over production for interchangeable work. The Law Books enumerated those parts that needed to be interchangeable as well as the gauges available for this

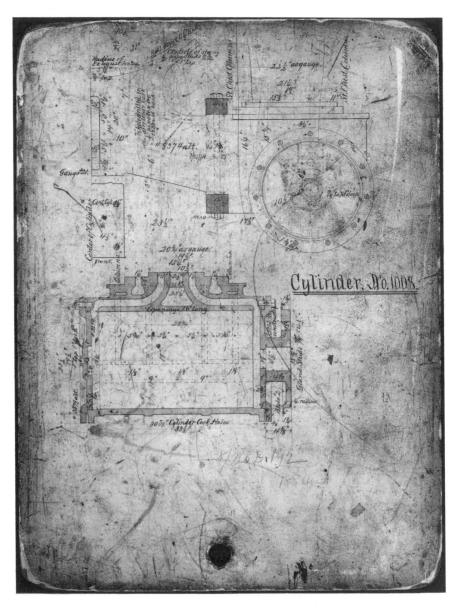

Dating to the 1880s, this well-used shop card for a cylinder shows both the mature Card System and the use of gauges. At three places in the drawing, dimensions were mandated by gauges. Cylinder number 1,008 could power engines in various designs. STANFORD UNIVERSITY LIBRARIES

work.[48] When preparing shop cards for these parts, draftsmen wrote the gauge numbers on the cards, or they gave nominal dimensions and indicated that workers should use gauges. When a foreman received his weekly Shop List outlining upcoming work, he requested the appropriate shop cards from the drawing room, and by reference to the cards he ordered the necessary gauges from the standard-gauge shop.[49] He then distributed the cards and gauges to specialist piecework machinists.

The burden of achieving the precision necessary for interchangeable work lay in the variously skilled hands of Baldwin's hundreds of machinists. To lessen the chances for deviation in measuring, the machinists were required to use special calipers, rods, and templates made in the standard-gauge shop.[50] But accuracy depended on the skillful use of these tools. Compared with the production methods and standards of the early 1850s, this work put higher demands on machinists' dexterity. To achieve interchangeable work, they had to meet the standard of precision in part after part, with only a small margin of error allowed. Given the high cost of such semifinished materials as forged axles, Baldwin could not afford sloppy final machining work that failed to meet its tolerance needs.[51] Such concerns further heightened its reliance on skilled men.

On the other hand, the reach for interchange standards fundamentally altered the character of machinists' work. Piecework, the Card System, and the new gauging techniques all standardized production methods, stemming the variability in machining and finishing that had barred the path to interchanging parts. These standardized methods also had "the tendency . . . to work the intellectual element out of the shop and confine it to the draughtsman's room."[52] Common policies established in the drawing room now guided work far more than did the separate judgments of skilled hands. The central authority required to maintain interchange standards sapped skilled workers' autonomy and discretion. Conversely, the actual achievement of those standards in hundreds of locomotive components in engine after engine increased Baldwin's dependence on skills.[53]

Baldwin's achievement of interchange standards in the 1860s required a substantial increase in power tooling as well as a heightened reliance on hand skills. No precise enumeration of its tooling survives, but Baldwin's fixed capital in plant and equipment grew from $471 per employee in 1860 to $1,051 ten years later.[54] Plant horsepower per man also increased, again suggesting more use of power tooling. Evidently the new interchange standards were frequently achieved by means of machine tools, rather than by hand skills alone. The new tools included general-purpose lathes, slotters, planers, and drill presses, all adaptable to a variety of parts.[55] Whether parts were finished by hand methods, machine tools, or a combination, only highly skilled machinists could ensure the precision required for duplicate, interchangeable parts. Although interchangeable production transformed their work, Baldwin's machinists acquiesced in these changes, and they were essential to the success of the new methods.[56]

The company sought a technical ideal in interchangeable parts, but the ideal served economic ends. How interchangeable were locomotive parts in reality, and what benefits accrued to Baldwin and its customers from the new production system? These questions were closely related, since neither Baldwin nor the railroads wanted to pay for a level of precision beyond what would benefit the production or operation of engines. Determining how interchangeable locomotive parts were is a difficult question; unlike guns, clocks, sewing machines, or other American System mechanisms, it is impossible to disassemble surviving locomotives to see if the parts interchange.[57] So we are left with documentary evidence, some of which is conflicting. Baldwin claimed that it had achieved parts interchangeability by 1865. Its 1871 catalog reiterates the claim, and it seems improbable that the locomotive builder would or could have deceived its customers.[58] Knowledgeable observers who visited the plant, such as Professor Robert Thurston and Charles Fitch, also described Baldwin's production as interchangeable.[59] On the other hand, the company employed two fitters in 1876.[60] This suggests that some of the parts made in the separate shops required hand-fitting during erection. The presence of fitters seems to defy the interchangeability claim. This seeming contradiction is resolved, however, when we consider interchangeability in locomotives within the technological and economic contexts of the 1860s and 1870s, rather than viewing it as an absolute concept to be evaluated by mechanical tests.

Baldwin had originally sought interchangeable production to meet the repair-parts needs of its customers. Railroads with engines sidelined by broken connecting rods, crossheads, or other parts wanted to receive new parts from Baldwin quickly, or to be able to take needed parts from an engine already undergoing repairs or from a stock of spare parts on hand. In any of these cases the replacement part or its mating surfaces required some adjustment to compensate for wear that had already occurred during the engine's service life.[61] Such adjustment was straightforward, since locomotives' moving parts were designed with adjustments in mind, and engineers routinely trued up many bearing surfaces before every run. Unlike many American System mechanisms, this adjustability allowed for looser tolerances in locomotive parts and minor fitting adjustments in the final assembly of engines. The crucial issue for Baldwin and its customers involved holding tolerances to the necessary limits; further precision served no purpose in adjustable parts.

Baldwin's customers viewed these parts as interchangeable, and they derived great benefits from the new production systems. On the Erie Railroad the cost of locomotive repairs fell from 14 cents per mile run by each locomotive in 1870 to 3.9 cents in 1880 as engines with interchangeable parts came into the system.[62] Following an 1877 accident to the Cincinnati Southern Railway's engine number 3, the line ordered 112 repair parts including a complete cab, new pumps, springs, and driver boxes. Baldwin shipped out the entire order three days after receiving it.[63] In the preinterchange era three weeks would have been a wonderfully fast turnaround for this job. Finally, in 1877 the general manager of the Memphis and Little Rock Railway wrote to the firm, describing its experiences with a batch of ten Baldwin engines made with interchangeable parts in 1872:

> The Road until recently has been in very bad condition, and every one of the engines has been off the track and turned over. . . . A great advantage of these engines is that all parts are interchangeable, thus enabling us to use portions of the machinery of one damaged engine to keep another one at work, and I am satisfied that there is not one of the lot that has the same parts now with which it came out of your shop.[64]

On this line and others, interchangeable parts kept engines at work, although such parts perhaps fell short of the mechanical ideal of absolute precision.[65]

In new construction Baldwin found interchangeability an unalloyed success. The new production standards streamlined erecting work in particular—heretofore a notable bottleneck—by curtailing the need for extensive hand-fitting. America's railway age flourished after the Civil War, just as Baldwin switched over to interchangeable production. The productive efficiencies of the new system greatly aided the firm's ability to garner a growing share of this enlarged market, as shown in table 6.2. The table shows a steady advance in Baldwin's market share after the 1850s, an advance arising from the company's concurrent expansion in product variety and its improvements in productivity. The data cover twenty years because Charles Parry laid much of the foundation for interchangeable work in the 1850s. In that decade Baldwin concentrated on

TABLE 6.2 Indices of Production and Productivity at Baldwin: 1850, 1860, and 1870

Year	Locos. Built	Number of Classes	Avg. Weight per Loco. (lbs.)	Fixed Capital per Man	Pounds of Product per $ of Capital	Pounds of Product per Worker	Market Share by Decade
							21% (1840s)
1850	37	12	44,661	$625	6.61	4,131	
							16% (1850s)
1860	79	17	51,812	471	12.85	6,064	
							24% (1860s)
1870	273	22	64,513	1,051	11.52	12,104	
							32% (1870s)

Source: Data on market share from White, *Short History,* p. 21. Sources for all other categories are given in appendix B.

Note: Capital ratios here are based on constant 1850 dollars.

such organizational reforms and derived great benefits. Its capital productivity increased by almost 100 percent and its labor productivity by almost 50 percent. The decline in capital per man in the 1850s provides indirect evidence that Baldwin's main productivity-enhancing efforts first centered on such low-cost organizational reforms.

During the 1860s Baldwin encountered real challenges to improving its efficiency. By 1870 the variety of classes built had increased, as had the average weight of each engine. Baldwin met these developments with great increases in plant and equipment, as capital per worker more than doubled. This investment and its standardizing reforms doubled labor's productivity over the decade. It is impossible to tell how much of this improvement resulted solely from interchangeable production, but it would be a mistake to consider interchangeability in isolation. The specialization under piecework, the simplification of redesigned locomotives, the standardization of detail-parts designs, and the creation of uniform interchangeable parts under Armory practice were complementary efforts that Baldwin refined and extended during the 1860s. All contributed to improving productivity.

Interchangeable production had little discernible effect on the selling price of locomotives, either up or down.[66] The productive efficiencies shown in table 6.2 suggest that production costs declined substantially during the 1860s, but there is no indication that Baldwin lowered prices. For a given design, railroads ordering a batch of engines paid roughly the same per unit as customers buying only one.[67] It appears as though Baldwin took larger profit margins (per unit) in batch orders while such customers gained the operational benefits of interchangeable parts at no extra cost.

Although interchangeability left locomotive prices unchanged, its advantages in repairs gave Baldwin a powerful marketing tool. As railroads grew larger over the nineteenth century, they placed larger batch orders for new motive power. Such customers had the most to gain from interchangeability in locomotive parts, particularly when engines needed quick repairs. In 1850, 59 percent of Baldwin's total output was sold in batch orders, rather than orders for single engines, with an average order size of 2.6 locomotives. These numbers increased slowly at every decennial, and by 1900, 86 percent of output was in batch orders, averaging 6.9 locomotives each.[68] This trend did not depend directly on the availability of interchangeable engines. But as an early interchangeability leader in the locomotive industry, Baldwin offered this feature to purchasers while other builders struggled to catch up.[69] This competitive advantage, as well as the internal production benefits of interchangeability, helps explain how Baldwin doubled its market share between the 1850s and the 1870s, becoming the dominant power in the American locomotive industry.[70]

All studies to date consider interchangeability solely among American System

manufacturers of consumer products.[71] Baldwin shows why a batch producer
building capital equipment to order also sought such standards to improve
customer service and promote internal efficiency. Other noteworthy builders
of heavy machinery also followed this course. Two Philadelphia machine tool
makers, William Sellers and Bement and Dougherty, began making tools
with interchangeable parts in the 1850s.[72] Machine builders like Sellers and
Baldwin did not have the challenges in volume production of American System
manufacturers like Colt and Singer to push them to interchangeability. But
they made complex, heavy machinery that they sold directly to customers across
the nation. Such builders and customers had far closer and longer relationships
than did the seller and buyer of a sewing machine or bicycle.

The fruits of this closeness are evident in mid-nineteenth-century American
industry's most important effort in interchangeability: standard sizes for nuts,
bolts, and their threads. This idea was proposed in 1864 by William Sellers
and pursued by a committee of the Franklin Institute composed entirely of
representatives from Philadelphia's capital equipment builders.[73] Baldwin
provided two members, Charles Parry and Edward Longstreth, who created
its interchange system. Sellers, Parry, and the other builders on the committee
operated in national markets; they knew the advantages uniform threads would
provide to their own firms, their customers, and the nation. They supported
Sellers' recommended system, adopted it as their own, and saw it become the
national standard. The benefits of interchange standards, and efforts to achieve
them, were hardly the sole province of the federal Armories and American
System manufacturers.

PRODUCTION IN THE 1870S AND 1880S: CONSOLIDATION
AND A NEW APPROACH TO GROWTH

The decade of the 1870s saw a slowdown in the pace of change in Baldwin's
production processes. In a sense the company was catching its breath and
consolidating its gains after the extensive reordering of 1855–72. The dismal
state of the locomotive market after 1873 also deflated the impetus for further
changes. Baldwin made a record 437 engines in that year, but the depression
that followed the Panic of 1873 severely wounded American railroading. Annual
production would not surpass the 1873 peak until 1880. During most of this
interval Baldwin's prices, profit margins, workforce size, and wage rates all fell
back as well.

The decade started on a promising note, however, and gains made during
the prosperity of the late 1860s through 1873 aided the company's survival
thereafter. Demand was robust in the five years before the panic, with output
increasing by over 250 percent. The company used the profits of those boom
years to expand its facilities to allow further growth. With the construction

Baldwin's 1869 erecting shop filled half a city block on Broad Street. A variety of engines and parts are shown under construction, and above the boiler to the right are pulleys and ropes used to lift heavy parts into place. The long row of benches and machinists by the windows demonstrates how the production of interchangeable parts rested upon hand skills.
HAGLEY MUSEUM AND LIBRARY

of a new erecting shop in 1869, Baldwin covered most of four city blocks with its buildings, three fronting on Broad Street. In 1873 the company took over the old Norris Locomotive Works shops, three blocks to the west on Seventeenth Street, for use as a tender shop. In these expanded shops workers wielded impressive new tooling. The blacksmith shop used a 3,000-pound steam hammer for forging work, a Sellers steam-powered bull riveter speeded boilermaking, and a locomotive crane carried completed boilers into the erecting shop. All of these new buildings and tools cost money, of course, and Baldwin's overall capitalization grew to $3 million by 1873.[74]

By the early 1870s Baldwin had become one of the largest metalworking factories in the nation, if not the largest. In the locomotive industry of 1870 Baldwin had almost three times the capital, roughly double the workforce, and more than twice the output of its nearest rivals. To improve their control over this sprawling operation, Baldwin's partners extended and refined their 1860s reforms—standard parts, the shop cards, and interchange standards— and they added new controls in the early 1870s, chiefly the List System and inside contracting. Baldwin also created an extensive departmental subdivision of its workforce. The 1873 force of twenty-seven hundred men worked in thirty-five separate shops.[75] Mostly these shops took the names of the parts they made, rather than the trades they employed. Machinists worked in the frame, connecting-rod, wheel, valve-motion, and other shops. This orientation of work by components helped define Baldwin's skilled men as industrial workers rather than practitioners of traditional crafts. In these changes the partners sought to improve their control and promote the efficiency of the huge workforce. But had Baldwin grown too big to be an efficient producer?

Recent findings, based on data from Philadelphia in the mid-nineteenth century, argue that there were diseconomies of scale in metalworking plants like Baldwin. Having measured how efficiently firms of various sizes used their raw materials, labor, and capital, Bruce Laurie and Mark Schmitz find the large metalworking firms were "among the least productive" compared with smaller firms in their sector or with companies in other industries. They argue that this inefficient use of resources, in a blind pursuit of growth and profits that ignored a declining rate of return on investment, was "wrong-headed and misguided."[76] Do the proprietors of the Baldwin Locomotive Works, America's largest metalworking plant in the early 1870s, deserve this verdict?

It is impossible to do the same sort of productivity analysis for the locomotive industry that Laurie and Schmitz conducted for all Philadelphia industries, so the question of the relative efficiency of Baldwin's production processes cannot be answered in this direct fashion.[77] But Baldwin's growth in market share from 16 percent in the 1850s to 32 percent in the 1870s challenges the notion that this was an inefficient producer, particularly given that the firm also achieved great increases in its labor and capital productivity during the 1860s.[78] The data in table 6.2 suggest that growth in fact brought enhanced efficiency—that Baldwin gained economies of scale. But the most important argument against characterizing Burnham, Parry, Williams & Co. as "wrong-headed and misguided" relates to survival rather than efficiency per se.

When the Panic of 1873 struck, it hit the capital goods sector quite hard indeed. Baldwin's output fell by over 70 percent in two years (1873 vs. 1875). With a plant running at only one-third of capacity, it must have been inefficient. But Baldwin remained open and in business. Such smaller builders as Schenectady, Hinkley, Pittsburgh, Grant, and Rhode Island shut down entirely, if temporarily, during the worst of the depression. All of these firms lacked the capital base and market strength to slog through the worst years. Grant and Hinkley fell into bankruptcy. Baldwin made locomotives and profits throughout the depression.[79] The partners knew that size and market power were essential to surviving in this industry of regular booms and busts.[80] Survival might seem a prosaic concern compared with productive efficiency,

TABLE 6.3 Indices of Production and Productivity at Baldwin: 1870, 1880, and 1890

Year	Locos. Built	Number of Classes	Avg. Weight per Loco. (lbs.)	Fixed Capital per Man	Pounds of Product per $ of Capital	Pounds of Product per Worker	Market Share by Decade
							24% (1860s)
1870	273	22	64,513	$1,051	11.52	12,104	
							32% (1870s)
1880	517	54	64,176	1,074	11.67	12,530	
							31% (1880s)
1890	946	91	92,636	1,269	15.37	19,504	
							39% (1890s)

Source: Data on market share from White, *Short History,* p. 21. Sources for all other categories are given in appendix B.
Note: Capital ratios here are based on constant 1850 dollars.

but in the capital goods sector it was often a more pressing issue. Even in good times capital equipment builders necessarily sought growth by broadening their markets, seeking to optimize their plants, workers' skills, and design capacities. Such firms continually had to balance desires for efficiency with preservation of optimal productive capacity. Baldwin would soon encounter productivity limits in its growth, but it would also surmount those limits and grow even larger.

To survive the six bad years following the panic, Baldwin seized on every available or potential market for locomotives. While many of its traditional customers, U.S. mainline railroads, fell into receivership, Baldwin courted industrial customers, the new narrow-gauge carriers, and the export market. It built custom engines for urban elevated lines and developed new steam streetcars. Most of these new products were smaller than regular mainline engines and offered lower profit margins, but they kept the factory going. The standardizing reforms and managerial controls of the 1850s and 1860s were essential to Baldwin's ability to produce for these various market niches. This policy of building power for all markets also caused the variety in designs to mushroom, as shown in table 6.3.

The depression was lifting by 1880, and in that year output finally surpassed the 1873 peak. But as the table indicates, all was not well despite the returning demand. The rate of improvement in capital and labor productivity achieved during the 1860s had largely stalled in the 1870s. Baldwin's policy of building for all markets had caused the variety of classes to jump, which hindered further productivity gains. But the chief problem in 1880 was Baldwin's need for new facilities and machinery to improve productivity. The company's capital expansion of the 1860s had halted in the following decade as the partners diverted funds to pay Baird's interest, and austerity reigned after the panic. By 1880 Baldwin had exhausted low-cost avenues to improving productivity. Further efficiency gains required extensive capital improvements.

During the 1880s the partners followed this course by reinvesting profits in the firm. Thanks to prosperous demand and good profits over most of the decade, Baldwin's fixed capital grew from $2.8 million to $5.7 million between 1880 and 1890, while capital per man increased almost 20 percent. Much of this investment went into new buildings, such as an 1880 office building with two floors for the growing corps of draftsmen needed as new designs multiplied. In 1884 the original machine shops at Broad and Hamilton Streets, built by Matthias Baldwin fifty years earlier, burned down. The firm immediately replaced the old three-story structure with a new four-story building, crowded with tooling. In 1890 Baldwin built a new erecting shop, large enough to erect seventy-five engines simultaneously.[81] Beyond this investment in buildings, most of Baldwin's spending went into tooling. Starting in the 1880s the company largely reordered its production processes with a new generation of machine tools.

The mechanization of production is apparent in the growth of plant horsepower from 0.17 hp per man in 1880 to 0.56 hp ten years later. This increased power drove a variety of new special-purpose machine tools developed to heighten throughput. Charles Parry and Samuel Vauclain designed most of the new tools, which were registered for patents and then built either across the street at the William Sellers Company or seven blocks away at Bement. They included a gang drilling machine, which bored twenty accurately spaced holes simultaneously in boilerplate, large-capacity boilerplate shears, an automatic feed table for punching machines, and an automated mechanism for turning tapered bolts.[82] Before 1881 no known member of the Baldwin company had patented any new production mechanism; between 1882 and 1891 Parry and others took out fourteen such patents in the drive to mechanize operations. The new machinery boosted speed and accuracy, particularly in boilermaking, which was notably labor-intensive. Although the new tools primarily enhanced productivity, some also took on a measure of the skill that heretofore had rested in workers' hands. For the first time in locomotive building, machines were displacing skilled men.

When he visited Baldwin in the mid-1880s Charles Fitch commented on this new generation of tooling: "As the conditions of fire-arms manufacture introduced the interchangeable system and improved machinery into a great range of small manufactures, the conditions of locomotive building are exercising a like influence in the introduction of uniform and labor-saving methods in the manufacture of marine engines and other heavy work."[83] Parry's ideas

This massive machine was a frame slotter, most likely built across the street from Baldwin at the William Sellers Company circa 1900. The slotter is machining four frames (i.e., for two engines) simultaneously. Compared with finishing each frame separately, this method assured that all were interchangeable, and it saved considerable time and expense in the setting up and machining of such large and unwieldy parts. HISTORICAL SOCIETY OF PENNSYLVANIA

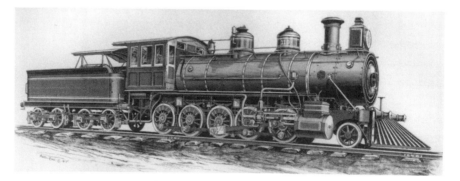

This 1885 engine for the Dom Pedro Segundo Railway of Brazil marked the beginning of a period of substantial growth in locomotive size and power. A 2-10-0 type, or Decapod, for freight service in the mountains, the 5-foot-3-inch-gauge engine weighed 141,000 pounds. The first to this overall design, it sold for $18,500. Five years later Baldwin's record weight reached 187,000 pounds for an 0-10-0 locomotive. Such large machines taxed the facilities and tooling of all builders, causing some to shut down entirely. SMITHSONIAN INSTITUTION

were filtering out from Baldwin, through the medium of Sellers and Bement, and into the factories of capital equipment builders in other industries and places.[84] As the largest firm in the locomotive industry, Baldwin had the capital necessary to innovate in these new techniques. Its growth in market share (1870s) and output (1880s) further aided its ability to adopt special semiautomated tooling, which was also an attribute of contemporary American System manufacturing.[85]

New machine tools improved the productivity of Baldwin's three thousand workers during the 1880s. In turn, raising the productivity of this new capital required it to be used more intensely. Baldwin had utilized gas lighting since 1850 to allow overtime work during periods of strong demand. But as a machinist wrote in 1880, even with gas light "men can't do much work after night. If you go into a shop at night you see a black immensity with little spots of light in it. In the middle of each of these little light spots you find a little machinist trying to do a little work."[86]

To remedy this problem and improve the productivity derivable from its capital expenditures, Baldwin installed electric lighting in its shops circa 1881.[87] As Baldwin's capital growth and its lighting program accelerated in the late 1880s, the company switched over to double shifts for round-the-clock operations. Floods of incandescent and arc lighting allowed this intensified use of Baldwin's extensive plant and equipment.[88] Output surged ahead in the late-1880s, with production climbing to a record 946 locomotives in 1890. Productivity had also improved substantially, as seen in table 6.3. But new barriers to further improvements loomed ahead. The average weight of Baldwin-built engines grew by 44 percent during the 1880s. Late in the decade big freight locomotives regularly surpassed 140,000 pounds, a size that greatly strained all Baldwin's operations. Such behemoths called for more and larger tooling, but this created another problem, because the firm's network of stationary steam power plants and line shafting to drive the tools could not be extended indefinitely. By 1892, 80 percent of the stationary steam horsepower generated in Baldwin's powerhouses was lost to turning the shafting, leaving a mere 20 percent to drive the tools.[89] To cope with the growth in engines and the need for more usable power to make them, Baldwin turned to a novel energy source: the electric motor.

ELECTRIC DRIVE PRODUCTION, 1890–1900

By 1890 Baldwin's shops filled four Philadelphia city blocks and occupied portions of three others (see map on page 190). Running such a sprawling operation in the middle of a congested industrial district was difficult, but the company had little choice. The partners were fiscally conservative and unwilling to abandon their investment in Bush Hill for a new site in the open suburbs.

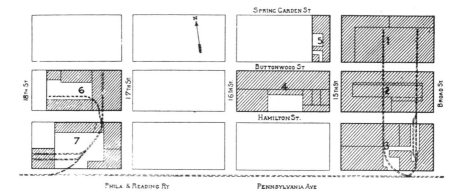

1. Main Office, Drawing Room, Erecting Shop, Cylinder Shop, Paint Shop, and Store Room.

2. Boiler Shop, Machine Shop, Brass Machine Shop, Brass Foundry, and Wheel Shop.

3. Machine Shop, Blacksmith Shop, Hammer Shop, and Power Plant.

4. Iron Foundry, Repair Shop, and Flange Shop.

5. Superintendent's Office, Laboratory, Pattern Shop, and Electrical Department.

6. Machine Shop, Tender Shop, and Sheet Iron Shop.

7. Spring Shop, Drop Hammer Shop, Hydraulic Smith Shop, Wood Shop, and Power Plant.

NOTE.—The shaded spaces indicate buildings occupied by the Baldwin Locomotive Works.

Baldwin's shops in the 1890s. HAGLEY MUSEUM AND LIBRARY

The city's great reservoir of skilled metalworkers also tied Baldwin to the district it had occupied for over fifty years.

As demand increased, the company added floors to its shops or wedged new departments into its crowded property, but these were compromise solutions at best; it was far better to increase throughput in existing structures. Here too the dispersed facilities posed difficulties. Higher output required more powered tooling, but the separate powerhouses required for each building's line shafts sapped overall efficiency. In the 1870s and 1880s Baldwin fought this problem with successive generations of steam prime movers.[90] Generated horsepower grew from 400 in 1870 to 2,500 in 1892, but frictional and shafting losses robbed much of the increase. So when the first generation of electric motors began to appear in the 1880s, the new plant superintendent, Samuel Vauclain, had reason to pay attention.

In 1889 Vauclain drew up plans for the erecting shop that would replace the 1869 facility. Over that twenty years Baldwin's annual production had grown by over 250 percent, with the average weight of engines increasing 44 percent.

The new erecting shop, with the Sellers overhead crane demonstrating its advantages. Despite electrification, the builder made ample provision for natural light.
HISTORICAL SOCIETY OF PENNSYLVANIA

These mounting demands gave Vauclain reason to notice that the E. P. Allis Company of Milwaukee, another capital equipment builder, had just installed the first electrically driven overhead traveling crane in America.[91] Vauclain approached the Sellers company, and when the new erecting shop opened in December 1890, two 100-ton-capacity Sellers electric cranes traversed the two erecting bays. These could lift the largest locomotive, carry it over engines in progress on the nineteen erecting tracks, and place it on the tracks leading to its buyer. Although costing $30,000 apiece, the cranes speeded production, simplified coordination, and allowed the firm to cut sixty laborers from the force that muscled parts around the shop.[92] This labor savings alone repaid the cranes' cost within two years.

In 1893 Baldwin applied electric power on a larger scale in the wheel-lathe shop, where its full advantages became apparent. Before electricity came to this shop, overhead shafts and belting drove its large wheel lathes. The jungle of shafts, pulleys, and belts hanging from the ceiling blocked much natural light and prevented the use of overhead cranes. Forty laborers shoved wheel sets

The wheel shop had an overhead electric crane and motor-driven wheel lathes. The crane plucked up wheel sets awaiting turning from the foreground and delivered them to the lathes. The two men working in the aisle indicate the size of the lathes and the room. AL GIANNANTONIO

weighing up to 3 tons down wide aisles, taking half an hour to mount each set on a lathe for turning. Sets awaiting machining needed to be carefully ordered to flow through the wheel shop in the order required in the erecting shop. But "with the electric drive all this is changed. The floor is literally filled with wheel lathes [driven electrically] while a traveler [electric crane] overhead serves them all without interference from belts or countershafts; the number of laborers employed has been cut down to eight or ten; the lost time between jobs does not exceed five minutes; and the wheels can be picked up as wanted."[93]

Natural light flooded the shop, aiding the accuracy of work, additional lathes filled the aisles, and the new drive system achieved a 50 percent savings in power costs.[94] This example convinced the partners, and Baldwin went headlong into electric-drive tooling and cranes. Despite a soft locomotive market after the 1893 panic, the company expanded its electrification program, convinced that such countercyclical spending would secure long-term advantages for the firm. During the 1890s Vauclain installed over 320 direct-current electric motors, totaling 3,500 horsepower and making Baldwin a national leader in this new field.[95]

In all its effects, electric power greatly intensified the pace of production. Beyond the changes noted in the wheel shop, electricity allowed higher speeds for tooling, promoted flexibility in plant layout by freeing machines from line shafts, and facilitated the running of selected tooling when desired.[96] The new power source brought with it such novel tools as compressed-air-powered nut drivers in the erecting shop, molding machines in the foundries, and hydraulic flanging machines for the boiler shop. As the company struggled to stay abreast of swelling demand after 1897, electricity's enormous advantages were duly appreciated by management. In 1901 superintendent Vauclain noted, "If we should abandon electric driving our manufactured product would now cost us from 20% to 25% more for labor . . . [and] were it not for electric driving, the Baldwin Locomotive Works would have to cover 40% more floor space." [97]

Baldwin's workers probably did not share Vauclain's enthusiasm. Electric drives intensified the pace of work for such skilled men as the wheel-lathe operators. Materials handling by power cranes gave the machinists far less idle time between jobs, and the higher speeds of electrically driven tooling forced operators to work faster. Since piece rates applied to most tasks and electricity boosted output, they probably took home more pay in return for

This pressman stands in front of a hydraulic flanger in the boiler shop. Until the adoption of such machines, Baldwin's boilermakers had to form flanges and steam domes largely by hand, heating an iron or steel plate in a furnace, then hammering it over a form using large wooden mallets. Besides being labor-intensive, this method often weakened the plates at just those bends and flanges that most required strength. The hydraulic press exerted up to 365 tons of pressure, forming the desired shaped between steel dies in a single heat of the plate. The press offered speed and better accuracy compared with hand-forming. CARL HOOPES

TABLE 6.4 Indices of Production and Productivity at Baldwin: 1890 and 1900

Year	Locos. Built	Number of Classes	Avg. Weight per Loco. (lbs.)	Fixed Capital per Man	Pounds of Product per $ of Capital	Pounds of Product per Worker	Market Share by Decade
							31% (1880s)
1890	946	91	92,636	$1,269	15.37	19,504	
							39% (1890s)
1900	1,217	118	129,521	1,501	12.79	19,204	
							39% (1900s)

Source: Data on market share from White, *Short History*, p. 21. Sources of all other categories are given in appendix B.

Note: Capital ratios here are based on constant 1850 dollars.

these exertions. But the company no doubt reserved for itself the lion's share of the productivity benefits. Unskilled men were the primary casualties of the new system; laborers were fired by the score as cranes took over their materials-handling jobs. Between 1888 and 1904 Baldwin spent almost $0.5 million on materials-handling devices like cranes, hoists, and elevators.[98] Without electrification the company simply would have sunk beneath the long-term increase in demand and the growing size and weight of road engines during this period.

Baldwin made other refinements in its production processes at the century's end. Railway master mechanics increasingly specified cast steel for many components previously made of forged iron. Axles, crossheads, connecting rods, and entire locomotive frames came to be made of the new material, which was stronger and less likely to contain the hidden flaws often found in wrought-iron forgings worked up under a steam hammer. The switch to cast steel arose largely from the need for stronger parts as engines grew. But as steel replaced wrought iron, the need for skilled smiths and hammer men fell off, and Baldwin cut the size of its blacksmith shops in half while doubling its output.[99] On the other hand, casting parts in steel required additions to the force of skilled patternmakers and molders. One trade's loss was another's gain.

So where did all these efficiency improvements leave the company in 1900? Surprisingly, Baldwin was little better off than it had been ten years earlier. On the positive side, total output—pounds of product built in a year—was up almost 80 percent, comparing 1890 with 1900, and market share had also increased. But as table 6.4 shows, labor and capital productivity had fallen, despite electrification. The productivity decline of 1900 had nothing to do with diseconomies of scale. Baldwin's factory was not too large; its products were too big, and the plant was too small to cope with them. The increase in average locomotive weight between 1890 and 1900 was the largest decennial jump ever. The 1900 output of 1,217 units was another record.

Baldwin's situation in 1900 arose from a dynamic common to the capital equipment builders and unknown in mass-production manufacturing. Unlike the manufacturers, Baldwin had no use for a corps of middle-level managers to set production quotas, plan orderly transitions in product design, and mesh projected demand with product design and productive capacity. The locomotive builder had to await the market for signs of its future; customer demands directly determined both the volume of production and the character of locomotive designs. Despite all its efficiency drives during the 1890s, Baldwin was playing catch-up in 1900 as output leaped by 35 percent compared with the previous year. Thus even with its new cranes, elevators, motors, and hoists, the company's 1900 wage bill for unskilled labor was higher than ever before.[100] With the largest engines regularly reaching over 200,000 pounds in 1900, the factory was swamped in a flood of orders. Baldwin moved quickly to cope with this rising tide; in 1900 the three-story Broad Street machine shop dating from 1863 — the lower right structure in the map on page 190 — was converted to six stories. While new walls and floors went up around them, the nine hundred men in this shop continued their work machining parts.[101] Baldwin had to await market demands, but it could not lag far behind them either.

As it turned out, the first decade of the twentieth century marked the apogee of American locomotive building. The industry built almost forty-three thousand engines in the decade, with Baldwin taking 39 percent of the total.[102] To meet this boom market, Baldwin further expanded its facilities, first in the constricted urban district of Bush Hill, and then in 1906 striking out for open land in the suburbs. After 1900 Baldwin broke with its past in a number of ways, so the watershed of the new century is an apt moment to summarize what managers and workers had accomplished at the Broad Street factory between 1850 and 1900.

EFFICIENT AND OPTIMAL PRODUCTION OF LOCOMOTIVES

While Baldwin continuously reordered its production methods, the basic attributes and constraints it had confronted in 1850 remained essentially unchanged in 1900. Locomotives were massive, complex, and costly to make, requiring a large capital investment and an extensive labor force with intensive skills. Engines were made to order, generally to custom or semicustom designs, and were quite literally "built" — assembled individually, piece by piece. Such dynamics remained true at all locomotive firms throughout the nineteenth century; in large measure they were inherent in the entire capital goods sector. Over the same period the most fundamental changes confronting locomotive builders were long-term growth in demand, an accelerated multiplication in locomotive designs, and a general trend of increasing size and complexity in engines.

While these various attributes and trends operated on all firms in the industry, Baldwin proved uniquely successful in recognizing them and adapting its operations to accommodate them. Before 1850 Matthias Baldwin had attempted to gain economies in production by marketing a line of standard engines that were built to order, one at a time. When the master mechanics blocked this course with their desire for custom products, Baldwin, Baird, and Parry realized that building custom engines efficiently required using as many standard components as possible. Piecework and interchangeable parts represented an attempt to gain some of the advantages of manufacturing in capital equipment building. Having developed these efficiency-enhancing measures for its major market, U.S. mainline railroads, Baldwin then promoted the optimal use of its systems, facilities, and workers by cultivating broad markets with a variety of motive power needs.

In improving efficiency and optimization, Baldwin's partners reordered production in a discernible pattern. They first focused on managerial and organizational reforms in the 1850s and 1860s. In the latter decade these planks were complemented by capital expansion focusing on general-purpose machine tools. The depression-ridden 1870s blocked further investment, so the company extended its production controls to allow expansion into new markets. Renewed prosperity in the 1880s brought more investment in plant and equipment, including a new generation of special-purpose tooling. Here too Baldwin was building upon its past. Its ability to amortize the new specialized tooling probably depended on the earlier broadening of markets that had been manageable only through the systems and standards developed circa 1855–73. The new steam riveters, plate drillers, and other tools further promoted accuracy and standards, but at some cost to skills in the labor force. The partners sought productivity enhancement, but skilled men were displaced. Baldwin's example and the tools it developed passed through such intermediaries as Sellers and Bement and served as models of, in Charles Fitch's words, "uniform and labor-saving methods" for other capital equipment builders.

As demand mounted late in the 1880s, the company encountered a new limitation to further increases in capital intensity: power transmission losses. With locomotives growing steadily in size and weight while Baldwin's urban location proved increasingly constricting, the company seized upon electric power in 1890. Daniel Nelson, historian of the "new factory system," calls Baldwin's "the first important application of electric power to drive machinery" in America.[103] Desperation drove Baldwin to national leadership in this embryonic field. Existing methods could not make the products in the size and quantities demanded. Electric drives intensified production, increasing the pace of work for skilled men and temporarily cutting the ranks of unskilled laborers. Electrification also allowed Baldwin to remain in the heart of the city—for the time being.

By 1900, after fifty years of extensive changes in production, Baldwin had become an international industrial giant, with eighty-two hundred workers and $20.5 million in sales.[104] Over the period from 1850 to 1900 the firm improved its labor productivity by an average of 3.1 percent each year—an impressive accomplishment, given that workers in the American industrial sector as a whole averaged 1.9 percent annual productivity growth.[105] Despite all the changes of this half century, however, Baldwin's essential constraints remained untouched. In 1850 the company required sixty days to build a 22-ton locomotive; fifty years later its engines weighed 65 tons on average, and they still took sixty days to make.[106]

7

TRIUMPH AND ECLIPSE

1900–1915

On the evening of February 27, 1902, 250 captains of American railroading, industry, and finance attended a celebratory dinner at the Union League, the private club at the pinnacle of Philadelphia's social and industrial elite.[1] They came to honor the Baldwin Locomotive Works on its seventieth anniversary, an occasion also marked by the company's completion of its twenty-thousandth locomotive. Among the guests were the presidents of the Philadelphia and Reading Railway, the Illinois Central, and the mighty Pennsylvania Railroad. Industrialists such as George Westinghouse and John Fritz also attended. They took their port and cigars along with leading members of the nation's financial community, including Edward Stotesbury of Philadelphia's Drexel & Co.; John Crosby Brown of Brown Brothers, the New York investment banking house; and John G. Johnson, the nation's top corporate lawyer. This was heady company for their hosts, the Baldwin partners, who had begun their careers as teenage clerks, draftsmen, or machinist apprentices.

Although railway presidents came to cheer Baldwin that evening, in the normal course of business the builders were the lines' junior partners—subordinate to their needs in demand and to their desires in design. The Baldwin company's success grew largely from its astute adaptations to that status. Baldwin had prospered within the bounds of this patron-client relationship, and by 1902 its partners—George Burnham, John Converse, William Henszey, William Austin, Alba Johnson, and Samuel Vauclain—were acknowledged captains of industry themselves. In that year their firm's sales exceeded $24.5 million, and its average weekly payroll totaled 12,100 employees.[2] During its seventy years Baldwin had exported its products to over fifty-five countries or colonies, and since the 1860s the firm had accounted for roughly 35 percent of all locomotive production in America. By 1901 its competitors had realized that their only hope to compete with the Philadelphia giant lay in combination, so eight smaller builders merged to form the American Locomotive Company (Alco).[3] Once united, they achieved rough parity with Baldwin. Alco represented a new challenge, one that Baldwin met with further growth. While it had required seventy years to build its first twenty thousand engines, Baldwin doubled that total in the eleven years following the 1902 dinner at the Union League. The celebrants at that dinner were backing a thoroughbred of American industry.

Baldwin's impressive output in the decade 1898–1907 resulted from extending and elaborating on its proven capabilities; the turn of the century saw no fundamental changes in the company's structure or in locomotive construction. Baldwin's particular success was in meeting robust demand with a combination of strengths in management, labor, and production while adapting to the evolving structure of its industry and to customers' design influence—themes treated in the five previous chapters. These themes are woven together here to show how Baldwin had fully refined its techniques of building custom capital equipment by 1900. This account also details changes within the firm that arose from its great expansion. During this period Baldwin finally incorporated, the flood of orders created new difficulties in control and coordination, the company began construction of a suburban satellite factory, and its fifty-year heritage of labor-management peace was shattered.

Such developments as incorporation and the new Eddystone plant mirrored trends common in American industry of the period. But as a machinery builder Baldwin's ability to modernize its plant and tighten its managerial controls was circumscribed by the particular dynamics of the capital equipment sector. As always, the manic-depressive demand cycle of expensive capital equipment still ruled Baldwin's sales, requiring a cautious approach to plant modernization and investment. Circa 1900, moreover, leading carriers greatly accelerated the technical flux in locomotive designs. At the time these realities caused little concern among the partners. Indeed, the essence of Baldwin's strength lay in its accommodation and adaptation to this reactive posture. Building only to order, the firm had a worldwide reputation for the rapid delivery of well-crafted products, competitively priced and fashioned to suit customers' particular needs. Baldwin's success arose from a longstanding partnership with its railroad customers, a bond celebrated at the 1902 Union League dinner. As American railroading prospered, so would the locomotive builder.

This relationship carried Baldwin to its height in the first decade of the twentieth century as American railways' demand for motive power reached its long-term peak in 1905. For many years after the Panic of 1907 Baldwin's officers waited for the market upturn surpassing the 1905 plateau that their experience suggested was inevitable. It never came. Baldwin's changed fortunes after 1907 were rooted particularly in the mounting debility of its major market segment, American mainline railroads, as a newly hostile regulatory

Baldwin construction number 20,000 went to this 4-6-0, or Ten-Wheeler, for the Chicago Short Line in 1902. With 73-inch drivers, it was destined for fast passenger service.
RAILROAD MUSEUM OF PENNSYLVANIA (PMHC)

climate at the Interstate Commerce Commission humbled the carriers. But the locomotive builder's inability to respond effectively to this challenge also signified broader transformations in American industry. The nineteenth-century attributes of industrial strength that Baldwin typified with its skilled, labor-intensive production of custom products increasingly became liabilities to the professional managers who dominated the corporate, capital-intensive, high-volume manufacturing economy of the twentieth century. Baldwin's apogee between 1900 and 1915 also marked the passing of an entire order of industrial production.

A HOTHOUSE OF GROWTH: AMERICAN RAILROADING, 1898–1907

If only a single decade in American history could be called "the railway age," there is a strong case for selecting the ten years that started in 1898. In that year of economic recovery from the post-1893 depression, the national railway network totaled 245,300 miles after seventy years of construction. Ten years later, track mileage reached almost 328,000, a 34 percent gain on top of a substantially mature base. While the network grew, many smaller carriers merged, and by 1906 seven interest groups controlled two-thirds of the American railway system.[4] Consolidation improved efficiency, productivity, and service. As the nation's agricultural and industrial production surged in this period of general prosperity, the volume of rail-borne freight and passenger traffic increased by 107 percent.[5] Behind such numbers were countless streams of freight and passenger cars, heavier and more numerous every year. Hauling them and their burden of the nation's business required ever expanding fleets of the ponderous machines of steel that Baldwin men created at Broad Street.

In 1897 the national locomotive fleet totaled thirty-six thousand engines,

This massive 355,000 pound 2-6-6-2 built for the Great Northern Railway was known as a Mallet compound, a type named for its Swiss inventor. The rear cylinders drove the rear six drivers with high-pressure steam from the boiler. Upon exiting those cylinders, the steam was piped forward to a pair of larger low-pressure cylinders that drove the six forward drivers, with the steam then exhausting up the stack. The front set of drivers ran in an articulated frame to allow the engine to negotiate curves despite its long wheelbase. With their massive boilers and quantities of drivers, Mallets particularly suited slow, heavy freight service. They provided the power of two locomotives in one, requiring only a single engine crew. RAILROAD MUSEUM OF PENNSYLVANIA (PMHC)

The broad cultural movement of Progressivism swept through American politics, industry, and engineering circa 1900. Progressives valued efficiency, rationalization, and professional expertise as tools to order and improve society. Progressivism even had an impact on locomotive design, as railways around the nation took an increasing interest in standardizing their motive power to secure lower engine prices, improve operating efficiencies, and cut maintenance costs. Thus in 1903 Baldwin prepared standard designs in a variety of wheel arrangements for use on E. H. Harriman's Associated Lines: the Union Pacific, Southern Pacific, Chicago and Alton, and two smaller lines. This Union Pacific 4-4-2 or Atlantic type followed standards that mandated common designs for all Harriman's lines and common parts for use across different engine types. But as trains grew heavier and faster, the standards quickly became a hindrance, not a help. Soon the Atlantics gave way to Pacifics (4-6-2s) and other types in fast passenger service. RAILROAD MUSEUM OF PENNSYLVANIA (PMHC)

The years after 1900 saw an unprecedented proliferation of new locomotive wheel arrangements, including the Pacific (4-6-2), Mikado (2-8-2), Mountain (4-8-2), and various Mallet designs—all resulting from the railways' pressing need to improve productivity. Shown here is a Santa Fe type, or 2-10-2, a wheel arrangement Baldwin originated in 1903 for that line. This 1907 engine for the Pittsburgh, Shawmut, and Northern was built to novel specifications and bore Baldwin construction number 30,000. RAILROAD MUSEUM OF PENNSYLVANIA (PMHC)

with roughly one-third carrying Baldwin builder's plates. Ten years later the national fleet included fifty-five thousand engines.[6] These new locomotives were of unprecedented size and power. While the largest Baldwin product of 1890 weighed roughly 187,000 pounds in working order, in 1906 the Great Northern Railway took delivery of five Mallet compound engines that each pushed the scales past 355,000 pounds.[7] Locomotive power capacity was proportional to weight, and these new giants provided a key contribution to improving railway productivity.

The railways retained primary control over the pace and character of technical change in locomotives after 1900. But the rise of performance specifications gave Baldwin and other builders a close collaborative role in innovation. On the major consolidated lines, the master mechanics of the Gilded Age now reported to such higher executives as the chief mechanical officer (CMO) or the superintendent of motive power.[8] As the titles imply, these men headed large, bureaucratized mechanical departments covering vast systems. Unlike the old master mechanics, the CMOs appreciated the role of technical standards in bringing order to their far-flung operations. The general Progressive-era concern for efficiency and rationalization also ignited a deeper interest in standards among railway officers. But in locomotive design, the fifty-year tradition of railway-imposed innovation actually accelerated after 1900.[9] To cope with the growing flood of traffic, the CMOs used performance specifications to order powerful new engines to improve speeds and hauling capacity. Working closely with such customers, the locomotive builders developed these new designs, which rendered much of the railways' existing motive power obsolete as the major carriers embraced technological innovations to improve productivity. In the past the builders' domestic sales had been driven by railroads' needs to add motive power as track mileage grew and to replace worn-out models. The accelerated pace of technical innovation after 1900 added a third component to locomotive sales: replacements for engines made obsolete by customer-sponsored technical change.[10] With the additional needs of industrial and foreign buyers, these were halcyon times for American locomotive builders.

EXPANSION AT THE BROAD STREET FACTORY, 1898–1907

The mid-1890s were years of national depression. In 1897 Baldwin employed 3,200 men and made 501 engines, 40 percent for export. Every depression in the past had been followed by record-breaking demand for engines, and the Baldwin partners no doubt expected that the hard times of the mid-1890s would pass eventually. But they could hardly have foreseen the mountain of orders the new century brought. Chart 7.1 shows annual employment and output in this era. While 1890 had set sales and employment records, Baldwin's performance after 1900 was of an entirely different order of magnitude. The company employed 18,500 men in 1907, placing it among the largest American industrial employers. These workers labored in two shifts, with locomotives rolling from the erecting shop twenty-four hours a day.

In meeting this crescendo of demand, Baldwin mostly relied upon controls and procedures dating back to the 1850s and 1860s, such as the Card and List Systems. The higher output of ever larger engines did require an extensive remodeling of the company's various shops, however. During this period American industrial companies were developing a "new factory system" that

CHART 7.1 Baldwin's Annual Employment and Output, 1890–1907

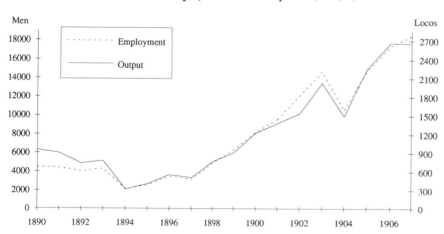

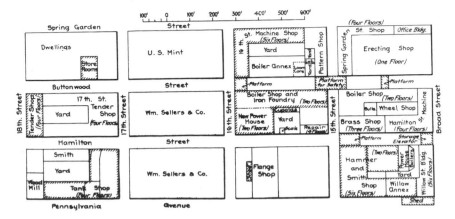

Plan view of Baldwin's shops during their 1902 remodeling (cross shading shows improvements made that year). Among other changes, the company took over portions of Hamilton and Buttonwood Streets to add further space in its constricted location. The firm also built an engine finishing shop uptown at Twenty-sixth Street. HAGLEY MUSEUM AND LIBRARY

rationalized flows of materials and increased managerial controls. To streamline operations, many firms found that the only way to circumvent plant bottlenecks and expand was to transfer all operations from constricted urban districts like Bush Hill out to new suburban factories. But the Baldwin partners remained unwilling to walk away from their seventy-year investment at Broad Street. Because horizontal expansion was impossible, the firm grew vertically, remodeling eight shop buildings in 1902. The choice of vertical expansion resulted in some seemingly anomalous operations. For example, a two-story structure housed the iron foundry at street level with a boiler shop above. In the Eighteenth Street tender shop, workers made tender frames on the second floor while the tenders were painted four stories above the street.

Although such high-rise shop buildings were unusual, the 1902 remodeling generally followed the rationalizing goals of the new factory system. Superintendent Vauclain grouped sequential operations in the same structure or adjacent ones where possible. Building on its pioneering work of the 1890s, Baldwin installed electric drives for tools, cranes, and elevators in great numbers. A new engine-finishing shop located uptown at Twenty-sixth Street allowed the 1890 erecting shop to concentrate on heavy work.[11] This eased the perennial bottleneck in erecting, where work remained comparatively labor-intensive, and improved throughput. To equip the enlarged factory, the partners purchased over $1 million in new tooling in 1902 and 1903.[12] All these efforts increased capacity by over 30 percent.

As so often in the past, Baldwin accomplished its 1902 expansion within an already congested plant. Because the firm had to build locomotives to order and "railroads wait until the last minute to place orders," the company could only

Locomotive tenders were riveted together in this large room, a straightforward task. But the room itself was three stories up in the Eighteenth Street tender shop. AL GIANNANTONIO

react after increases in demand.[13] Before the 1902 enlargements, delivery dates for new orders stretched out twelve months or more. With full order books and healthy profits beckoning, Baldwin's officers set about enlarging capacity. While a desire for profits drove this expansion, the company was also motivated by its general sense of partnership with its customers. In this industry, builders and buyers had long-standing and personal relations, and the builders prided themselves on meeting the railways' needs.[14] The American Locomotive Company consolidation of 1901 also played a role in Baldwin's frenetic expansion. Alco represented a real competitive threat to the Philadelphia firm, the first serious challenge in over thirty years. It too was enlarging its plants, heightening the pressure on Baldwin. Despite this invigorated competition, Baldwin maintained its 39 percent share of a greatly enlarged market (1890–99 vs. 1900–1909).[15]

All the changes in plant and equipment in 1902 could not alter the fact that locomotive building remained a massively labor-intensive effort. Armies of laborers, founders, machinists, and boilermakers were recruited to the task, and Baldwin's overall growth in employment between 1898 and 1907 was an accomplishment in and of itself. As the curves in chart 7.1 suggest, labor-intensity actually increased over much of this period as growth in the size of individual road engines outpaced productivity improvements from new plant and equipment.[16] To oversee this greatly enlarged workforce, Baldwin heightened its reliance on inside contractors, using them even to supervise unskilled laborers' work.[17] Their technical capabilities also remained important in particularly demanding tasks. For example, contractors divided up the work in the erecting shop, with their gangs specializing in particular jobs: setting valves, erecting piping, installing lagging and jacketing, and so on.[18] Each gang worked speedily under the direction of a contractor whose pay depended directly on the number of engines completed. Among high-volume manufacturers contracting was a dead letter by this time, but it remained irreplaceably valuable to a job-lot locomotive builder.

Although contractors were vital for the high-speed production necessary

The Baldwin partners had grandstand seats when a parade passed down Broad Street early in the century, most likely celebrating the July 4 holiday. From left, the three men in the window were George Burnham Jr., John Converse, and William Henszey. Together they owned roughly half of the huge company outright, but Converse and Henszey were getting along in years. DEGOLYER LIBRARY

to meet demand, their autonomy came at some cost to quality. In the sales boom after 1900 poorly constructed engines became a sufficient problem for comment in the railway trade press. One author called for "Premiums for Good Workmanship on New Locomotives," while another wanted "Acceptance Tests for Locomotives."[19] Insofar as Baldwin was building some substandard engines, the fault lay largely with contractors who rushed their work. But as one correspondent to the *Railway Gazette* noted, the root cause lay in the motive power purchasing habits of the railroads, which all crowded into the market at once "and almost demand immediate delivery." Large volume, high speed, and high quality simply became incompatible at times.

Baldwin's growth in the decade after 1897 also heightened the demands on its management systems. In general this period saw the extension and modernization of management practice in American industry. Partnerships gave way to incorporated firms run by managerial hierarchies. Companies engaged in high-volume manufacturing pioneered many new systems that sought increased control over workers, technology, and markets.[20] But the

Baldwin partners believed that this managerial revolution offered little of value to their company. Their resistance to innovations in management forms and practices arose from more than mere stubbornness; they believed that many trends current in American industry were inapplicable to their operations or even counterproductive to their purpose and conception of the company.

The leading instance of such resistance to change was the partnership itself, the owners' unwillingness to incorporate. The trend toward incorporation was so common in American industry circa 1900 that Baldwin's status as a very large privately held firm had become unique. But in 1901 and again in 1907 the partners chose to reorganize their firm rather than incorporate. As partner Alba Johnson explained, the partnership gave "a permanency in the conduct of the business," with ownership and management in the same hands.[21] Johnson also saw Baldwin's avoidance of hierarchical bureaucracy as a great advantage, since the owners were active managers who had a personal interest in the firm's success. In maintaining the partnership, Johnson was arguing for the value of Baldwin's old producer culture. The company and everyone in it had one clear purpose: building locomotives. The partners remained convinced that a bureaucratic management structure would dilute that goal.[22]

This conviction helps explain the partners' antipathy to the Scientific Management movement then rising in American industry. The firm could ignore Taylorism in part because Baldwin's pioneering efforts of the 1860s in systematic management had proven well adapted to growth. For example, the company maintained its eight-week production schedule under the List System from 1872 down to 1910. While the average size of Baldwin-built engines more than doubled over the period, managers and workers were simply expected to accomplish more work. No wonder the partners saw little need or value in Taylor's system with its bureaucratically supervised workforce. The company saw only one major managerial development before 1909. When the workforce swelled after 1905, Baldwin created a new employment department, which centralized hiring, timekeeping, and payroll functions.[23] Most other changes sweeping American business management at the turn of the century bypassed the office building at 500 North Broad Street. The partners were assured by their past and confident of their future. But time revealed rather quickly how vulnerable Baldwin was to changes in the national economy and polity.

Baldwin's growth during the 1898–1907 boom provided ample confirmation that the policies it had developed in the mid-nineteenth century remained viable after 1900. In the ten years after 1897 the company broke its own production records six times, output climbed by over 250 percent, and annual revenues often exceeded $30 million. This expansion was not trouble-free, however. One would expect healthy profit margins at the end of this decade-

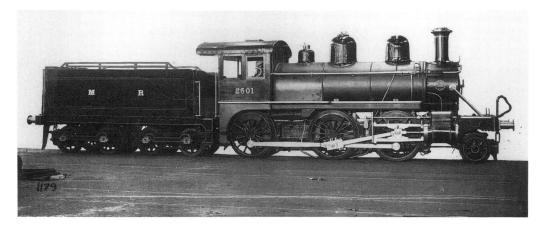

In 1899 the Midland Railway of England ordered a batch of thirty locomotives from Baldwin, including this 2-6-0. In the same year the builder sold forty more engines to two other English carriers. Such railways far preferred to buy their own domestic products but found them unavailable. In 1897–98 a six-month lockout of engineers—*machinists,* in American parlance—at all British engineering firms paralyzed the heavy machine business. Afterward, pent-up demand kept those builders occupied, sending orders across the Atlantic. RAILROAD MUSEUM OF PENNSYLVANIA (PMHC)

long runup in demand, yet Baldwin's 1906 profits on locomotives amounted to only 7.3 percent of sales.[24] This was an adequate but hardly impressive performance for a company in an oligopolistic industry with record sales and customers patiently waiting out eighteen-month delivery schedules. Behind the limited profit margin were a number of issues that illustrate the constraints on Baldwin's managers and the challenges of rapid growth in the capital equipment sector.

While Baldwin's expansion was astounding, advance planning and calculation by management had nothing to do with it. One overriding force drove the firm: railway demand for motive power. When it exploded, the factory quickly became choked with materials and men. Although the company sought to relieve the pressure with plant expansions in 1900, 1902, and 1906, such remodeling amid congested conditions only increased the disruption. Then the new economic reality of inflation took its toll on profits. As Alba Johnson later wrote of this boom period, "Notwithstanding large investments in labor-saving machinery and improved appliances, the cost of labor and administration constantly mounted higher, so that . . . profits were found to be lower rather than higher."[25] The great increases in locomotive size and weight in these years heightened the labor-intensity of production, adding further costs and eroding the productivity gains that Baldwin's managers and workers had so laboriously earned during the nineteenth century. Wage inflation and the growing labor-intensity of production caused the firm to reverse its Gilded Age policy of increases in the real pay of its workers. Between 1898 and 1906 Baldwin's men suffered a 15 percent decline in the purchasing power of their wages.[26]

In years past the company might have simply increased prices to relieve the pressure on profits and wages, but two new factors prevented that course. With Alco's formation Baldwin faced a truly powerful competitor for the first time in almost forty years. The two giants accounted for 85 percent of U.S. locomotive production in the decade after 1900, an oligopoly that could easily have renewed the price-fixing tactics so common in the nineteenth-century industry. Instead Baldwin and Alco embarked on a bidding war. This seems like irrational behavior for the two companies, since they both faced record demand for their products, substantial costs of plant expansions, and declining labor productivity. Their price competition seems to have arisen in part from Alco's need to promote sales volume to service its consolidation debt.[27]

The highly dynamic pace of technical change in locomotives of the period also contributed to the competition in bids. Engines were growing in size, power, and complexity so rapidly that the builders no doubt found it difficult to bid accurately on the new giants the railways were ordering. Also, the large new engines simply cost more to make. The railways were determined to have this new motive power but objected to the builders' increasing profit margins in direct proportion to their cost increases.[28] So the carriers solicited competitive bids from Baldwin and Alco, making it "impossible for the [builders] to increase prices at a rate commensurate with the growth in the size of . . . locomotives and the increases in the cost of manufacture." Baldwin's partners could afford to shave per-unit profit margins, since their aggregate business was booming. A 7.3 percent profit on locomotive sales was hardly a stellar performance, but that translated into $2.29 million—a sufficient sum divided among the six owners.

Notwithstanding such difficulties, the firm brought its techniques of custom-building capital equipment to high refinement in this era. In 1899 Baldwin achieved the coup of securing three large contracts to build seventy locomotives for English railroads.[29] Such poaching in the land where the railway age began shook John Bull badly. In 1905 the Pennsylvania Railroad offered Baldwin a contract for five hundred big freight engines if the locomotive builder could achieve rapid deliveries. Despite sizable orders already on hand, Baldwin took the contract, making deliveries at the specified rate of twenty locomotives a week in addition to its other business.[30] Two years later the firm closed a deal with a French line, the Paris-Orléans Railroad, for twenty engines built to the company's blueprints. The particular challenge here lay in the drawings themselves. The five hundred sheets detailed over ten thousand parts for each locomotive—all noted in metric dimensions. Accurate conversion of these plans to English measure would have been prohibitively costly and time-consuming.

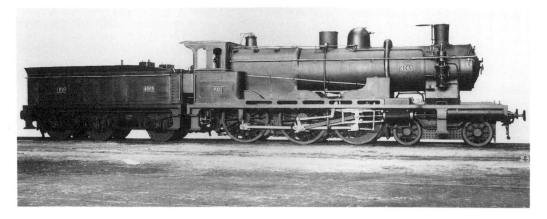

Baldwin made twenty of these 4-6-0 types to metric dimensions for the Paris-Orléans Railway in 1907. It was the firm's first experience applying metric standards throughout an order. The job proved less difficult than it first appeared, however, since machinists generally worked to gauges and templates rather than blueprints. RAILROAD MUSEUM OF PENN-SYLVANIA (PMHC)

So Baldwin ordered metric standards, made up new gauges and templates to the print dimensions, and built the twenty engines to metric dimensions throughout.[31]

As these examples suggest, Baldwin's essential strength lay in its ability to marshal a skilled, labor-intensive workforce to build custom products rapidly, on demand, and at competitive prices. Baldwin's challenge came in meeting its markets; directing them was impossible. Therefore the partners believed that preemptive managerial control, then rising in the mass-producing industries, was an empty promise in their industry. Given this decision, it made sense to rely on the company's producer traditions, which stressed individual effort, initiative, and responsibility. But this raised a new problem. These traditions were relatively easy to foster in Baldwin's limited managerial ranks, but would conceptions of individual effort and reward continue to motivate a workforce that after 1900 generally exceeded ten thousand men?

THE COMPANY AND ITS WORKERS

During the nineteenth century Baldwin's basic policy in employee relations had been to treat workers as individuals in a partnership sharing the firm's purpose of building locomotives. The firm expected men to work to their capacity; in return they received good wages. Those demonstrating untapped abilities were promoted up through the innumerable gradations of skill required to build engines. Given the volatility of the locomotive market, layoffs were common, but when demand again picked up, old hands received preferential treatment in hiring, and they flocked back to their jobs. These basic tenets defined the company's producer culture. For forty years it had secured peace between labor and management.

After 1897 Baldwin faced the difficult problem of promoting individual initiative in a workforce that had swollen to mammoth proportions almost overnight. Growing from 3,200 men to 18,500 between 1897 and 1907, the force quickly came to be dominated by newcomers unused to the company's traditions. Most of these new men were outright machine tenders and laborers, for the fivefold increase in employment over the decade had accelerated Baldwin's forty-year trend of employing industrial operatives. Because Philadelphia's reservoir of skilled workers was inadequate to provide for the company's expanding needs, Baldwin had to subdivide tasks rigorously so that new men needed only a narrow competency.[32] Some evidence of this transformation of work is evident in the company's own definition of *skilled labor*. In 1903 a Baldwin officer wrote, "A skilled workman is . . . one who is familiar with the use of a tool, a machine, or a process."[33] Compared with the graduates of Baldwin's 1860s apprentice program, an employee competent to operate a single machine tool hardly qualified as a skilled worker. But the company's loose definition of skill accorded with general trends in American industry, trends so pervasive that some craft unions would grudgingly accept such operatives as members.

Though Baldwin needed these new men, their vast numbers threatened to overwhelm the company's producer culture. Their highly specialized skills were also incompatible with the firm's tradition of internal promotions; many of these new men were simply machine tenders, exchanging their time for a wage. The company made no secret of the fact that if workers began to band together seeking better wages, they would be instantly dismissed.[34] But in addition to this threat, the partners sought some positive means to revitalize the labor-management partnership of earlier years, so in 1901 they inaugurated a new apprenticeship program.

Baldwin's old apprenticeship program had died in the late 1860s, a casualty of workers' specialization under piecework. The company began the new program in part to counter the narrowing of skill that piecework had fostered. Baldwin's new system of worker training was an example of the "new apprenticeship" movement in American industry that dated to the late 1890s.[35] Like traditional apprenticeships, these programs aimed to train young men broadly in their chosen craft. The novel aspect was an additional component of classroom education, generally in vocational subjects. Unlike Baldwin, most machinery builders had never discontinued apprenticeships during the Gilded Age, so in this sector "renewed apprenticeship" would be a more accurate description.

Baldwin's superintending partner, Samuel Vauclain, developed the program, and he commented frequently on its goals and benefits. Chiefly he argued that by instilling in these boys the "traditions of a plant," the company created a cadre of workers who would "form the backbone . . . for the development of a good, loyal body of men." Vauclain also saw the program as an antidote to the " 'race suicide' of the trades," which had occurred as industrial employers killed off craft skills by piecework and specialization.[36] Broadly trained program graduates would become leading workmen, contractors, and foremen. In renewing apprenticeship Vauclain sought to provide in the impersonal, teeming factory the same chance for individual recognition and advancement that he and his fellow partners had enjoyed in the nineteenth century.[37]

To accomplish this variety of purposes Baldwin created a three-tiered apprenticeship program.[38] First-class graduates would become top mechanics, tool room hands, and layout men. Contractors and subforemen came largely from the second class, while the third class was on a trajectory leading to foremanships or the executive ranks. A superintendent of apprentices, Nathaniel W. Sample, oversaw the program. He had indentured as a machinist at Baldwin in 1859, so the new program had a direct tie to its predecessor. Foremen and contractors in each shop were responsible for the boys' training, and the apprentices rotated through different departments to acquire broad experience. First-class machinist boys started out on planers, passed through nine intermediate types of machinist's work, and ended up in the demanding task of setting locomotive valve gears in the erecting shop.[39] After the six o'clock whistle blew, first- and second-class apprentices had to attend night classes in a variety of technical subjects at one of Philadelphia's many vocational schools. Such classwork in algebra, geometry, and mechanical drawing was a key part of the "new apprenticeship," which sought to impart theoretical competency in addition to manual skills.

Diaries kept by a drafting room apprentice, Vernon Gotwals, from 1906 to 1911 describe one young man's training.[40] Gotwals began at simple jobs but was quickly brought along. Working under the tutelage of senior men, the young apprentice was soon turning out quite demanding drawings such as designs for new boilers and erecting plans for complex Mallet designs. The pay was meager, and his workday often stretched to ten P.M. on the nights he did not attend electrical-engineering or physics classes at Drexel Institute. Despite the low pay and long hours, he loved his job, often spending his lunch hour in the erecting shop watching as skilled erectors rendered his plans in steel.

The evidence suggests that Gotwals's experience was typical and that apprenticeships were in high demand. From the company's viewpoint the program proved an incomplete success, however. After seven years Vauclain reported that the training system "has been highly beneficial to us. We do not now need to consider going outside to hire skilled men." [41] While true enough, it also appears that most apprenticeship graduates soon left the company's employment. From

The occasion is unknown, but sometime after 1902 a photographer found these Baldwin apprentices congregating in front of the massive plant on Broad Street. LIBRARY OF CONGRESS

1901 to 1910, 1,300 boys entered the program, but in 1910 only 250 apprentice graduates were working at Baldwin.[42] This poor retention suggests that the program largely became a pipeline to the managerial ranks rather than a system of worker training.[43] Graduates not tapped for management positions evidently obtained better-paying or less specialized jobs outside locomotive building.

Although the program fell short of Vauclain's stated purpose of creating a large cadre of loyal workers, it did provide noteworthy benefits. During their terms the apprentices provided an inexpensive pool of increasingly skilled labor.[44] No doubt great numbers of boys were the sons of Baldwin workers, and these fathers must have appreciated the training given their sons. The new training system also represented a modest effort by Baldwin's partners to bridge the gap with their workers that had widened with the company's overall growth and division of labor. Finally, the program provided good public relations.

The Baldwin company received frequent press coverage in the early years of the twentieth century, and the apprentice program contributed to the attention. For forty years editors in search of good copy had been drawn to this firm, a giant of American industry making a mysteriously powerful product that symbolized man's technological triumph over nature. Circa 1900 Baldwin's officers began to contribute to the coverage, writing so many articles as to suggest a purposeful public relations campaign.[45] Although the general public seldom purchased locomotives, there was a widespread Progressive-era concern with "the labor question." The partners took pride in their company's forty-year heritage of labor peace, in contrast to the strife so common during the Gilded Age. As Progressives turned to the mediation of such conflict, Baldwin seemed to have some answers.[46]

In 1902 the National Civic Federation sponsored an "industrial conference" to examine opportunities for a rapprochement between American labor and capital, and at that meeting Baldwin was cited for its tranquil labor relations. To examine this success a reporter, E. A. Bingham, visited the firm, writing up his findings in an article called "The Labor Situation at the Baldwin Works."[47] Bingham repeatedly cited examples of Baldwin's producer-culture traditions. He led by describing the firm's status as "an old-fashioned partnership [in which] from the front office to the ash heaps the old idea of partnership and reward and promotion is realized in a fashion that makes the works unique." Even a new apprentice "knows that if he is faithful and bright . . . he may be a partner someday." Ordinary workers were motivated by the chance for promotion and by the piece-rate pay system, since "no limit is put upon a man's skill. . . . The more he earns for himself, the more he makes for his employers."

Bingham's article did not describe actual conditions at Baldwin in 1903 so much as summarize the old nineteenth-century ideology that management sought to perpetuate. But the company's expansion after 1897 had rendered much of its culture anachronistic even as an ideology, let alone as a statement of actual conditions. By 1903 trusts and corporations, not "old-fashioned partnerships," ruled American business. Since 80 percent of apprentices left Baldwin's employment, evidently few believed that their own careers could parallel the rise of partners like William Austin or Samuel Vauclain. These owners may have thought that "no limit is put upon a man's skill," but eight out of ten apprentices evidently decided that Baldwin had little to offer them.

The Baldwin partners clung to this culture, despite its increasing irrelevance, because it had sustained and justified their own careers while serving their company well for a half century. During the Gilded Age Baldwin backed up its individualist rhetoric with high pay and wage increases. As the industry leader it could afford such policies, which simultaneously blunted workers' desires for union representation while securing competitive advantages for the firm. After the late 1890s the partners still clung to the old ideology, but they had to abandon the wage policies that gave force to the rhetoric. Philadelphia was nationally known as an open-shop city hostile to unionism, which gave the Baldwin partners more latitude in their internal policies. But the firm's overall growth also increased the likelihood that at least some workers would eventually challenge management.

CLIMAX AND CHANGE, 1906–1909

In 1906 the Baldwin company employed 17,400 men, and it made 2,666 locomotives and other products worth over $46 million.[48] It was a banner year, thanks to the financial strength of American railroading. Business was so good that a congestion of parts and workers crowded the factory, despite the expansions of only four years earlier. With no further remedy available in center-city Philadelphia, the partners finally decided to strike out for the suburbs. In May 1906 they purchased a 185-acre site at Eddystone, just south of the city. By year's end, foundry operations had been moved down to Eddystone, where a thousand Baldwin workers toiled.[49] The Pennsylvania Railroad and the Baltimore and Ohio traversed the site, which also fronted on the Delaware River. With plenty of land available for growth, it was the ideal location for an entirely modern locomotive plant capable of meeting the demands of the future.

Eddystone was Baldwin's response to the evident financial strength of its U.S. railroad customers. But by 1906 the great consolidation systems that dominated American finance and commerce had become targets of public hostility, largely because of that domination. President Theodore Roosevelt, the Congress, and much of the country saw these powerful corporations as exploitative monopolies, requiring close regulatory oversight for the national good. Their chosen instrument was the Hepburn Act, a law enacted in June 1906, which removed ultimate review of railway rates from the Supreme Court and gave that power to the Interstate Commerce Commission. The commissioners would wield their power vigorously.[50] At first, however, there was little outward evidence of any fundamental change in either railroading's financial health or Baldwin's flush order books. While 1906 saw record output for the firm, 1907 promised even more business. Then October saw a sharp fall in the value of shares traded on the New York Stock Exchange. The bears' stampede set off the Panic of 1907 and a subsequent depression. The economy had been overheated for months, but financial analysts "attributed the Wall Street slump to unfriendly legislation, particularly the Hepburn Act."[51]

The 1907 Panic, often called "the bankers' panic," did not have the devastating effect on the general economy that followed the 1893 stock market sell-off. For example, thanks to deep price cuts U.S. Steel ran at 75 percent of capacity in 1908, and a year later its mills faced a backlog of orders.[52] But in the neighborhoods around Baldwin's Bush Hill factory, the panic was a disaster. As 1908 opened, pay cuts and layoffs swept the plant that only five months earlier had employed twenty thousand men working two shifts. Even as it slashed the workforce, Baldwin took contracts at prices below cost, trying to keep some portion of the organization alive.[53] By July the force totaled forty-six hundred men on a twenty-five-hour workweek. During 1908 only 614 new locomotives rolled out of the erecting shop, the lowest sales year since 1897. The partners were sobered by the low volume of business but expected sales to rebound in time. Instead the next two years brought a number of internal challenges.

On March 23, 1909, one of Baldwin's six partners, William P. Henszey, died. Hired originally by Matthias Baldwin, Henszey had been a member of the partnership since 1870. He held a 20 percent interest in the firm, worth over $6 million. His will contained the usual terms forbidding his executors to press for payment of his ownership interest, but Henszey's death came at a difficult time in Baldwin's finances.[54] The firm had borrowed heavily to finance plant expansions in the boom just past, and business was then in a slump. Further financial demands loomed in that two partners, George Burnham and John Converse, were ninety-three and sixty-nine years old. So the partners finally decided to incorporate the firm.[55] For years they had sought to maintain the managerial autonomy the partnership form provided. But once American Locomotive had gained access to the public capital markets by incorporating in 1901, it was inevitable that Baldwin would follow sooner or later. Under its 1909 incorporation Baldwin remained privately held, thus preserving the unity of ownership and management. Incorporation allowed the company to enter the bond market, however, and within a year Baldwin issued $10 million of first mortgage bonds.

This bond issue was a fundamental departure for the company. In the past the partners' capital expenditures had been limited by the funding sources available: cash on hand, ongoing profits, and limited short-term borrowing. The firm's capital expansion had therefore been reactive; it had grown if and when market conditions demanded. Now the new corporation could attempt to plan its growth in anticipation of the market. Samuel Vauclain particularly desired to take such a proactive approach in 1909. Output in that year was only 38 percent of the 1906 total, but Vauclain believed that sales would soon rebound. Despite the low volume of orders, he was finding the Broad Street plant constricting. The problem was the inexorable growth in the size of road locomotives, which by 1907 regularly exceeded 300,000 pounds. To get such behemoths out of the 1890 erecting shop at Broad Street, "wheels had to be jacked up, doorway frames cut out, and switch lamps taken off the locomotives." Railroads would certainly require even larger engines in the future, and Vauclain saw only one solution: a new erecting shop at Eddystone. The company's conservative financial chief and new president, John Converse, opposed such a large expenditure with Baldwin's markets so unsettled, but Vauclain lobbied hard and won the other stockholders' approval.[56] The bond issue would be used in part to finance the new shop.

LINES OF DIVISION, 1910

Baldwin's first foray into the bond market required careful handling to launch the firm as a blue-chip investment. Negotiations to place the issue through the investment banking houses of Brown Brothers and Kuhn, Loeb & Co. proceeded in February and March 1910. While the financiers were hammering out the details, bedlam broke out in the streets of Philadelphia. On February 19, 1910, over five thousand union motormen and conductors struck the Philadelphia Rapid Transit Company (PRT).[57] The strike quickly turned ugly, with a number of public disturbances. On February 23 a riot broke out in front of Baldwin's works on Broad Street, Philadelphia's main thoroughfare. The trouble seems to have started between picketing PRT strikers and police assigned to protect the trolleys running down Broad Street. Baldwin workers in the street during their lunch hour began jeering at the bluejackets. Never known for brilliance or restraint, the Philadelphia police responded to this provocation by firing on two locomotive workers, wounding one in the leg.[58] The commotion drew more of Baldwin's men outside. Vice president William Austin described what happened next:

> The police ran the men back into the shop and then trouble began. Before anyone knew what started it we heard pistol shots. I looked out of the office window and saw a long line of policemen, about 2 dozen, lined up in front of the Willow Street Shop actually firing into the second and third story shop windows. They were answered by a volley of nuts, bolts, washers, shaft hangers, iron rods, etc., some very heavy, and all calculated to kill if landed in the right spot. The police shot at least 200 shots into the shops. Fortunately no one was hurt on either side but for a while it was very serious business.[59]

This incident became front-page news around the country; the Duluth (Minn.) *News Tribune* headlined it as the "Battle of the Baldwin Works." The company's management no doubt hoped that calm would soon return, but Baldwin's workers were enraged by the police action, described even by the sober Philadelphia *Evening Bulletin* as "Russian rule."[60]

With the power of the city, the state, and the PRT all allied against the striking carmen, Philadelphia's labor leaders turned up the heat. On March 4 an umbrella group for organized labor, the Central Labor Union, declared a general strike in all industries throughout the city in sympathy with the carmen. It was a desperate gamble, of dubious legality, which American Federation of Labor president Samuel Gompers refused to endorse; but by the second day up to one hundred thousand workers around the city had quit work. In its second week the general strike began losing strength as sympathy with the PRT strikers gave

"BATTLE OF THE BALDWIN WORKS," AN INCIDENT OF PHILADELPHIA STRIKE

ARRESTING A BALDWIN WORKMAN

MAKING ARREST IN KENSINGTON

BATTLE AT THE BALDWIN WORKS

BATTERED AND ARRESTED

When the history of the great Philadelphia trolley strike is written the "battle of the Baldwin works" will figure prominently in the narrative. The sympathies of the great body of the men employed in the huge works were with the striking street car employes from the outset of the trouble, and when the tide of affairs turned toward the shops a serious riot took place at Thirteenth and Spring Garden streets. A crowd of workmen attacked the police de-

The "Battle of the Baldwin Works," as the Duluth *News Tribune* headlined the story, was only one of many bitter incidents in Philadelphia's 1910 streetcar strike, which also saw the bombing of a car barn, incessant stoning of cars run by replacement motormen, and repeated battling between police and large crowds of strike sympathizers. But newspaper editors across the country gave wide play to wire-service photos of police firing at random into Baldwin's fortresslike factory—images symbolizing the vitriolic passions that ruled the entire city. MINNESOTA HISTORICAL SOCIETY

way to the hard reality of food and rent money.[61] But Baldwin's workers were just beginning their fight.

Baldwin's employees were unwilling to join the general strike en masse, but a substantial number had grievances against their own employer, grievances that drew them to evening rallies and meetings.[62] Thanks to the accounts of informants who reported to superintendent Vauclain, we can see how the sympathy strike turned into an action against the Baldwin company itself. On March 9, five days into the general strike, a rally held at Philadelphia's Labor Lyceum "was crowded, mostly with men from the Baldwin plant."[63] Speakers exhorted the men to join unions, arguing that they were underpaid and urging them to attend a special meeting, for Baldwin men only, to be held the next day. At that gathering an erecting shop worker argued that Vauclain was guilty of "bad faith" for not restoring wage cuts exacted during the 1908 recession.[64] By March 12 thirty-five hundred Baldwin employees had banded together in an organization known simply as the "Baldwin Workers" to petition the company.[65]

It is unclear and ultimately unimportant how much of this protest arose spontaneously among Baldwin men rather than having been elicited by labor organizers. National unions like the International Association of Machinists and others quickly sensed their opening. The IAM had begun to organize Alco's workers in January 1910, and preserving that advance alone offered a sufficient argument for hurrying organizers into the Baldwin plant.[66] But groups of Baldwin men also acted purely independently; for example, men from one machining department petitioned Vauclain for the return of pay cuts, a Saturday half-holiday, and time-and-a-half wages for overtime. In the workweek ending Saturday March 19 roughly twenty-five hundred employees out of twelve thousand did not report for work.[67] Most evidently returned the following week, but organizing continued, as did petitions seeking Saturday half-holidays with no pay cut. Workers also complained about piece-rate cuts, the narrowing of skills under piecework, and unfair treatment by contractors.[68]

Samuel Vauclain was not about to tolerate unions in his plant, but a number of factors forced him to take a conciliatory stand, at least publicly. Negotiations to place the $10 million bond issue with the bankers were almost complete, and an open breach with the workforce could only hurt bond sales.[69] Also, the locomotive market was finally picking up in the spring of 1910 after two bad years. At least for now, Vauclain preferred making engines to breaking unions. Ascribing some minor troubles to the sympathy strike, he announced on March 12 that new orders would shortly bring an increase of four thousand men in the workforce, and he urged all to roll up their sleeves and get to work.[70] His employees did just that, union and nonunion alike; by August, sixteen thousand men filled the shops at Broad Street and Eddystone.

As spring gave way to summer, Baldwin's labor-management relations remained outwardly quiet. The company avoided provoking an open break, while workers concentrated on organizing. Evidence suggests that at the contest's opening, anticompany sentiment was strongest among immigrant workers in semiskilled or unskilled positions, although a number of skilled men also joined the union cause.[71] After three months of organizing, Philadelphia Locomotive Lodge No. 466 of the IAM claimed two thousand Baldwin machinists as members.[72] Other trades joined these men, and by 1911 a total of thirteen different craft unions, all American Federation of Labor affiliates, were represented in Baldwin's two locomotive plants.[73] They included locals of the patternmakers, molders, blacksmiths, boilermakers, asbestos workers, and metal polishers unions. To create internal cohesion and present a united front, representatives of each Baldwin local sat on the "Locomotive Builders Council," which sought to coordinate the actions of all trades.[74]

Vauclain was also busy wooing workers that summer. In response to their petitions he granted a paid Saturday half-holiday. He also moved to bring the contractors under tighter rein, issuing new Shop Rules in July that increased financial controls over the contractors.[75] Private shorthand notes kept by Baldwin's president, William Austin, over that summer also show a determined effort to root out contractors' abuses.[76] The officers' main concern was contractors' exploitation of the company, not workers. But by curbing the autonomy of these production bosses, Austin and Vauclain reached out to many workers whose dissatisfaction was primarily directed at the contractors who oversaw their workday. To secure workers' loyalty even further, Vauclain raised wages by an average of 7 percent over the level paid during the boom of 1907.[77] As 1911 opened, a curious calm had fallen over Baldwin's labor-management relations as the company and union men each vied for the allegiance of uncommitted workers. While neither side would outwardly provoke the other, both sought to build strength for the coming showdown.

MANAGEMENT STRIKES BACK, 1911

In the spring of 1911 Baldwin's officers were again juggling financial affairs with issues in the workforce. The bond issue of the previous year had sold well, but the company was still hungry for cash. The officers decided to reincorporate to allow a public stock offering. Vice president Alba Johnson opened negotiations with Philadelphia's Drexel & Co. and White, Weld & Co. of New York to take the company public. At the same time, Samuel Vauclain decided to break the unions' grip before it got any stronger.

Vauclain moved with exact calculation at precisely the moment of his own choosing. On June 1, 1911, he announced a new "Employees' Beneficial Association."[78] The association was a novel departure for Baldwin, which had generally avoided such programs of welfare capitalism. It included an employee payroll deduction savings plan, injury and illness benefits, and death benefits for the heirs of men killed on the job. These provisions were available only to members of the association, but the firm placed few restrictions on eligibility. Finally, association members enjoyed preferential rehiring in the event of layoffs. While aspects of the plan were undeniably generous, altruism alone hardly motivated it.[79] Vauclain created the association to hold the loyalty of his nonunion men and perhaps win back some who had gone over to the union side.

Just before producing this carrot, Vauclain wielded his large stick. The 1910 upswing in orders had proven brief, and with sales off the superintendent seized the moment. On May 26 the company announced an impending layoff of twelve hundred men. Calculated to produce a showdown, the layoffs plucked union workers from all departments.[80] After a pro-union boilermaker was discharged, the boilermakers walked off the job, and men in all the organized trades quickly followed. None of the national unions involved had authorized a strike, but the strikers could take heart from the evident solidarity of trades in the plant. Thanks to the industrially organized Locomotive Builders Council, the thirteen striking unions spoke with one voice. By June 8 the council claimed that 10,700 men out of a force totaling 13,000 had laid down their tools.[81]

In fact the unions were putting up a brave but false front. Baldwin's payroll records for the first full week of the strike reveal that slightly more than half of the workers refused their call and stayed on the job.[82] The company said little in response to the walkout, announcing simply that strikers seeking to return could apply only as individuals; Baldwin refused to deal with unions or union members per se. On June 24 the company knew it would ultimately prevail, for

Shall Morgan Own This Country?

Would you like to be the property of J. Pierpont Morgan?

If you pride yourself on being a free man, you will answer: NO!

But how are you going to stop it, if Morgan gets hold of all the land and industries?

Morgan has just got hold of the Baldwin Locomotive Works. He has inflated its capital many millions of dollars, and he has made up his mind to squeeze additional dividends, on the watered stock, out of the men who work at the Baldwin plant.

The workers object. They are out on strike. They do not intend to be starved into submission. And they will not surrender, if you will help them.

Do you want them to be Morgan's slaves? Do you want to be one yourself?

If you do not, then help the men who are out on strike at Baldwin's.

Tomorrow an authorized collector, with proper credentials, will call at your home. Give whatever money you can to the strike fund. Every penny counts; every dollar will tell.

STRIKERS RELIEF COMMITTEE,

232 North 9th St.

SIMON KNEBEL, Chairman,
JOHN J. MILLER, Secretary.

To rally financial support, the strike leaders circulated handbills playing on J. P. Morgan's involvement in Baldwin's 1911 public stock offering. For many working men and women, Morgan personified the merger or trust movement, which in their view had resulted in labor's enslavement to capital and Wall Street interests. SMITHSONIAN INSTITUTION

on that payday and every one thereafter, employment slowly swelled as strikers returned and the firm hired new men.[83] As the strike crumbled, the company turned away hundreds of reinstatement applicants whom it considered labor militants. All returning workers had to turn in their union membership cards to get their jobs back.[84] With some two thousand men still holding out, the Locomotive Builders Council finally declared the strike over on August 29, 1911. Almost thirty years would pass before organized labor returned to the Baldwin plant, overcoming management's continued resistance with the intercession of the federal government during the New Deal.

While Vauclain made no public comment either during or after the strike, he had provoked it, and the outcome clearly suited him. By breaking the unions the firm had maintained its "right to manage." This was no small matter for capital equipment builders. Given their circumscribed influence over marketing and innovation, these companies generally believed it essential to retain control over production.

But the strike also represented a failure for Vauclain, one that he felt deeply.[85] His career had revolved around work and workers since the age of sixteen, when he first took up the chisels and files of a machinist. He identified with these men who extracted a livelihood from the skills in their hands, and he thought he knew their concerns. While business advantage motivated his abhorrence of unions, he sincerely believed in the virtues of individual effort, initiative,

Sam Vauclain (*left*) stands on the steps of the Broad Street office building during the 1920s. At right is the boxer Gene Tunney. RAILROAD MUSEUM OF PENNSYLVANIA (PMHC)

and personal advancement, and he saw organized labor as antithetical to those principles. Such ideas had guided his own rise from apprentice to partner; surely their merit was evident to all. During his twenty-five years as general superintendent he had tried to foster this producer culture throughout the company. Baldwin's production records and its heritage of labor peace again seemed to confirm his beliefs.

Although he had precipitated the 1911 strike, it came as a personal blow that six thousand men had cast their lot against him, his company, and his ideals. During the strike Vauclain went to four old employees whose tenure dated back to Matthias Baldwin's era and asked them to write down their recollections of the 1860 walkout.[86] Perhaps something of value could be learned from that incident when workers of another era had broken with the company's culture, Matthias Baldwin, and Charles Parry. On both occasions strikers had argued for time-and-a-half wages for overtime. Shortly after beating back organized labor in 1911, Vauclain granted that fifty-year-old demand. After the strike he also ended the 1901 apprentice program. Most graduates left Baldwin after completing their terms, and clearly the program had failed to foster "a good, loyal body of men."[87]

The walkout demonstrated that the company's traditions and its workplace culture provided no shield against labor unrest. But slightly more than half of Baldwin's workers had spurned the unions' call to strike. To acknowledge their loyalty, Vauclain created the Baldwin "Loyal Legion." Membership was limited to those who had worked throughout the strike. Members wore a solid gold badge that indicated their years of service with the company. As of February 1912, 42 percent of Baldwin's eighty-five hundred men were in the legion. Over fifteen hundred members had seniority ranging from eleven to sixty-two years.[88] Although the company's interests, traditions, and culture had been shaken by the strike, they still received the support of many employees.

HOLLOW TRIUMPH, 1911–1915

While Vauclain was besting the unions in the summer of 1911, Alba Johnson was managing Baldwin's reorganization as a publicly held corporation. The firm finally joined the modern corporate world in 1911, with its stock traded on the New York and Philadelphia exchanges, with consolidated executive powers in the company president, and with oversight by a board of directors that for the first time included outside directors.[89] With its up-to-date financial structure and its triumph over union labor, Baldwin's officers felt poised for further growth and a bright future, but that future proved difficult to manage.

The 1909 incorporation had given Baldwin a capitalization of $20 million. In 1911 three brokerage houses working in concert restructured the firm's finances to create a new corporation possessing a capital base of $40 million, equally divided between preferred and common stocks.[90] When word of this overnight doubling in capitalization came out during the strike, union leaders pointed to stock watering at the hands of J. P. Morgan, but the charge lacked merit. Under the regular financial practice of the day "preferred shares represented a firm's tangible assets, whereas common shares had as their basis good will and earning power."[91] Baldwin's top officers retained control through a majority stock interest while the investing public avidly bought shares that came on the market. Savvy investors like Pierre du Pont added Baldwin equities to their portfolios, seeing a profitable company with a promising future.[92] Indeed, Baldwin generally earned comfortable profits every year until World War I, when locomotive and munitions sales to the Allies boosted its margins even higher.[93]

Baldwin's officers were certainly optimistic about the future. They had sought the 1911 restructuring largely to provide capital for expansion. Given the competition with Alco, Vauclain in particular was determined to take advantage of this newfound ability to expand in anticipation of renewed demand. Although Broad Street and Eddystone ran at only 75 percent of capacity in 1911, Baldwin laid plans for a new locomotive plant in the West, purchasing 370 acres of land in East Chicago, Indiana. Sales were no better in 1912, but Vauclain completed a new erecting shop with 7 acres under roof at Eddystone during the year.[94]

The irony of plant expansions during a period of overcapacity underscores a unique problem facing the locomotive builder. Because the railroads, rather than Baldwin, directed the overall character of innovation in locomotives, the builder could have a plant that was simultaneously too small and too large. In 1912 the carriers did not buy in large quantity, compared with sales before the panic, but they did order even bigger locomotives. Despite the idle capacity of 1912, meeting the railroads' technical requirements required a larger erecting shop. This capital burden came on top of inexorable increases in labor-intensity, with the growth in road locomotives finally forcing Baldwin in 1911 to lengthen its List System production timetable for engines from eight to twelve weeks.[95]

Despite such problems, Alba Johnson, Samuel Vauclain, Wall Street financiers, and the investing public all saw a promising future for Baldwin after its incorporation. But in the years after 1906 a fundamental challenge was shaping up for the builder and for American railroading itself. Since the passage of the Hepburn Act America's railroads had become increasingly debilitated. Their sickness arose from a combination of inflation, which pushed up their costs, and regulation, which froze the rates they could charge shippers. In this squeeze, railroad profits and purchasing inevitably suffered.

The Hepburn Act of 1906 had signified a new regulatory climate for the railroads, a determined resolve to rein in these powerful corporations. That determination became apparent in 1910, when the Interstate Commerce Commission rejected the carriers' petition for a rate increase. Comparing 1910 with the flush times of 1907, railway traffic had increased by 10 percent while earnings had actually declined. But the regulators were unmoved. The ICC rejected renewed requests for rate relief in 1914 and again in 1915. By that time the carriers were confronting a grim financial situation. Profit margins steadily declined between 1907 and 1914, although railway traffic increased by almost 25 percent over the same period. The lines faced record demands for transportation services, but without adequate profits they could not modernize their systems to meet that demand efficiently. Albro Martin estimates that in the seven years following the passage of the Hepburn Act, American railroads suffered a capital investment deficiency of $5.6 billion.[96] In 1907 the nation's railroads stood on a pinnacle of profits and achievement. By 1917 these same companies were so broken down that the federal government had to nationalize the system as an emergency measure during World War I.

Few observers in 1907 could have predicted that the ICC would cripple the nation's largest industry over the next decade. Even those Wall Street financiers who held a direct stake in railroading's financial health did not comprehend the dimensions of the problem until after 1912.[97] Baldwin officers also failed to realize that the boom conditions of 1898–1907 would never return. In January 1913 president Alba Johnson announced expansion plans to give

Baldwin an annual capacity of over five thousand units.[98] Hindsight shows that such projections were scarcely more than dreams, bearing no relationship to true market conditions. Domestic orders for locomotives had peaked in 1905 at sixty-three hundred units for all builders, and Baldwin's output for U.S. customers reached its all-time high in 1906.[99] American mainline carriers—known as Class 1 railroads, defined by the ICC as those having annual revenues of over $1 million—provided Baldwin's major market. After 1906 their aggregate orders for new engines would never again surpass four thousand units a year.[100] While Johnson and Vauclain saw growth and a bright future in 1912, their company's major market segment was six years into a long-term decline that would only worsen. What had happened, why was Baldwin so wrong in its projections, and why did the company largely fail to realize its error until it was too late to recoup?

Three factors explain the reversal in Baldwin's sales. The ICC's denial of rate relief was of primary importance. As profits narrowed, the operating companies curtailed capital expenditures wherever possible. Locomotives were expensive items that the lines had always forgone at the slightest hint of financial uncertainty. Because of the regulatory climate after 1906, uncertainty became the reigning condition, and Baldwin's sales suffered as a result. The ICC's intransigence was only part of the problem, however. Absent any regulation, it appears unlikely that Baldwin could have ever regained the sales pace of the 1905–7 era, let alone advance to five thousand units annually.

Between 1900 and 1907 a key purchasing goal of the railways was to acquire modern, efficient motive power. These purchases to replace existing and serviceable locomotives came on top of the carriers' regular buying to meet the needs of expansion and to replace engines at the end of their normal service life. In 1901 one-third of the national locomotive fleet had been built in the previous decade—a predictable proportion, given a thirty-year service life.[101] In 1907 almost two-thirds of the fleet was that young. In other words, in the years before the panic the major carriers were replacing engines faster than they were wearing out. This buying spree had to end sooner or later, even without ICC interference.[102] A perceptive young college student working for the summer as a machinist's assistant in Baldwin's erecting shop accurately described the problem in a 1904 letter to his father: "The time is not far off when the number of steam loco's will reach a maximum production, as in the case of bicycles. Ten years I should say at the outside. Baldwin's do not seem to be providing against that time, by extending the field of usefulness [i.e., diversifying]."[103] Although Baldwin's officers were aware of this accelerated buying pattern before 1907, they failed to realize that a correction was inevitable.[104]

A third factor also explains Baldwin's post-1907 sales slump: the rapid pace of technical change in locomotives. Beginning around 1900 the railroads had begun purchasing increasingly powerful engines—qualitatively better products. That dynamic pace of innovation carried an eventual penalty for locomotive builders, however. With engines of better quality, the quantity of locomotive purchasing would necessarily decline at some point. Railway demand for stronger locomotives predated the ICC's regulatory stranglehold; thereafter the lack of rate relief heightened the carriers' desire to accomplish more work with fewer engines.[105] The pace of innovation in locomotive design accelerated even as total sales fell off. Baldwin could do little about this problem. Indeed, the company prided itself on fulfilling the technical requirements of its customers.

These factors explain Baldwin's sales decline while also suggesting why the company remained passive for many years after this fundamental reversal. Quite reasonably Baldwin's officers failed to conceive that an agency of the U.S. government would create such hardship for its major customers and its own operations.[106] Beyond this issue, hindsight suggests that the Baldwin company operated in a sector that became particularly vulnerable to disruptive changes and market reversals as the century advanced and the national economy passed through fundamental changes.

When a large enterprise fails to adapt, one is tempted to blame its managers as shortsighted, complacent, or lacking in professional vision. Baldwin's officers partly deserve such a verdict. Under the producer culture espoused by Alba Johnson and Samuel Vauclain, the company had a single purpose—to build

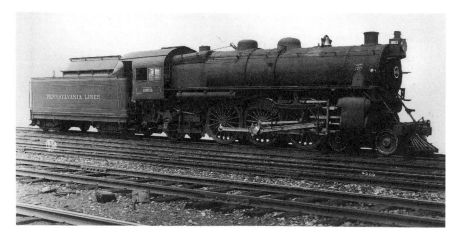

Around 1900 the Pennsylvania Railroad began replacing its 4-4-0 types with 4-4-2 engines to accommodate heavier passenger trains. As steel took the place of wood in passenger-car construction, the 4-4-2 design soon proved inadequate. Baldwin had originated the 4-6-2 or Pacific type in 1901, and in 1910 the Pennsylvania selected this configuration to supplant its 4-4-2 locomotives. Here is a PRR class K-3 Pacific, designed by the carrier and built by Baldwin in 1913. This engine bore Baldwin construction number 40,000. While a PRR 4-4-2 had 27,400 pounds of tractive force, the Pacifics exceeded 38,000 pounds. RAILROAD MUSEUM OF PENNSYLVANIA (PMHC)

locomotives. It was not a property to be "managed" for its own sake. They scorned professional managers, relying instead on internal promotions. Because of the strength of their convictions and the limitations of their backgrounds, Baldwin's top managers were painfully slow to realize that the company needed to diversify into other product lines. When World War I intervened Johnson and Vauclain were understandably distracted from the sickness in American railroading, since the war spurred export sales and Baldwin profited greatly in the lucrative munitions market.[107] Wartime patriotism and profits provide only a partial excuse for their blindness, however. While the domestic locomotive market peaked in 1905, Baldwin began a serious program of diversification only in 1929.[108]

To modern-day readers that seems an inauspicious year for embarking on a corporate expansion. At the time its significance was that Samuel Vauclain finally stepped down as president after a decade in the job and a forty-six-year career at Baldwin. Vauclain stands as a towering figure in the company's history, on a par with Matthias Baldwin and Charles Parry, although they lacked his gift for self-promotion. As superintendent, Vauclain accomplished miracles in extracting record output from the old Broad Street plant. But if he is due that credit, he also presided over the company's decline in the 1920s. In that decade president Vauclain completed his personal monument: a vast and expensive factory at Eddystone. With its completion in 1928, the firm abandoned all the old Philadelphia shops, leaving Bush Hill after ninety-three years. In the new suburban plant Vauclain was determined to break through the constraints that had repeatedly bound him at Broad Street in every sales boom since the 1880s. While he built this superb factory, however, engine orders simply evaporated. Throughout the 1920s the Eddystone plant generally ran at less than one-third of its annual capacity of three thousand engines. And while his sales fell off by 70 percent, Vauclain and his board of directors doubled Baldwin's annual dividend for most of the decade, draining the firm's cash surplus. Had the company still been a partnership, men like John Converse or Alba Johnson could have reined in the headstrong and charismatic Vauclain. But he was alone at the top now, with his old partners either dead or retired. Vauclain's subordinates owed their advancement to him, and they would not cross the boss.

Not all Baldwin's problems were internal, however. American Locomotive had a very different corporate structure and different managerial talent, yet it too failed to adapt to the challenges of the new century.[109] Other major capital equipment firms making machine tools, ships, and stationary engines had also stumbled by the 1920s. While their products varied, these firms mostly took a reactive posture in innovation and market demand that mirrored Baldwin's stance. In these crucial areas the locomotive builder had an eighty-year policy of following the lead of its railway customers. Largely because of this reactive

The Eddystone plant circa 1918. At the time Broad Street remained open and Eddystone was incomplete. Nonetheless, it was a huge facility, with its own ocean shipping docks and two railway lines traversing the site. When fully operational in 1928 it included almost 600 acres, with 100 acres of floor space. Undoubtedly the best-equipped steam locomotive plant in the world, it seldom exceeded one-third of its production capacity, except during World War II.
RAILROAD MUSEUM OF PENNSYLVANIA (PMHC)

heritage, Baldwin's managers remained blind to the need for technical leadership and diversification into new product lines until their problems had become nearly insurmountable. Once Baldwin finally began diversifying, its acquisitions of other capital equipment firms making complementary products appeared to be a sound policy, but the firm acted at the wrong time, 1929. When the Great Depression struck shortly thereafter, capital goods sales plummeted, pulling Baldwin into voluntary bankruptcy in 1935.[110]

THE LONG DECLINE FOR BALDWIN
AND AMERICAN CAPITAL GOODS

Historical accident, government regulation, managerial failure, and the company's heritage of awaiting its markets all played a role in the Baldwin company's decline after 1907. The builder would have a long history after that pivotal year, making locomotives until 1956 and other lines of capital equipment through 1972.[111] During the century's two world wars Baldwin served as a prominent arsenal of the Western alliances. But the challenges, mistaken policies, and missed opportunities of 1905–29 sent the company into relative decline, both on the national industrial scene and in the locomotive industry it had dominated for so long. Beyond the specific problems discussed already, the locomotive builder's waning fortunes also arose from larger economic trends and transformations. Baldwin did not change so much as the American economy and industrial capitalism did.

After 1900 and increasingly after World War I, the companies and industries of the first industrial revolution gave way to those of the second as the United States built upon the past or cut its encumbering embrace. The nineteenth-century age of coal, iron, steam, and textiles had done its work of creating an industrial nation. Now oil, steel, electricity, chemicals, and automobiles would increasingly sustain the mature economy as a consumer society. Inevitably many capital equipment builders, the creators of smokestack America, waned in the rising consumer-oriented society.[112] While Baldwin had exemplified American industrial strength in the 1880s and 1890s, the icon of the 1910s was the Ford Motor Company, whose Model T liberated millions of Americans to travel wherever they wished.

The changes in energy, materials, and products allowed transformations in the organization and structure of industrial companies. Incorporated and integrated manufacturing giants, like U.S. Steel and American Tobacco, overseen by a professional managerial class became the dominant economic power. This first occurred in industries where high-volume production or distribution was technologically feasible—chiefly in materials processing and those lines of standard consumer products that could accommodate the technology of volume production. Wherever possible the new managerial class integrated raw materials processing, mechanized production, and national distribution within one firm. The managers of these modern business enterprises originally sought the control of integration to capitalize on new technologies in production and distribution. But the new giants disliked tolerating any uncertainties, disruptions, or risks. So control and standardization became valued goals in and of themselves. These firms particularly sought to market patented product technologies, necessarily to standard designs. Such a patenting strategy provided competitive advantages while focusing customers on quality rather than price. For many decades this strategy also served to bypass antitrust objections, allowing the large manufacturers to dominate their markets by controlling technology.

The Baldwin Locomotive Works stood as far from these developments as it possibly could—even in 1915, after eighty years of development. By that time many integrated manufacturers exerted control over product innovation with in-house industrial research laboratories dedicated to developing new patents; Baldwin had to await the railways' lead. The new corporate giants were resolute in making only standard products; Baldwin turned out an endless variety of designs day after day. The big manufacturers influenced the business cycle itself, using advertising to create demand and easing sales downturns by price cutting

During World War II Baldwin strained to build the machines required for total war. This 1944 photo shows an M-4 Sherman tank with a 2-8-8-4 engine for the Duluth, Missabe, and Iron Range Railroad. This was the most powerful steam locomotive ever built, exerting 140,000 pounds of tractive effort. RAILROAD MUSEUM OF PENNSYLVANIA (PMHC)

and by making standard products for inventory; Baldwin's expensive, made-to-order locomotives placed it at the mercy of capital equipment's mercurial demand cycles. The automated production technology used by high-volume manufacturers largely determined their workers' skills and the pace of effort; Baldwin's labor-intensive operations remained heavily dependent on the hand skills and exertions of its workers. For most of a century Baldwin had refined its techniques of building custom capital equipment, but by 1915 powerful new giants had risen up around it by relying on entirely different strengths and strategies.

The contrasts between old and new even penetrated the locomotive industry. As American railroading began its long slide into decline after 1907, Baldwin responded to falling sales by broadening its product line, building on its strength in custom work. The company's 1908 catalog listed 379 different sizes and types of engines, while the 1915 edition offered 492 varieties.[113] If none of these five hundred–odd types suited a customer, Baldwin would build to its drawings or create a new design on demand, guaranteed to perform on any of the world's fifty-seven different railway track gauges. This impressive flexibility in production and products sought to capture any category of locomotive demand the markets offered. But the decline in railway purchasing continued into the 1920s and accelerated during the depression of the 1930s. Even before this Baldwin and Alco had resorted to price fixing, a token of their resignation to diminishing sales. In 1933 Baldwin made twenty-three engines—its lowest annual total since 1848.[114]

Into this total collapse in railway equipment purchasing came General Motors. Its Electro-Motive subsidiary sought to revitalize this anemic market by introducing diesel-electric motive power. Baldwin and Alco had experimented with diesels in the 1920s, producing a handful of prototypes. The new technology appeared to offer limited advantages, particularly for switching engines and in urban jurisdictions with smoke abatement regulations. Beyond that, however, its potential for mainline railway operations seemed problematic—at least in the 1920s and the early 1930s. While its thermal efficiency far exceeded steam power, the diesel engine required highly advanced metallurgy, and it simply weighed too much for most locomotive applications. Adapting diesel power for mainline freight or passenger trains would require a massive effort in research and development and an entire reordering of the railways' steam-based operations. Neither Baldwin nor Alco had the capital for a full-fledged developmental effort, and in any event both companies saw little reason to walk away from the steam technology which they and their customers knew so well.[115]

On the other hand, General Motors had the capital, internal combustion experience, research and development capacity, and managerial strength to take the initiative in diesel development.[116] Armed with these tools, executives at GM's Electro-Motive subsidiary could also perceive an opportunity in diesel locomotion where Baldwin, Alco, and many railways only saw problems. Through the 1930s, the latter group correctly viewed diesel locomotives as being largely incompatible in railway operations if used to supplement steam power, requiring entirely different fuels, parts, and maintenance techniques. But top executives at Electro-Motive had little interest in supplementing steam. They wanted diesels to replace it entirely.[117] Given the infancy of railway diesel technology, however, replacing the nation's 50,000 steam locomotives required an incremental approach. So in 1935 the automaker made a firm commitment to enter the diesel switcher market, just as Baldwin entered bankruptcy court.

Before building a single new switcher GM established "one fundamental policy . . . Electro-Motive Corporation will build a standardized product and [will] not undertake to build to the many different standards and specifications on which each railroad demands to purchase."[118]

The policy made sense; the automaker had great experience in the mass production of standard products, which offered cost savings over custom designs. But more than economy was at issue in this directive. Electro-Motive wanted control over locomotive innovation.[119] It would give the railroads the savings from production economies in exchange for a power they had exercised for eighty years. Its efficiency in production was real, but GM used it to achieve a larger goal: control over its markets, a fundamental desire of manufacturers.[120] GM's insistence on control through standard design was entirely antithetical to the customer-service philosophy that had guided Baldwin since the 1850s. Electro-Motive would require the railroads to adapt to its structure and philosophy. And while Baldwin's five hundred standard designs of 1915 had failed to reverse the slide in locomotive sales, GM's new technology would do just that, despite its inflexible policy.

A detailed account of how Baldwin and the other steam locomotive builders failed to meet this new competitive threat is beyond the scope of this study. Suffice it to say that Baldwin hardly knew what hit it. During the 1920s and 1930s, the general decline in sales had impelled established builders to focus on innovations in the design and production of their steam products. This strategy made a virtue out of necessity. The railroads could not purchase in quantity, but they desperately needed quality improvements to bolster their operating efficiency. By providing close design collaboration, the builders fulfilled their most pressing market demand, lessened price competition, and cemented relations with particular carriers. By incorporating a vast range of incremental design improvements, Baldwin doubled the tractive effort of its largest freight engines between 1906 and 1941—an impressive accomplishment for such a mature technology.

But Baldwin's efforts in incremental changes in steam technology so preoccupied the firm that it scarcely acknowledged or comprehended the revolutionary diesel. Although GM's new technology quickly made market inroads after 1935, Balwin's vice president and director of sales, Robert Binkerd, wrote in 1937 that the steam locomotive would reign into the foreseeable future.[121] Binkerd could not see very far. Just a year later, U.S. railroads ordered more diesels than steamers. Late in the decade Baldwin began competing with GM in diesel switching engines, but its 11 sales through 1939 barely compared with Electro-Motive's output of 301 units (1936–39).[122] As Baldwin tentatively ventured into switchers, in 1939 Electro-Motive made its first road diesel for freight service, freight power accounting for most of the American locomotive market. Baldwin and Alco failed to produce competing models before Pearl Harbor, and thereafter the War Production Board required them to concentrate on steamers and diesel switchers while allotting the market for big road diesels to GM.[123]

Even though World War II greatly stimulated the petroleum economy, in 1945 many major carriers remained skeptical of the diesel, a costly innovation that would require revolutionary changes throughout their operations.[124] But the railways' steam locomotive fleets had steadily aged since 1907, after decades of austerity arising from the snowballing problems of hostile regulation, the rise of auto and truck competition, the depression, and wartime production controls. Their old, depreciated, and worn-out motive power was ripe for replacement after 1945, just as the diesel's advantages in operating and maintenance costs became undeniable. Ironically, the new technology won the

railways' favor partly for the same reason—improved operating efficiency—
that had motivated their continuous design experimentation in steam power.
Beyond that advantage, the diesel offered economies throughout railroad
operations, serving as a retrenchment technology for an industry whose decline
accelerated greatly in the postwar period. Baldwin finally launched a line of
diesel products after 1945, but GM's locomotives were technically superior
while its experience in volume manufacturing provided other key first-mover
advantages.[125] In little more than a decade following the war, Electro-Motive
largely completed its rout of Baldwin and the other established builders as
American railroads undertook a wholesale conversion from steam to diesel
power.

In 1956 Baldwin capitulated to the new reality and ceased locomotive
production after 125 years of continuous operation. The company's failure
arose largely from its history. In the middle of the nineteenth century the
railways had demanded that the locomotive builders take a supporting position
in locomotive innovation. The carriers would set the pace and direction of
technological change, leaving the locomotive builders to give tangible form to
these initiatives. As long as Baldwin's markets were growing, the peril implicit in
this inability to control the character of its own products remained manageable.
In these years the company became accustomed to its reactive posture. When
American railroading began its long-term decline after 1907, Baldwin's heritage
of reaction and its lack of design control became liabilities. Having ceded its
destiny to the railroads in the 1850s, the company could never regain control of
its own future.

CONCLUSION

BALDWIN, THE CAPITAL EQUIPMENT SECTOR, AND THE NINETEENTH-CENTURY ECONOMY

THE BALDWIN COMPANY AND MANY OTHER CAPITAL EQUIPMENT BUILDERS FELL into obscurity in the mass production, corporate giantism of the twentieth century, prime examples of capitalism's creative destruction. This is one reason for their neglect in industrial history. In their nineteenth-century heyday, however, companies like Baldwin exemplified the nation's industrial strength. Locomotives, machine tools, stationary engines, generators, and mill machinery symbolized the rising technically driven society. Beyond their standing as powerful symbols, such complex mechanisms created all the products and services that revolutionized modern life. High-volume manufacturers provided uniform products in endless quantities, but—like individuals and the nation as a whole—these manufacturers depended on capital goods as the mainspring that drove productivity growth in their own operations and throughout the national economy.

This study of the Baldwin Works has attempted to show how such complex mechanisms were sold, designed, and created in the first place. Baldwin enjoyed unique success for over eighty years, although its operations exemplified the entire capital goods sector. Its record of constant design collaboration with its customers boosted locomotive power output from 30 to 1,500 horsepower between 1832 and 1900. Those years saw no fundamental innovation in engine design; instead such figures underscore the productivity benefits to railways and shippers that accrued from constantly incorporating a range of minor innovations. In the last half of the nineteenth century Baldwin's product variety jumped from 12 to 118 classes built annually, while its labor productivity improved by 360 percent. Again, no fundamental innovations, like assembly-line technologies, reordered its production techniques. But through continuous organizational and technical changes, Baldwin's workers lifted their productivity by an average of 3.1 percent a year, compared with the 1.9 percent rate for workers nationally. This achievement also had broader ramifications for shippers and consumers, lowering the cost of transportation services and thus the prices of all products shipped by rail. By 1906 finished locomotives, weighing anywhere from 4 to 175 tons, rolled from Baldwin's erecting shop at the rate of one every three hours, twenty-four hours a day, for almost two years. Over the preceding seventy years the factory at 500 North Broad Street had built more than one-third of the entire American motive power fleet. Further afield, from Ireland to Japan, locomotives chuffed into stations and freight yards bearing a nameplate that read simply, "Baldwin Locomotive Works Philadelphia U.S.A."

How can we account for this impressive, influential, and durable success? Baldwin lacked many of the tools modern corporations have developed to lessen risks and promote stability, yet the business environment of the nineteenth century offered up far more instabilities and challenges than are presented by our modern economy.[1] Recent work by business historians has emphasized the adaptations made by one group of companies—incorporated mass producers—to enhance their internal efficiency and external control. But as this account has emphasized, Baldwin spurned their bureaucratic managerial structures, standard products, mechanized production methods, and mass-marketing strategies. The roots of its success and growth lay elsewhere.

Instead Baldwin responded to the demands of its sector and more generalized risks in the economy by developing a distinctive business strategy that possessed its own internal coherence and logic. That comprehensive strategy encompassed related policies in product innovation, management, production, and labor relations. Mirroring other capital equipment firms, Baldwin learned that growth derived from the broadening of markets, product lines, and productive capabilities, while long-term success depended on workers' skills, quality products, close relations with customers, and continuous technical support. Because the bulk of recent scholarly attention has focused on mass-production manufacturers, it may be worth stating explicitly that this strategy of capital equipment firms did not represent a variant on the manufacturers' models of organization and operation. The builders practiced a distinct alternative format of industrial production, one that minimized risks whenever possible while capitalizing on opportunities for growth.

That format was the machinery builders' response to challenges of a character or extent unknown in mass-production manufacturing. Compared with American System manufacturers, the builders had to marshal large, skilled workforces and substantial capital to construct complex, heavy engineering products that saw continuous technical evolution. These firms faced a notably perilous business cycle and ever shifting terms of trade with their customers. They marketed their expensive machines in national and international markets to buyers who often insisted on entirely unique designs. Before building such tailor-made products, the machinery makers generally had to sell their products through competitive bidding, often with extensions of credit. All of this took place in an economy where the flow of information was uncertain, financial markets imperfect, and periodic depressions inevitable. The owner-managers of these firms operated in a world of risk and uncertainty. Although the rewards could be great, theirs was a difficult route to wealth.

Two issues in particular, boom-or-bust sales and customers' design influence, impelled Baldwin to take a variety of steps that minimized risks. Risk was unavoidable, but it could be contracted out or shared with other parties to lessen its potential effect on the firm. Working with its competitors, Baldwin periodically reached industrywide price-fixing agreements that relieved the squeeze between materials costs and engine prices. Relying on suppliers for just-in-time parts inventories freed working capital. Technical collaboration with the leading carriers cemented close relations with important customers while continuously updating Baldwin's standard product line with modern designs for other buyers. Inside the firm, the builder lowered the costs of continuous technical change by using standard parts wherever possible in fulfilling railroads' custom design needs. It also forced much of the burden of labor cost containment onto inside contractors. Such policies helped ensure Baldwin's survival while also providing the basis for long-term growth.

Capital equipment builders prospered by optimizing their productive capacity. This strategy secured customers in a range of different sectors in America and around the world. Baldwin's decision to broaden its productive capacities was not simply a response to an external technical reality—the master mechanics' design power. Such optimization became a positive choice, resulting from the firm's close and dynamic interaction with customers in a variety of markets. In fulfilling their varied motive power needs, Baldwin derived both long-term expansion and some short-term protection during depressions in the mainline American market. Its policy of broadening markets also depended upon a series of informal strategic alliances with a range of parties. Outside the firm, Baldwin built relationships with leading American carriers. These included a de facto partnership with the Pennsylvania Railroad, technical collaboration with lines like the Santa Fe, and actual equity interests in the Kansas Pacific and Northern Pacific. The locomotive builder also drew growth from other informal relationships with the "Philadelphia interests" and with investment bankers like Drexel and Morgan. By striking similar alliances with commission merchants around the world, Baldwin unlocked the export market.

Delivering the goods in all the design variety required by customers called for more informal alliances inside the firm. So Baldwin relied upon inside contractors for flexible production management. It also reached out to establish a partnership with skilled workers. Individual workers and contractors came and went—although turnover was low, thanks to high pay—but the firm preserved its alliances with these groups for half a century. By promoting cooperative relations with contractors and skilled men, the builder protected its chief asset in flexible production: its human capital. Efficiency in production was also important, so the company utilized piecework and standard parts designs where possible. But Baldwin always had to balance the promotion of efficiency with the preservation of optimal productive capacity.

Baldwin's alliances to key groups inside and outside the firm worked to lessen risks and achieve growth. Rather than emulating the high-volume manufacturers' strategy of trying to exert managerial control over their environments, Baldwin grew to great size largely by sharing powers and risks among its partners and with suppliers, commission merchants, competitors, customers, and workers. Few of these relationships were formal or contractual, but such flexibility offered real advantages. The locomotive builder operated in highly variable economic and technical environments where adaptability was vital because control was impossible. These policies propelled Baldwin to the top ranks of American industry from the 1860s to the 1910s, and overall they sustained the company for 125 years.[2] Such a collaborative business strategy clearly provided a viable formula for enduring success. When studies of other capital equipment builders follow this examination of Baldwin, we will learn more of the character of this sector and the role these collaborative practices played in sustaining the nineteenth-century economy.

In the twentieth century, Baldwin's eclipse by General Motors' Electro-Motive subsidiary would seem to suggest that corporate managerial capitalism had become a better organizational form than the older collaborative format. While Electro-Motive did have its day in the sun, this view appears indefensible on historical terms. Each format—custom building of heavy machinery and mass production manufacturing—rose to meet circumstances within its own originating sector. These sectors in turn were defined by unique constellations of economic realities, social relationships, and cultural beliefs as they coalesced at particular moments and evolved over time. A judgment in favor of one or the other format necessarily discounts the historical context that gave each its vitality.[3]

From the present vantage point, it is apparent that corporate managerial capitalism carries its own liabilities, whether in heavy machine building or in volume manufacturing. Managerial hierarchies and control strategies also rose at many capital equipment builders after 1900. Like Baldwin, these bureaucratic firms found it difficult to survive in the consumer-driven economy and the post-industrial society that followed. In the machine tool industry, the putative advantage of making only standard products proved to be a liability, hindering American firms' ability to compete with more innovative and adaptive firms abroad. The short-term, risk-averse, and earnings-driven perspectives that grew in American managerial capitalism after 1945 also contributed to the postwar decline of the U.S. capital equipment sector. Finding its endemic uncertainties and constant innovations intolerable, the professional managerial class at these machinery builders often chose to exit the sector entirely, redeploying assets to achieve more predictable returns. For example, General Motors tried to sell its Electro-Motive subsidiary in the early 1990s. No buyer was interested.[4]

The capital equipment builders' history offers other insights for modern industrial policy. The upheavals of recent decades in American volume manufacturing have demonstrated the hollowness of American managers' fetish for control and the pitfalls of placing quantity production ahead of quality products. Concurrently Japanese firms frequently garner benefits from cooperation and coordination—between managers and workers, and beyond the boundaries of the firm. In response to inroads made by such foreign competitors, American volume manufacturers have sought to reinvent themselves in the last decade with a range of policies known collectively as "agile manufacturing." That rubric encompasses goals that Charles Parry could have authored: collaborating with suppliers, enlisting workers' abilities, cutting lead times in production, curbing stultifying managerial bureaucracies, and responding to evolving customer demands with continuous product innovation. Collectively these policies represent a turn away from the single-minded pursuit of efficiency in favor of adaptations that lessen risks and promote firms' adaptability to seek out new market opportunities. Although the capital equipment sector has declined in recent decades, its nineteenth-century strengths are being rediscovered by high-volume manufacturers seeking to reverse the historical limitations of their own format.[5]

Furthermore, close observation of the modern industrial scene suggests that many of the dynamics described in this study remain common in today's capital equipment sector. Modern makers of heavy machinery like Boeing Aircraft encounter highly variable demand; they cultivate direct relations with customers; they must sell their products before building them, often through competitive bids and with extensions of credit; and they rely on skilled workers in production. Finally, their customers still have important influence over design issues, an influence frequently exerted through performance specifications. While custom or customized capital goods are relatively rare today, examples do survive in supercomputers, ships, industrial robots, locomotives, and machine tools.[6] The rising concern for quality among modern American builders and manufacturers has even begun a shift from standard to custom designs as firms court customers with made-to-measure products.[7] Clearly, those factors that distinguished the nineteenth-century capital equipment sector remain important today, even though the sector itself has declined in prominence.

Recovering the history of the machine builders should also cause fundamental reevaluation in many aspects of American industrial history. Although mass production and its American System antecedents dominate in that literature, at best they account for only a portion of the nation's nineteenth-century industrial prowess. The capital equipment firms' unique model of industrial production suggests the need for some revisionism in a range of historical fields. Consider some of the implications of just one issue: Baldwin's pursuit of broad markets possessing varying motive power needs, rather than the American System strategy of using a standard product to exploit intensive demand. By optimizing its productive capacity in this fashion, Baldwin used its remarkable flexibility to create an alternative route to scale economies that economic historians should find noteworthy.

Once the locomotive builder had developed this flexibility in production, the partners perceived that custom locomotive designs suited Baldwin's long-term interests, even if they were nettlesome in the short run. Technological historians have detailed the standardizing efforts of American System firms, but they have largely overlooked the costs of such standards and the benefits of incremental change as demonstrated in the capital equipment sector. Continuous innovation in capital goods boosted sales as quality improvements—frequently advanced by buyers—rendered older models obsolete.

In the 1850s and 1860s Baldwin found that it needed new managerial controls to meet the differing needs of customers while securing efficiency in production. Therefore the firm established a battery of oversight systems twenty to thirty years in advance of the systematic management movement. Beyond this record of innovation, business historians will also note the importance of actors outside the firm—Baldwin's controls were internal tools to reach the external ends of fulfilling the master mechanics' requirements while securing new market niches.

Baldwin's management systems suggest that Frederick Taylor was a late entrant into the field of managerial reform—a revision that is long overdue, particularly in labor history. But in building a variegated product line, Baldwin remained far more reliant on a core group of skilled workers than were contemporary high-volume manufacturers. This reliance suggests that labor historians may find new insights by looking beyond the all-too-common assumption that managers had one overriding goal: to improve profits by controlling and exploiting workers. Baldwin's managers long sought a partnership with skilled men whose talents provided a key competitive advantage for the firm. For many decades its workers responded favorably to these overtures. Labor historians may also find that sensitivity to sectoral differences and to the differing demands of efficient and optimal production helps explain the uneven advance of organized labor in the nineteenth century.

As these examples suggest, the general contours of industrial history should change substantially after historians place the unique strategies and structures of the capital equipment builders alongside those of volume manufacturers. At this early stage in examining the capital goods sector, this history of a single firm cannot authoritatively predict the new interpretations that historians will offer of our industrial past. But this study joins a growing literature on custom and batch producers in other sectors, making goods ranging from ceramics to textiles. Collectively these accounts underscore the vitality of American industrial history and the differing character of firms as a result of variations in market demands, customers' design influences, managers' cultural beliefs, workers' skill endowments, and governmental policies. On its own, this study shows that even the largest American industrial companies circa 1900 could achieve their stature by choosing from a range of very different tactics. Such a finding challenges Chandler's powerful synthesis; indeed it suggests that synthesis of any sort may prove difficult to achieve. But this account also suggests the promise of a new industrial history in which variety, contingency, and choice replace economic, organizational, and technical determinisms. Much work remains before that new narrative can be written. But as Baldwin's operations show, achieving a comprehensive understanding of industrial history requires incorporating the capital equipment firms and the unique methods they employed to build the infrastructure of industrial America.

APPENDIX A

BALDWIN'S ANNUAL OUTPUT AND EMPLOYMENT, 1832–1932

Year	Locos.	Men	Year	Locos.	Men	Year	Locos.	Men
1832	1	30*	1861	40		1890	946	4,493
1833	0		1862	75	675*	1891	899	4,440
1834	5		1863	96		1892	731	4,039
1835	14		1864	130	900*	1893	772	4,313
1836	40	240*	1865	115		1894	313	2,150
1837	40	300*	1866	118	1,000*	1895	401	2,551
1838	23		1867	127		1896	547	3,490
1839	26	250*	1868	124		1897	501	3,191
1840	9		1869	235	1,700*	1898	755	4,888
1841	8		1870	280	1,455	1899	901	6,336
1842	14		1871	331	1,900*	1900	1,217	8,208
1843	12		1872	442	2,500*	1901	1,375	9,595
1844	22		1873	437	2,478	1902	1,533	12,150
1845	27		1874	205	1,300*	1903	2,022	14,720
1846	42		1875	130		1904	1,485	10,573
1847	39		1876	232	1,697*	1905	2,250	14,811
1848	20		1877	185	1,106*	1906	2,666	17,432
1849	30		1878	292	1,861	1907	2,655	18,499
1850	37	400	1879	298	2,000*	1908	617	4,600*
1851	50		1880	517	2,648	1909	1,024	5,788*
1852	49	400*	1881	554	2,889*	1910	1,675	14,553
1853	60		1882	563	2,943*	1911	1,606	13,116
1854	62	500*	1883	557	2,920*	1912	1,618	13,631
1855	47	430*	1884	429	2,377	1913	2,061	15,814
1856	59		1885	242	1,563	1914	804	6,710
1857	66	600*	1886	550	2,411	1915	867	7,717
1858	33	400*	1887	653	2,879	1916	1,989	
1859	70	600*	1888	737	3,329	1917	2,737	16,785
1860	83	675	1889	827	3,579	1918	3,580	

Year	Locos.	Men	Year	Locos.	Men	Year	Locos.	Men
1919	1,722	16,945*	1924	535		1929	446	
1920	1,534	15,758*	1925	721		1930	439	5,515*
1921	969	6,100*	1926	836		1931	87	
1922	684		1927	654		1932	65	610*
1923	1,696		1928	375	7,523*			

Sources: Because surviving internal records on Baldwin's employment are scattered and incomplete, this table was constructed from a variety of sources. Workforce data for 1884 to 1906 are from "The Development of the American Locomotive," *Journal of the Franklin Institute* 164 (Oct. 1907): 266 and 269. Other entries on employment are from more than thirty different sources: letters, magazine profiles, government documents, and so on. Output data for 1832 to 1923 are from Baldwin Locomotive Works, *History of the Baldwin Locomotive Works,* p. 182.

Notes: Baldwin's workforce fluctuated greatly from year to year, even from week to week. Entries marked with asterisks are measures of the workforce at a given moment in time; all other figures are the average size of the force for that year.

Output totals during World War I include an undetermined number of gun mounts (to which the company assigned construction numbers).

Any entry left blank is unknown.

APPENDIX B

DATA ON MAJOR AMERICAN LOCOMOTIVE BUILDERS

Firm/Location	Capital	Workforce Size	Output (no. of locos.)	Cost of Raw Materials	Payroll Costs	Value of Products
			1850			
Baldwin/Philadelphia	$250,000	400	37	$102,700	$144,000	$300,000
Norris/Philadelphia	200,000	350	36	141,225	105,600	270,000
Taunton/Taunton, MA	75,000	150	20	104,500	54,000	165,000
Hinkley/Boston	(400,000)	250	40	88,400	81,600	300,000
Wilmarth/Boston	50,000	90	NL	23,840	30,000	100,000*
Souther/Boston	40,000	100	NL	102,700	44,400	166,000*
Lowell/Lowell, MA	600,000	500	NL	143,700	180,000	380,000*
Rogers/Paterson, NJ	(350,000)	604	46	221,900	180,000	460,000*
Grant/Paterson, NJ	80,000	180	10	89,500	45,360	180,000*
Schenectady/Schenectady, NY	65,000	200	12	(49,800)	60,000	130,000
Cincinnati/Cincinnati, OH	200,000	NL	NL	70,000	NL	150,000*
			1860			
Baldwin/Philadelphia	($900,000) 350,000	675	83	$369,000	$270,000	$750,000
Norris/Philadelphia	(750,000)	580	79	317,500	194,880	670,000
Taunton/Taunton, MA	218,500	175	23	100,500	72,000	180,000
Hinkley/Boston	(25,000)	80	10	29,450	31,200	80,000
Mason/Taunton, MA	200,000	425	NL	161,000	144,000	330,000*
Souther/Boston	100,000	85	3	67,075	42,960	135,500*
Cooke/Paterson, NJ	150,000	250	36	NL	(168,000)	315,000*
Rogers/Paterson, NJ	300,000	720	90	418,900	288,000	765,000*
Grant/Paterson, NJ	150,000	260	30	150,000	144,000	300,000
Manchester/Manchester, NH	100,000	24	NL	16,754	9,000	37,500*

Firm/Location	Capital	Workforce Size	Output (no. of locos.)	Cost of Raw Materials	Payroll Costs	Value of Products
			1870			
Baldwin/Philadelphia	($12,500,000)	1,445	259	$1,655,132	$1,276,590	$2,988,463
	2,000,000					
Pittsburgh/Pittsburgh	300,000	375	60	410,000	240,000	860,000
Taunton/Taunton, MA	(218,500)	350	34	232,825	156,925	478,894
Dickson/Scranton, PA	(600,000)	500	NL	500,000	400,000	1,200,000*
Mason/Taunton, MA	(700,000)	658	46	450,000	430,000	1,000,000*
Schenectady/Schenectady, NY	130,000	600	74	658,487	336,685	959,248
Cooke/Paterson, NJ	500,000	760	55	(236,500)	423,544	(1,397,000)*
Rogers/Paterson, NJ	500,000	928	118	900,000	700,000	1,675,000
Grant/Paterson, NJ	300,000	500	105	(445,700)	350,000	1,400,000
Manchester/Manchester, NH	150,000	400	60	378,600	240,000	690,000
Rhode Island/Providence, RI	500,000	850	NL	(1,475,300)	379,312	NL
			1880			
Baldwin/Philadelphia	$3,000,000	2,648	NL	$1,976,153	$1,364,964	$3,800,599
Brooks/Dunkirk, NY	151,000	445	NL	414,101	194,417	524,534
Taunton/Taunton, MA	(218,500)	362	NL	466,256	222,678	729,263*
Mason/Taunton, MA	600,000	742	NL	223,000	287,000	551,000*
Schenectady/Schenectady, NY	350,000	725	NL	613,200	302,200	945,000
Cooke/Paterson, NJ	700,000	660	NL	538,036	260,000	921,129*
Rogers/Paterson, NJ	450,000	1,000	NL	(300,000)	342,572	630,844
Grant/Paterson, NJ	300,000	486	NL	575,000	242,000	875,000
Manchester/Manchester, NH	125,000	307	NL	264,305	106,205	407,889

Source: U.S. Manufacturing Census schedules.

Notes: Parentheses indicate dubious data. Italicized figures for Baldwin are the author's alternative estimates.

Asterisks indicate entries known to include the value of non-locomotive products; other entries may include such products as well.

NL = not listed in census schedules.

Data are for the fiscal year ending June 1, not the calendar year.

These data from the U.S. Manufacturing Census schedules provide a portrait of the locomotive industry for the period 1850–80. The information is useful in a general sense, but specific entries are often grossly inaccurate.

The footnotes to the table indicate some of the liabilities of reading these data as hard facts. There are others: In many cases information on firms is simply unavailable, either because the enumerator missed that company or because the records have been destroyed. Moreover, firms often did not know their own costs in detail. In other cases capital figures clearly include working capital even though enumerators were supposed to list only fixed capital. Some firms made non-locomotive products—using the listed capital, payroll, and raw materials— that are not included in the value of products. Some of the manufacturing census data are used at various points in this study because they are the only information available, but the figures, particularly the dollar amounts, are often quite suspect.

Because the census information is so often incorrect, it was inadvisable to use these data to gauge changes in Baldwin's productivity. Even if the data were accurate, a productivity analysis based on dollar values of inputs and products would still confront the problem of finding accurate constant-dollar indices to allow comparison over time. The episodes of price fixing in the industry would also throw off such an analysis. Finally, Baldwin's products were in constant evolution, they were made for a variety of markets, and they were priced according to what those markets would bear. Therefore another route to analyzing Baldwin's productivity seemed required.

In chapter 6 I use a data series based on weight of products, rather than dollar values, to gauge the firm's productivity. Such a technique was used, for many of the same reasons, by Albert Fishlow in "Productivity and Technological Change in the Railroad Sector," and it offers noteworthy advantages. The issue of changing money values is minimized, although not eliminated. Such a weight index homogenizes Baldwin's quite varied output into a single measure while also accounting for quality improvements.

To construct the index I derived lists of Baldwin's output by class designations for each decennial year from 1850 to 1900; these years were chosen because of the availability of horsepower and employment data from the manufacturing census, data that are presumably more accurate than the dollar figures therein. The weight of each class at each decennial was then determined from Baldwin's engineering data at the DeGolyer Library and at the Smithsonian. The resultant totals of pounds of output appear in the text.

The data should be seen as an index of changing productivity rather than a precise measure of all output. The weights are based on the weight of engines in working order. Not included are engine tenders, replacement parts, non-locomotive products (1850 only, proportionately a small figure), and electric trucks (1900 only, and again relatively small in proportion to total output).

Baldwin also devoted a portion of its workers and plant to locomotive repair jobs throughout this half century, work not included in the index. I believe that these omissions do not detract from the validity of the overall portrait of improving productivity. Some may question the index—for example, arguing that one engine cylinder that weighed 50 percent more than another perhaps did not require half again as much time to machine. While such ratios are debatable, growth in product size and weight did require more workers and larger machine tools, cranes, and shop space.

This weight index also accounts in part for improvements in product quality. Quality encompassed two essential issues: hauling power and repair costs (since fuel costs were a secondary concern and locomotive service life remained relatively constant at twenty-five to thirty years). The index cannot account for repair issues, although such costs fell substantially during the period, but it does include power improvements, which were proportional to weight. In fact locomotive hauling capacity increased at a faster rate than did engine weight.

In chapter 6 I use the weight index to gauge improvements in Baldwin's productivity over time. Labor productivity is a straightforward question, since worker-hours were essentially unchanged in each of the years sampled (at each point the contemporary trade cycle for locomotives was either well into a market advance or peaking). Measuring capital productivity required the use of an index to transform Baldwin's fixed capital figures from current to constant dollars. I used the Davis and Gallman index for the historic cost of capital stocks in the United States, found in Peter Mathias and M. M. Postan, eds., *The Cambridge Economic History of Europe* (Cambridge: Cambridge Univ. Press, 1978), vol. 7, pt. 2, p. 27. Baldwin's current-dollar capital figures and their sources are given here:

1850: $250,000 (U.S. Manufacturing Census)
1860: $350,000 (BLW Ledger, 1859–63, BLW-HSP)
1870: $2 million (estimate—based on the R. G. Dun credit reports on Matthew Baird's sale of ownership in 1873)
1880: $3 million (U.S. Manufacturing Census)
1890: $5.7 million (estimate—based on an R. G. Dun report of 1888 on the firm)
1900: $11.5 million (estimate—based on the valuation of Edward Williams's interest in the firm given in his 1899 will)

Any effort to measure the productivity of firms in the capital equipment sector is a daunting task. While Henry Ford would not let buyers of his Model T pick even the color, Baldwin's customers had a vast range of choices. This variety in the product line is what makes a productivity analysis interesting and worth undertaking, but in deriving the data required, I have gained immeasurable respect for the managers and workers who created such diversity in the first place.

ABBREVIATIONS AND ORIGINAL SOURCES

Because much of the Baldwin company's extensive archive was destroyed in the mid-1950s, this account necessarily draws on a wide range of primary and secondary sources. Individuals who provided original sources are noted in the acknowledgments; institutional repositories are listed below.

The abbreviations used in citations for major manuscript collections are given below. One other abbreviation form is used in the notes, a short-form citation for the Baldwin letterbooks (1844–66) in BLW-HSP. For example, a letter cited as IN 7/61 is found in the Incoming Letterbook dated July 1861; one listed as OUT 10/61 appears in the Outgoing Letterbook dated October 1861.

HISTORICAL SOCIETY OF PENNSYLVANIA, PHILADELPHIA

BLW-HSP: Baldwin Locomotive Works Papers. This collection of 225 volumes of papers and ledgers details Baldwin's history from 1836 to 1866. Donated to HSP by the company circa 1946, it includes 165 letterbooks, fifty volumes of financial data, and approximately ten miscellaneous books, including three detailing the company's apprenticeship program. The letterbooks, which include incoming and outgoing correspondence, contain approximately one hundred thousand letters, bills, contracts, engine specifications, telegrams, and receipts. Each letterbook was selectively reviewed and every fifth book was read entirely.

SMV-HSP: Samuel Matthews Vauclain Papers. This collection of correspondence, reports, and memoranda covers the period from 1905 to 1931, during which Vauclain was successively a partner, vice president, president, and chairman of the company.

John Clayton Papers: A small collection of the personal and business papers of Matthias Baldwin's son-in-law, John Clayton, who served as the company's legal counsel from circa 1845 to 1866.

HAGLEY MUSEUM AND LIBRARY, WILMINGTON, DELAWARE

WLA-HAG: William Liseter Austin Collection. This collection includes drawings, internal memos, photographs, and reports kept by Austin during his career (1870–1932) at Baldwin, where he rose from a draftsman to chairman of the board. The collection includes shorthand notes taken by president Austin in the summer of 1910 that detail inside contracting at the firm. There is an additional Austin collection at Hagley—cited in the notes as WLA(B)-HAG— consisting of a number of notebooks outlining sales and design negotiations with customers.

The Reading Company Collection: This large corporate archive of the Philadelphia and Reading Railway provides data on the carrier's frequent purchases of Baldwin engines (1840s–1920s).

SMITHSONIAN INSTITUTION, WASHINGTON, D.C.

BLW-NMAH: The Archives Center at the National Museum of American History holds a collection of Baldwin's Order Books (eight volumes, 1854–1900) and its construction lists or Registers of Engines (nine volumes, 1832–1956). These sources provided much of the data on the variety in Baldwin's products, noted in chapter 3, and on the weight of output, discussed in chapter 6. The Division of Transportation at NMAH also holds four scrapbooks that were created at Baldwin between 1870 and 1930 to preserve historical material. They mostly contain clippings from the technical press on Baldwin engines, but one volume was particularly valuable. It includes a sample of every printed document Baldwin used in the 1870s and 1880s, including contract forms, List System blanks, employees rules, and instructions to foremen—many of which are pictured here.

DEGOLYER LIBRARY, SOUTHERN METHODIST UNIVERSITY, DALLAS, TEXAS

SMV-DEG: The Samuel M. Vauclain Collection at DeGolyer covers the period from 1882 to 1929, during which Vauclain rose from a foremanship to Baldwin's presidency. The collection includes Vauclain's Production Notebooks (twenty-seven volumes, 1888–1919), containing much detail on output, revenues, the workforce, contracting, and so on. There are also a host of other internal company documents from the period 1900–1929.

BLW-DEG: This repository also holds an extensive collection of Baldwin's engineering data, including over eight thousand drawings (1870s–1950s). Supporting volumes include the Law Books, Specification Books, and Weight Books (used to compile the weight index of output in chapter 6).

DEPARTMENT OF SPECIAL COLLECTIONS, STANFORD UNIVERSITY LIBRARY, CALIFORNIA

BLW-STAN: This is a small collection of Baldwin engineering data, covering the period 1854–1918. It provides detail on the evolution of drawing room practice in the mid-nineteenth century.

OTHER REPOSITORIES AND COLLECTIONS CONSULTED

Baker Library, Harvard Business School, Boston: library holdings

Historical Society of Pennsylvania, Philadelphia: Wallace Rogers Lee Papers, Herbert Welsh Collection, library holdings

Van Pelt Library, University of Pennsylvania, Philadelphia: Philadelphia Social History Project Papers

The Franklin Institute, Philadelphia: library holdings

The Library Company, Philadelphia: library holdings

Federal Archives Center, Philadelphia: Baldwin bankruptcy records (1935–38)

Office of the Registrar of Wills, City Hall, Philadelphia: wills of BLW partners

Urban Archives, Paley Library, Temple University, Philadelphia: Central Labor Union minutes, library holdings

Railroad Museum of Pennsylvania, Strasburg: BLW Board Meeting Minute Book (1910), BLW builders photographs, library holdings

The Pennsylvania State Archives, Harrisburg: BLW Order Books (1901–7), Contract Books (1871–1930), engineering data

Pattee Library, Pennsylvania State University Library, University Park: Harrington Emerson Papers

Hagley Library, Wilmington, Delaware: Longwood Manuscripts, Sellers Company papers, R. G. Dun credit reports, Trade Catalogue Collection, library holdings

Library of the Association of American Railroads, Washington, D.C.: library holdings

National Archives, Washington, D.C.: Patent Extension Files, Papers of the U.S. Commission on Industrial Relations, manuscript schedules (microfilm) of the U.S. Manufacturing Census

Library of Congress, Washington, D.C.: photography department

Archives Center, National Museum of American History, Washington, D.C.: Pittsburgh Locomotive Works Papers

SERIALS

To supplement these original sources, the following serials were consulted (specific citations to major articles appear in the bibliography):

American Machinist, vols. 1–37

Annals of the American Academy of Political and Social Science, vols. 1–42

Annual Reports of the American Railway Master Mechanics Association, vols. 1–48

Baldwin Locomotives, vols. 1–19

Baldwin Locomotive Works, *Records of Recent Construction,* nos. 1–100

Cassier's Magazine, various

Engineer (Philadelphia), vol. 1

Engineering (London), various

Engineering Magazine, vols. 1–40

Fincher's Trades Review, vols. 1–2

Hillyer's American Railroad Magazine, vols. 1–6

Iron Age, 1880–90

Journal of the Franklin Institute, vols. 1–170

Machinists' and Blacksmiths' International Journal, vols. 8–9

Machinists Monthly Journal, vols. 9–27

New York Times, various

Philadelphia *Evening Bulletin,* various

Philadelphia *Inquirer,* various

Philadelphia *Public Ledger,* various

Proceedings of the Engineers Club of Philadelphia, vols. 1–33

Railroad Gazette (a.k.a. *Railway Age Gazette*), vols. 1–59

Railway Master Mechanic, vols. 27–28

Scientific American, vols. 4–14; new series vols. 1–23

Transactions of the American Society of Mechanical Engineers, vols. 1–20

Van Nostrand's Eclectic Engineering Magazine, various

NOTES

PREFACE

1. Details of the construction process drawn from BLW, *Record of Recent Construction* *#29.* For accounts of twentieth-century steam locomotive construction, see Weitzman, *Superpower,* and Trostel, *Building a Lima Locomotive.*

2. Edward H. Sanborn, "Locomotives," special report in *Twelfth Census of the United States* (1902), p. 3. Because Baldwin made custom products to order, it necessarily maintained a just-in-time inventory system for most parts and materials throughout its history.

3. For changes in erecting work, see "Erecting a Locomotive at Altoona," *Railroad Gazette* 20 (Aug. 31, 1888): 572.

4. Normally this was the first stage, but erecting could proceed in a variety of ways, depending on the order in which the main parts—wheels, cylinders, boiler, and frames—arrived at the erecting shop.

5. Steps described in BLW Test Department, memorandum to William Austin, May 7, 1907, folder: Technical Correspondence and Reports, 1902–6, box 5, WLA-HAG.

6. For a fine discussion of railroading's impact on American society and culture, see Douglas, *All Aboard!*

7. Karl Baedeker, *The United States* (Leipzig, 1909), p. 168.

8. A journalist who visited Baldwin in 1904 described the locomotive as "a living, almost a sentient organism, the crux of transportation which is so great a factor in civilization, a maker and breaker of fortunes, a gladiator in war and an emblematized cornucopia in peace" (Rogers, "Greatest Locomotive Works," p. 69).

9. Keith L. Bryant Jr., *History of the Atchison, Topeka, and Santa Fe Railway* (New York: Macmillan, 1974), pp. 212–16.

INTRODUCTION

1. *Capital equipment* is defined here as those mechanisms sold to secondary companies for use in their final production of goods or delivery of services.

2. Their tour is described in *Diary of the Japanese Visit,* pp. 9–11. Victor Clark notes Pennsylvania's national leadership "in industries that depended mainly on engineering and metallurgy" in the mid nineteenth century (*History of Manufactures,* 1:516). Philadel-

phia's status as "the largest tool building center in the country" is cited in Roe, *English and American Tool Builders,* p. 239.

3. "Local Affairs," Philadelphia *Ledger and Transcript,* p. 1.

4. For the literature on integrated corporations and modern management, see the numerous studies of Alfred D. Chandler Jr., as well as works by Lamoreaux, Litterer, McCraw, Yates, and Zunz. For the American System, see the writings of Fries, Hoke, Hounshell, Howard, Gordon, Mayr and Post, Rosenberg, and Smith (full citations given in the bibliography). American System manufacturers produced standardized light machines in large quantities, using special techniques to make them with interchangeable parts.

5. For twenty years American business history has been dominated by the writings and concerns of one man, Alfred Chandler, who has masterfully analyzed the rise of the integrated corporation. Chandler's work eminently deserves its praises, but it also has embedded presumptions, which this study will touch on. Chandler is unabashedly teleologic—he seeks the origins of "modern management." Premodern forms of managerial practice go largely unconsidered. While his work is often quite detailed, Chandler is so concerned with synthesis that he homogenizes very different firms, ignoring their fundamentally different attributes. For example, Baldwin achieved large size (number seventy-five in assets among Chandler's top two hundred firms of 1917), major market share (35–40 percent), and durable success (140 years) without making Chandler's key "three-pronged investment" in managerial hierarchies, volume production technologies, and marketing capacities. Indeed, Baldwin did not even incorporate until 1909. Finally, Chandler has chosen to focus his work sharply, a wise decision that has greatly increased his impact. That influence is deservedly large, but it has also left a vast range of unexplored topics in American business history.

6. Thomas Cochran writes that "a broadly competent machine tool industry appears to be essential to continuous industrial growth" (*Frontiers of Change,* p. 94). Note, however, that even Nathan Rosenberg's consideration of capital equipment firms ("Technological Change in the Machine Tool Industry") is motivated by his interest in their relationship to American System mass producers. Beyond their direct role in creating heavy machinery, capital equipment firms also became nurseries of technical knowledge and engineering talent vital to the entire economy. As Monte Calvert has noted, the profession of American mechanical engineering largely originated in the shops of the machinery builders. For example, Baldwin boiler shop foreman Elijah Hollingsworth became a co-founder of a leading American iron ship builder, Harlan and Hollingsworth. Alexander Holley, the progenitor of Bessemer steel in America, started out in the New Jersey Locomotive Works and the Corliss engine shops, where naval architect Nathaniel Herreshoff also worked as a young man. The architect of the American steam navy of the Civil War, Benjamin Isherwood, pursued his early training at the Novelty Iron Works, New York City's leading engine builder. The list could go on, although historians have yet to study how such graduates of heavy machine building diffused innovations throughout the industrial economy. In Britain the relationship between the capital goods sector and the mechanical-engineering profession was so evident that firms in the sector were simply called the engineering industries.

7. Rosenberg, *Technology and American Economic Growth,* p. 88. Rosenberg here defines capital goods (equipment) firms as "producers of the machinery and equipment which were used as inputs in the other sectors of the economy." I believe that Rosenberg's inclusion of farming equipment in this definition overly dilutes the industrial character properly associated with capital equipment.

8. There are good histories of textile machine builders by Gibb (*Saco-Lowell Shops*), Navin (*Whitin Machine Works*), and Lozier ("Taunton and Mason"). Morrison (*New York Shipyards*) and Tyler (*American Clyde*) give short general overviews of New York– and Philadelphia-area shipbuilders. To date we lack published secondary histories of nineteenth-century firms making machine tools, general machinery, bridges, stationary steam engines, or locomotives.

9. Philip Scranton's current research on batch producers, including a number of capital equipment firms, confirms "the existence of a spectrum of possible approaches to manufacturing," each involving "distinctive technical considerations, labor requirements, and marketing stances, hence different managerial challenges" ("Diversity," p. 28). See also Zeitlin and Sabel, "Historical Alternatives to Mass Production."

10. Nelson, *Managers and Workers,* pp. 7–8. Scranton notes the difficulty of calculating the numerical extent of batch producers like capital equipment firms: the industry classifications in the U.S. census are too broad, combining the output of batch and mass producers ("Diversity," p 32). I cite Nelson's table because it is one of the few ready measures available. Note that the preponderance of capital equipment firms in this list of large employers properly connotes the labor-intensity of heavy machinery building. But large size was not a prerequisite to success in this sector, which also included hundreds of smaller companies.

11. While these two production formats have received various names, their divergent nature has been acknowledged since at least 1832, when Charles Babbage described the *making* (building) system vs. the *manufacturing* system in metalworking (see Paul Uselding, "Measuring Techniques and Manufacturing Practice," in *Yankee Enterprise,* ed. Mayr and Post, pp. 103–26). Ca. 1900 authors in the technical press frequently commented on the different systems of management and production in the machinery trades (see Carpenter, "Jobbing Work and Efficiency"; Day, "Metal-Working Plants"; Richards, "Economic and Labor Factors"; and Smith, "Economics in Machine Shop Work").

12. Roe and Lytle, *Factory Equipment,* p. 1. In the modern era the term *manufacturing* has come to encompass all industrial production, including the format and practices described here as *building*. The two terms, however, are used here in their original meanings, as outlined in the text.

13. In the language of Thomas Hughes (*Networks of Power*), the buyers of capital equipment required designs that would mesh with their own technological systems of production or distribution.

14. The most crucial difference between builders and manufacturers centered on the postures they took in approaching their markets. In his influential studies of manufacturers, Alfred Chandler largely ignores market demands; his high-volume firms grow by investing in managerial bureaucracies, production technologies, and marketing arrangements. In this formulation customers and markets need not appear. Customers' vital innovative role in capital equipment meant that machinery builders seeking growth pursued entirely different tactics from the volume manufacturers'.

15. John James, Jeffrey Williamson, and other economic historians have pointed to improvements in capital goods production in the nineteenth century as a key factor driving increased capital-intensity, and hence productivity, throughout the industrial economy (see James, "Structural Change in American Manufacturing"). Indeed, Chandler cites improved production and distribution technologies (read capital goods) as causal factors in the rise of high-volume manufacturing, again underscoring this sector's primary role in promoting technical change and economic growth.

16. Baldwin used straight piecework for decades, beginning in the 1850s, and it was

common throughout the locomotive and machine tool industries. A few capital equipment firms—including Baldwin's competitor, American Locomotive—experimented with the labor management planks of Scientific Management, but such experiments were rare. See Scranton, "Diversity," p. 47.

17. By this time a number of specialist machine tool firms had opened in the Midwest, particularly in Cincinnati. Before then the trade's leading branch was in Philadelphia, where firms like William Sellers and Bement and Dougherty were classic builders of semicustom capital equipment.

18. While American mass production has preoccupied American historians, British historians have, logically enough, extensively studied that nation's large capital equipment sector. Their studies generally echo my description of the American sector's fundamental attributes. See Moss and Hume, *Workshop of the British Empire,* introd.

19. Baldwin continued to make locomotives until 1956 and other lines of capital equipment until the firm was liquidated in 1972. But this study terminates at 1915 for a variety of reasons. A focus on Baldwin's nineteenth-century heyday offered the most interesting historical contribution, the company's fortunes and the mainline American locomotive market both had peaked by 1915, and thereafter the firm shifted operations to a modern, qualitatively different plant. Finally, comparatively little documentation survives for the firm's more recent history.

20. As the nation's largest capital equipment company, Baldwin had a particular incentive to improve labor productivity, and the extent of its controls may have been exceptional, yet I suspect that other large builders established similar policies (many are noted in the text).

21. Chap. 6 might be thought of as a "labor process" study of locomotive building, although I have avoided that term because it tends to privilege a single aspect of production. Labor, technology, and management had a dynamic interplay.

22. Nelson, *Managers and Workers,* pp. 7–8.

23. As Scranton notes, "Batch firms could never sustain the illusion that they could control their market environments, manage technological change, or use formal rationality to make decisions" ("Diversity," p. 89). Their powers were reactive, not proactive.

24. John Staudenmeier notes this transformation from negotiation in technology and markets to control through standards and hierarchy in "The Politics of Successful Technologies," in *In Context,* ed. Cutcliffe and Post, pp. 150–71. Chandler's *Scale and Scope* (Cambridge: Harvard Univ. Press, 1990) has a useful appendix that charts the changing composition of the two hundred largest U.S. industrial companies between 1917 and 1948 (size measured by assets). The relative decline of the machinery builders is clearly evident in these data.

1: ESTABLISHING THE BALDWIN WORKS, 1831–1866

1. This community of mechanicians is described in Ferguson, *Early Engineering Reminiscences.* See also Wallace, *Rockdale,* chap. 5.

2. Thomas Cochran notes that industrialization crucially depended on "the establishment of the processes for building . . . machine[s] on a profitable, ongoing basis, the self-reenforcing process of machines building machines" (*Frontiers of Change,* p. 50). Like Matthias Baldwin, these men each worked on a variety of technologies. Nathan Sellers built papermaking machinery, Lukens made clocks and lathes, Lyon was known for his fire engines, and Mason created scientific instruments and lathe slide rests. Other members of the Philadelphia mechanical community of the 1820s included Alfred Jenks

(textile machines), Rush and Muhlenberg (steam engines), and Franklin Peale (machinery for the U.S. Mint).

3. The institute's history is superbly told in Sinclair, *Philadelphia's Philosopher Mechanics*. Matthias Baldwin was one of its founding members.

4. Coulson, "Some Prominent Members," p. 172. Jeweler Thomas Fletcher was another founding member of the Franklin Institute, further evidence of the catholic character of metalworking in this period (Sinclair, *Philadelphia's Philosopher Mechanics*, p. 31).

5. Brooke Hindle has described the habit of visual or spatial thought, as distinguished from verbal thought, that artists, artisans, and mechanicians shared (*Emulation and Invention*, pp. 16, 133–38). See also Ferguson, *Engineering and the Mind's Eye*.

6. Coulson, "Some Prominent Members," p. 173; Ferguson, *Sellers*, pp. 53 and 59. Mason was another founding member of the Franklin Institute.

7. The list of locomotive builders with antecedents in textile machinery production includes Baldwin, Mason, Taunton, Lowell, Rogers, and Cooke.

8. Ferguson, *Sellers*, p. 186, and Coulson, "Some Prominent Members," p. 174. Had Mason known that Baldwin would build roughly twenty stationary engines between 1830 and 1837, he might have continued with the partnership. Data on production from "Letter of the Secretary of the Treasury."

9. An 1830 Day Book shows that Baldwin continued to make the various tools described above. The entries also reveal strong ties within the Philadelphia mechanical community, with Baldwin numbering among his customers such noted mechanicians as Coleman Sellers, Samuel V. Merrick, John Agnew, and Franklin Peale. See pp. 1–16 in Day Book, Dec. 11, 1829–Aug. 16, 1836, BLW-HSP.

10. At the time of Baldwin's conversion a prominent Presbyterian evangelist, Dr. Ezra Styles Ely, was igniting evangelical fervor from his pulpit at Philadelphia's Pine Street Church. Baldwin became a staunch supporter of Ely's American Sunday School Union, and it appears that Baldwin's conversion accompanied Ely's proselytizing. For Ely, see Wallace, *Rockdale*, chap. 7; for Baldwin's religious beliefs, see Calkins, *Memorial*, pp. 39–45.

11. Calkins, *Memorial*, pp. 113–39. Modern historians often view the evangelical conversions and religious activities of members of the rising industrial elite as self-interested attempts to enlist Protestant belief in the tasks of sanctioning capitalist labor discipline and creating docile workforces. While this motivation may have applied to Baldwin as well, no documentary evidence supports such an assertion.

12. By the 1850s and 1860s Baldwin certainly enjoyed the life of a very wealthy man. He had a house in the city decorated with fine art, a country estate, many servants, and the carriage of a gentleman. But he seems to have believed that the accumulation of wealth was an unworthy goal of itself. Despite his personal yearly income of up to $210,000 (1864), Baldwin's estate in 1866 was valued at only $282,000 (not including his interest in the company). Although a vast sum for the period, this was far less than one might expect for a man who had earned almost as much in a single year. Evidently he had reinvested much of his profit in the firm while donating the bulk of his remaining wealth to charitable causes. For his estate valuation, see will of M. W. Baldwin, #456 of 1866, Office of the Registrar of Wills, Philadelphia City Hall.

13. After a visit from his close friend Henry Campbell, Baldwin wrote, "The stories you can tell and the hearty outbreak of merriment are entirely indispensable to my comfort. Since you have been away I have had the dispepsia continually." OUT 8/46: M. W. Baldwin to Henry R. Campbell, Jan. 15, 1847, BLW-HSP. For the UBA, see O. A.

Pendleton, "Poor Relief in Philadelphia, 1790–1840," *Pennsylvania Magazine of History and Biography* 70 (Apr. 1946): 170–71; for the Christian Commission, see Baldwin's entry in *Dictionary of American Biography* (hereafter cited as *DAB*), 1:542.

14. *DAB,* 1:541–42.

15. OUT 11/51: M. W. Baldwin to J. Roach, Jan. 8, 1852, BLW-HSP.

16. White, *American Locomotives,* pp. 240–41. The D&H engines mirrored an 1810-era design, first used in English collieries; Stephenson's *Rocket* was far more advanced.

17. Ward, *J. Edgar Thomson,* pp. 18–19.

18. White, *American Locomotives,* p. 13. The West Point Foundry Association of New York City built the first American engine, the *Best Friend,* in 1830 for the South Carolina Railroad.

19. Ferguson, *Sellers,* p. 70, and BLW, *History of the Baldwin Locomotive Works,* p. 9 (hereafter cited as *Baldwin History*).

20. White, "Old Ironsides," pp. 85–87.

21. White, *American Locomotives,* p. 270.

22. Sanford, "Pioneer," p. 20.

23. OUT 2/41: Baldwin and Whitney to Charles Moering, Aug. 18, 1842, BLW-HSP.

24. Sanford, "Pioneer," pp. 20–23.

25. In 1838 the Locks and Canals Machine Shop of Massachusetts made an average profit of $2,200 on each engine it sold. My figure of $60,000 (or $1,500 per engine) in profits for Baldwin is conservatively derived from that data. See Gibb, *Saco-Lowell Shops,* p. 95.

26. Letter Copy Book no. 1: M. W. Baldwin to George Schuyler, Mar. 12, 1836, and M. W. Baldwin to William W. Woolsey, Mar. 24, 1836, BLW-HSP. Bruce Laurie notes a wave of labor activism and strike actions in New York and Philadelphia in 1836 as workers sought wage increases to counter inflation (*Artisans into Workers,* p. 88).

27. As he wrote to a supplier in March, "Business is still brisk with me but [is] generally dull in Philada. on account of money being so scarce." Letter Copy Book no. 1: M. W. Baldwin to S. Vail & Son, Mar. 20, 1837, BLW-HSP.

28. Letter Copy Book no. 1: M. W. Baldwin to W. B. Fling, Feb. 8, 1837, BLW-HSP.

29. Calkins, *Memorial,* p. 76. No original sources documenting this crucial incident survive, so this section is largely based on Calkins's somewhat sketchy *Memorial,* written after Baldwin's death. Baldwin's commission-merchant creditors of this period included Anson Phelps of New York and Hendricks and Brother of Philadelphia.

30. In the early industrial period Vail's Speedwell Works of Morristown, N.J., was an important forge and machine shop, building much of the machinery in the pioneering Atlantic Ocean steamship *Savannah* (ca. 1819). Stephen Vail also backed the partnership of his son Alfred with Samuel F. B. Morse in developing the electric telegraph. See Buffet, "Vail Family," pp. 117–19. The estimate of the Vails' capital contribution to Baldwin is given in Clark, "Birth of an Enterprise," p. 437. The breakdown of ownership interest is from BLW Ledger, 1839–42, BLW-HSP. Beyond the Vails' actual cash contribution, they also provided a vital intangible asset to Baldwin: a creditworthy name to endorse the notes of the reorganized locomotive-building company.

31. Hufty's ownership share is in BLW Ledger, 1839–42, BLW-HSP. With Baldwin essentially in voluntary bankruptcy in 1839 and the locomotive market at a low ebb, it is understandable that Hufty made no capital contribution in return for his stake in the new firm.

32. Baldwin's full-time efforts in product development during 1839 and 1840 are described in "Petition of M. W. Baldwin for Extension of Patent," June 16, 1856, Patent

Extension Files, National Archives. The material in the text on the development of his flexible-beam design comes from this source.

33. This was the 4-4-0 or American type, which in fact proved a great success.

34. The company history takes over a full page to describe the mechanical motions involved in this design (*Baldwin History,* pp. 32–33), so a complete description will not be attempted here. Suffice it to say that Baldwin used connecting rods to power the front four wheels.

35. Clark notes ("Birth," pp. 438–39) that by 1840–41 George Vail was spending most of his time at his father's Speedwell Works, leaving his brother Alfred to oversee his interests at the locomotive company. The terms by which Hufty left the partnership are unknown, but he continued to work for Baldwin as general superintendent until 1854.

36. Vail's withdrawal is described in ibid., pp. 442–43.

37. Calkins, *Memorial,* p. 76; *Baldwin History,* p. 24; and OUT 3/44: Baldwin and Whitney to Stephen Vail, Oct. 15, 1844, BLW-HSP.

38. My list of seven majors includes Rogers, Schenectady, Taunton, Grant, Hinkley, Norris, and Portland (appendix B gives data on these and other firms in the industry). The figure of forty builders in 1854 is from "Locomotive Shops of the Country," *Railroad Advocate,* p. 1.

39. In 1849 a railway officer wrote to Baldwin, "Your engines stand deservedly high for their efficient and economical working . . . [but] I think it due to other establishments to give them a fair trial and thus aid in keeping up a wholesome competition." IN 1/49: William Mitchell (Western and Atlantic Railway, Atlanta) to M. W. Baldwin, Apr. 4, 1849, BLW-HSP.

40. See IN 7/44: R. R. Cuyler (president, Central Railroad and Banking Company, Savannah) to Baldwin and Whitney, Nov. 26, 1844, BLW-HSP. In this letter Cuyler also demanded (and got) a penalty clause in a contract for four engines in case of late delivery.

41. OUT 3/53: M. W. Baldwin (per George Burnham) to Indiana Central Railroad, May 11, 1853, BLW-HSP. By the 1850s such payments in securities were common, and they demonstrate how the railway supply industry served as a financing intermediary for railway development in America—quite the reverse of the conventional wisdom.

42. See performance table in OUT 3/44: Baldwin and Whitney to C. F. Hagedorn (Bavarian government consul), Apr. 13, 1844, BLW-HSP.

43. Baldwin, Vail, and Hufty, *Locomotive Engine Catalog* (1840). The 1840 product line included three models, all in the same basic configuration, varying only in power and weight.

44. OUT 8/46: M. W. Baldwin to Ferrier Wicksted, Oct. 22, 1846, BLW-HSP.

45. See his terms for wrought-iron tire purchases in Letter Copy Book no. 1: M. W. Baldwin to Rogers, Ketchum, and Grosvenor (Paterson, N.J.), Feb. 8 and 17, 1837, BLW-HSP.

46. IN 1/50: Abbott and Ferguson (Baltimore, makers of forged axles) to M. W. Baldwin, Feb. 25, 1850, and OUT 2/50: M. W. Baldwin (per George Burnham) to Parke and Brothers (Elkton, Md., boilerplate supplier), Mar. 1, 1850, BLW-HSP.

47. In 1845 Baldwin's paymaster borrowed $500 to meet the payroll from Orrick and Campbell, a boiler flue supplier. No doubt Baldwin reciprocated when other firms had such immediate needs. OUT 10/44: Baldwin and Whitney (per George Whitney) to Asa Whitney, Mar. 5, 1845, BLW-HSP.

48. Workforce data are incomplete, but extrapolating from known figures, Baldwin's payroll in 1840 probably averaged no more than one hundred men.

49. Matthias Baldwin had little interest in financial matters, but Burnham's com-

petence in this field greatly aided the firm's survival and success. As the founder wrote to a customer in 1849, "My money man [Burnham] tells me he is out of cash and I must get some for him forthwith." OUT 10/49: M. W. Baldwin to J. Edgar Thomson (Harrisburg), Nov. 23, 1849, BLW-HSP.

50. No clear enumeration of Baldwin's managers before 1859 is available. This total is derived from a layout map of the shops ca. 1850 in "Miscellaneous Records Book, 1846–75," p. 63, folder 7, box 9, BLW-DEG. The discussion in the text of foremen's duties and powers is drawn from over a dozen letters in BLW-HSP.

51. Because foremen hired most workers, one might argue that the historical record is insufficient to make this assertion. But the Baldwin letterbooks contain numerous letters written by non-Philadelphians seeking employment at the works. Almost without exception the company responded that it had no jobs available. According to these records, Baldwin was short of skilled hands only once in the antebellum period: in 1853, when the company needed patternmakers.

52. Most historians believe that skilled labor was in short supply during the antebellum era. In their study of industry in mid-nineteenth-century Philadelphia, Bruce Laurie and Mark Schmitz write that "workers were surely in short supply among industries still dependent on highly skilled labor . . . [i.e.] manufacturers of metals, machines, and heavy equipment" ("Manufacture and Productivity," p. 84). Very little in the pre-1860 Baldwin records supports this assertion, although the company did have to pay good wages to secure skilled men. Laurie and Schmitz's own citations for the shortage date to 1863 and 1864—hardly the best years for generalizations about the American labor supply, with over one million men in the armed forces.

53. OUT 3/44: Asa Whitney to M. W. Baldwin, July 24, 1844, BLW-HSP.

54. Baldwin wrote that an order had been delayed "in consequence of the influence of the holidays, as the men don't make as much time as they might." OUT 8/47: M. W. Baldwin to Benjamin H. Latrobe (chief engineer, Baltimore and Ohio), Jan. 6, 1848, BLW-HSP.

55. IN 1/44: A. G. Benner, Jos. E. Maguire, and Jas. Wray to Baldwin and Whitney, June 7, 1844, BLW-HSP.

56. "Whig Procession," Philadelphia *Public Ledger,* Oct. 2, 1844, p. 2; IN 7/44: Milton Mendenhall to Baldwin and Whitney, Oct. 14, 1844, BLW-HSP.

57. OUT 10/58: M. W. Baldwin & Co. to Israel Morris (Allegheny and Bald Eagle Railroad), Nov. 4, 1858, BLW-HSP.

58. Baldwin wrote to a customer in November, "I am daily expecting the return of my drafts, accepted for the amount of the engine. . . . I find it next to impossible to meet my engagements and can raise money only on the most undoubted paper." OUT 6/48: M. W. Baldwin to W. H. Bartless (Charleston, S.C.), Nov. 11, 1848, BLW-HSP.

59. IN 1/49: Hands Resolution to M. W. Baldwin, Nov. 15, 1848 (BLW-HSP; this resolution is filed out of place, following correspondence dated Jan. 30, 1849). The petition detailed workers' hardships and sacrifices over the month their wages went unpaid. While this incident suggests a peril of working in a large, impersonal factory, Theodore Hershberg notes in his study of Philadelphia industry in this period that over the long run, "the best work settings in terms of career opportunities, security, working conditions, and wages were large factories, not small shops" (*Philadelphia,* p. 41).

60. OUT 8/47: George Burnham to M. W. Baldwin (Cape Island, N.J.), Aug. 14, 1847, BLW-HSP. The time varied depending on the general volume of work in the plant. The figure of two months starts with the date raw materials were ordered.

61. Ratio determined from data on Baldwin in the manuscript schedules of the 1850 U.S. Manufacturing Census; see appendix B.

62. IN 5/50: M. W. Baldwin (Cape Island, N.J.) to George Burnham, Aug. 10, 1850, BLW-HSP. At this time Baldwin had to borrow money from a foreman, Matthew Baird, to meet the payroll.

63. In 1850 Baldwin required 10.8 man-years to build a single locomotive; by 1854 that figure had dropped to 8.06.

64. OUT 12/53: M. W. Baldwin & Co. to J. G. Rankin, Mar. 24, 1854, BLW-HSP.

65. BLW Ledger, 1849–52, p. 9, and 1854–58, p. 67, BLW-HSP.

66. OUT 2/58: M. W. Baldwin & Co. to John C. Cresson (president, Mine Hill and Schuylkill Haven Railroad), Mar. 4, 1858, and George Burnham to Matthew Baird (Charleston, S.C.), Mar. 31, 1858, BLW-HSP.

67. George Burnham wrote about this change in pay in March 1858, noting that "it enables us to keep more hands employed." See letter to Cresson cited in previous note.

68. Fincher, "Early History," pp. 521, 564. Barry had worked off and on for Baldwin since at least 1848, and he signed the workers' petition of 1848.

69. IN 1/59: Matthew Baird (Charleston, S.C.) to Charles T. Parry, Mar. 2, 1859, BLW-HSP.

70. The petition is in *Annual Report of the Secretary of Internal Affairs of the Commonwealth of Pennsylvania* (1882), pt. 3, *Industrial Statistics,* 9:278–79 (hereafter volumes in this series are cited as *Pennsylvania Industrial Statistics Report*). Overtime was compulsory (when needed by the company), which the union objected to as abridging the ten-hour day. The union also noted that foremen used the dual pay system (time and piece rates) to play favorites among the men. Men the foremen liked would get piecework (i.e., higher pay) while others had to work for time wages.

71. OUT 2/60: M. W. Baldwin & Co. to Thomas A. Scott (Pennsylvania Railroad, Altoona), Mar. 7, 1860, BLW-HSP.

72. George F. Johnson to Samuel M. Vauclain, June 27, 1911, M. Baird & Co. Scrapbook, p. 173, Division of Transportation, NMAH.

73. Curtis, "Organized Few," pp. 72–73.

74. OUT 2/60: M. W. Baldwin & Co. to Michael Kelly, Mar. 10, 1860, and to John Clayton, Mar. 27, 1860, BLW-HSP.

75. OUT 2/60: M. W. Baldwin & Co. to Dearborn Robinson & Co. (Boston), Mar. 5, 1860, and to R. C. Barkley (Charleston, S.C.), Mar. 1, 1860, BLW-HSP.

76. George F. Johnson to Samuel M. Vauclain, June 27, 1911, M. Baird & Co. Scrapbook, p. 173, Division of Transportation, NMAH.

77. Writing of the 1857 layoffs at Baldwin—although the firm is unnamed—Fincher notes that a key issue in the MBIU's formation was that 150 journeymen were laid off while 60 apprentices retained their jobs ("Early History," p. 521). Historians such as Julia Curtis ("Organized Few") and David Montgomery have interpreted this as meaning that the MBIU was established "primarily to stop the flooding of workshops by so-called apprentices" (Montgomery, *Fall of the House of Labor,* p. 185).

78. Some examples of Baldwin's dependence on craft skills (contra the viewpoint described in the previous note): The company rigorously maintained its traditional apprenticeship program, keeping the ratio of apprentices to the total force at roughly 9 percent, a comparatively low figure. Baldwin literally could not lay off its apprentices in 1857, having signed traditional indentures with the boys. When it discovered early in the 1860 strike that it was employing a runaway apprentice from another firm, the

boy was fired instantly, despite Baldwin's pressing need for labor. Finally, in recruiting replacements Baldwin wanted only those "who have served a regular apprenticeship to the trade," men who "command the highest wages." OUT 2/60: M. W. Baldwin & Co. to R. J. Hollingsworth, Mar. 13, 1860; to J. Kinsley Smedley, Mar. 12, 1860; and to John Stuart, Mar. 20, 1860, BLW-HSP.

79. Baldwin's workforce averaged 675 men in 1860, and it made eighty-three engines. The Rogers Locomotive Works of Paterson, N.J., had 720 on the payroll and built ninety locomotives.

80. OUT 7/61: "Memorandum of Notes and Accounts in the Hands of R. C. Barkley for Collection," July 20, 1861 (see also separate accounts listed on following page), BLW-HSP. As a point of reference, Baldwin's sales in 1860 totaled $750,000, so the loss of these Southern accounts was crippling.

81. Baldwin had a secret arrangement with the Pennsylvania to build engines for $250 less than any other builder's bid, terms granted "in consideration of the [PRR's] larger orders and prompt payments." OUT 9/63: M. W. Baldwin & Co. to J. Edgar Thomson, Oct. 10, 1863, and OUT 11/62: same to same, Jan. 13, 1863, BLW-HSP.

82. See terms in OUT 8/62, BLW-HSP.

83. Quotation from OUT 11/63: M. W. Baldwin & Co. to J. W. Jervis (Pittsburgh), Nov. 16, 1863, BLW-HSP. Baldwin's materials costs advanced 47 percent between November 9, 1863, and May 1, 1864, while wages went up 23 percent. OUT 3/64: M. W. Baldwin & Co. to Col. D. C. McCallum (general superintendent, U.S. Military Railroads, Nashville), May 6, 1864, BLW-HSP.

84. See Porter and Livesay, *Merchants and Manufacturers,* pp. 125–29, for a general discussion of these changes throughout the wartime Northern industrial economy.

85. E. Digby Baltzell, *Philadelphia Gentlemen: The Making of a National Upper Class* (Glencoe, Ill.: Free Press, 1958), p. 108.

86. "City Bulletin," Philadelphia *Daily Evening Bulletin,* Sept. 12, 1866, p. 5.

87. As George Burnham wrote to a customer, "The business will suffer no interruption as its continuance was provided for in the partnership agreement." OUT 9/66: M. W. Baldwin & Co. to J. H. Flynn (Western and Atlantic Railway, Atlanta), Sept. 9, 1866, BLW-HSP.

2: THE LOCOMOTIVE INDUSTRY, 1860–1901

1. Nelson, *Managers and Workers,* p. 6.

2. White, "Once the Greatest of Builders: Norris," pp. 17–56; Moshein and Rothfus, "Rogers Locomotives," pp. 78–79.

3. Taylor, *Transportation Revolution,* p. 74.

4. Fishlow, "Productivity and Technological Change," p. 585.

5. Lewis, *Iron and Steel in America,* p. 38. Robert Fogel dethroned the railroads as leading agents of nineteenth-century economic development. Albert Fishlow raised the carriers up again, while Alfred Chandler ascribes to them the central role in developing the structure of modern business enterprise.

6. White, *Short History,* p. 5.

7. Ibid., p. 3 (import figures); White, *American Locomotives,* p. 12 (tariffs).

8. Carriers in the 1840s and 1850s had no reason to build their own power when they could purchase locomotives on credit, enabling the engines to pay for themselves, as it were. Only one U.S. line, the Pennsylvania, built large numbers of its own engines, but even it frequently turned to Baldwin. Although many other American railroads built their own engines occasionally, their output was relatively small (White, *Short History,*

p. 15). Most carriers that made their own locomotives did so only to give their repair shops some useful work at times when they had excess capacity beyond their normal maintenance tasks.

9. By 1888 a railroad shop that built its own power needed to produce at least one hundred engines annually to achieve per-unit economies comparable to those secured in contract shops. "The Cost of Rebuilding Locomotives," *Railroad Gazette* 20 (Oct. 19, 1888): 689.

10. White, *American Locomotives,* p. 24. According to an 1853 estimate (p. 16), a locomotive shop capable of building thirty-six engines a year required a fixed capital base of $42,000. Many general machinery builders already had the needed tooling and shop space, making locomotive building a relatively easy field to enter. Surviving and succeeding were more difficult.

11. Nineteenth-century locomotive firms were often renamed as their ownership changed; their most common names are used here.

12. Hinkley had some good production years in the 1870s and 1880s, but because the firm shut down temporarily in 1870, 1880, 1885, and 1886, with its liquidation following in 1889, it was clearly waning for most of the period. The lack of a managerial succession probably accounts for Norris' closure ca. 1866.

13. White, *Short History,* p. 21. In 1901 many of Baldwin's smaller competitors were merged into the American Locomotive Company.

14. See output data for various industries in Burns, *Production Trends.*

15. "Giants at Eddystone," *Fortune*, p. 58.

16. As a 1913 article noted, "The locomotive industry . . . is as sensitive to the gyrations of the business world as the seismograph is to the tremors of an earthquake" ("Sensitive Business of Locomotive Making," *New York Times Annalist,* p. 10).

17. IN 4/51: Henry R. Campbell (Vermont Central) to M. W. Baldwin, June 10, 1851, BLW-HSP.

18. "Feasts and Famines in Car Shops," *Railroad Gazette* 38 (May 12, 1905): 449–50.

19. White, *Short History,* pp. 49, 53, 83, 103. Some of these companies were reorganized and resumed production.

20. Seven major locomotive builders had incorporated by 1870, a step Baldwin did not undertake until 1909. The scanty evidence suggests that these incorporated firms resembled partnerships in that their stock remained closely held and they saw little separation of ownership from management. The threat of bankruptcy probably provided the chief argument for corporate status. By incorporating, these builders emulated the example set by their customers—the major textile manufacturers and the railroads. For the subject of industrial incorporation, see Navin and Sears, "Rise of a Market," pp. 112–16.

21. Such policies are described in Lamoreaux, *Great Merger Movement,* p. 188.

22. *Railroad Gazette* 5 (Oct. 4, 1873): 406 and 6 (Oct. 24, 1874): 413, 177.

23. Philip Scranton has argued that most batch producers faced such "visible market power relations, in which influence over prices was situational, shifting erratically from seller to buyer and back" ("Diversity," p. 37).

24. Product quality and postsale technical support also played a role in competition among builders at all points in the business cycle. Generally it is unclear whether one builder held a particular advantage in these areas, but Baldwin probably garnered sales growth in the 1860s from its leadership in making engines with interchangeable parts.

25. Four major new firms entered the locomotive industry during the 1860s: Dickson (1862), Pittsburgh (1865), Rhode Island (1866), and Brooks (1869).

26. Baldwin's output fell from 437 engines in 1873 to 205 a year later. In contrast, the American System manufacturer Singer Sewing Machine achieved annual production increases throughout the depressed 1870s. Singer data from Hounshell, *American System to Mass Production,* p. 89.

27. "Locomotive Building," *American Machinist,* Dec. 31, 1881, p. 2.

28. "Locomotive Manufacturing in This Country," *Railroad Gazette* 13 (Mar. 11, 1881): 146. The carrier wrote to fifteen American builders. None could deliver the order before December 1882, and some named June 1885 for delivery.

29. "Locomotive Building," *Railroad Gazette* 13 (Feb. 18, 1881): 103. This source says that "beyond those at Chicago and St. Louis already noted, new locomotive works are projected at Indianapolis, Detroit, Cleveland, Wilmington, and Chattanooga."

30. Burns, *Production Trends,* pp. 296–97.

31. Scranton notes that the machinery builders "could hardly expect to manage demand," and the locomotive industry certainly bears this out. He goes on to argue, however, that such firms "aggressively attempt[ed] to influence its timing and extent through marketing and financial tactics" ("Diversity," p. 49). Evidence from Baldwin suggests that this is an overly sanguine view, implying that builders had more power to influence their markets than they actually possessed. Scranton cites the case of Baldwin sending sixteen engines to the 1893 Columbian Exposition. This came to naught as a marketing effort; Baldwin's output soon fell by 59 percent, comparing 1893 with 1894. But the evolving structure of markets did offer some latitude and promise to efforts in marketing.

32. As a point of reference, in 1880 a typical mainline freight locomotive sold for $10,500, while a skilled metalworker was paid $760 a year. A modern diesel locomotive can cost over $1.5 million.

33. John Holusha, "Locomotives in High Gear Again," *New York Times,* Apr. 5, 1989, p. D1.

34. Rogers, Taunton, Mason, and others continued to make non-locomotive products during and after the 1860s.

35. Laurie and Schmitz, "Manufacture and Productivity," p. 44.

36. Boston's early prominence in locomotive building—with three firms in the 1850s: Hinkley, Souther, and Wilmarth—derived from the 1840s railway boom in New England. When that passed, these builders became increasingly marginal. The Paterson firms of Rogers, Cooke, and Grant were long-lived, but the city itself was one-tenth the size of Philadelphia and lacked its strength as a financial and railroading center. Pittsburgh had a number of evident advantages for locomotive building: nearby steel and coal, a large population of skilled metalworkers, and comparatively good access to Chicago, America's railway center. Nonetheless, the city possessed only one major engine builder, the Pittsburgh Locomotive Works.

37. For a description of Philadelphia's wealth of specialty producers, see Scranton and Licht, *Work Sights.* Philip Scranton notes how batch output firms benefited by locating in such industrial districts "with densely concentrated specialist firms, each of which . . . intersect with others in fairly complex productive sequences" ("Diversity," p. 35). Conversely, the proprietor of the Manchester Locomotive Works of New Hampshire noted in the mid-1880s that his firm had "to be satisfied with less profit" compared with builders nearer the iron regions. See testimony of Aretas Blood (proprietor, Manchester Locomotive Works) in *Report of the Committee of the Senate upon the Relations between Labor and Capital,* 3:166.

38. See Chandler, "Anthracite Coal."

39. Quotation from John Clayton (Baldwin's son-in-law) to Mrs. John Clayton (Cape May, N.J.), Aug. 18, 1850, folder 10, box 1, John Clayton Papers, HSP. For the Baldwin-Thomson relationship, see Ward, *J. Edgar Thomson,* pp. 29, 162.

40. In 1869 the PRR ordered seventy engines from Baldwin, almost one-third of the builder's output that year. In the late 1860s the carrier began building its own power at its Altoona shops, but the PRR had such large needs that it remained Baldwin's most important customer into the 1950s. After Thomson and Baldwin died, many other figures contributed to the BLW-PRR alliance, including four Baldwin partners—Edward Williams, John Converse, William Morrow, and Samuel Vauclain—who had previously worked for the carrier.

41. Given its private ownership, the firm could not, until its 1909 incorporation, raise investment capital through bond or equity offerings. Nor could it secure long-term bank loans to finance additions to fixed capital, since nineteenth-century banks generally declined such transactions. Throughout its history Baldwin frequently met working-capital needs by borrowing short-term bank funds.

42. Because Baldwin's post-1866 business records were largely destroyed, it is impossible to document its relations to bankers fully. An agreement with Drexel & Co. from 1877 does survive, however, which indicates the importance of such ties to financial intermediaries. In that year Baldwin sold fourteen engines worth $208,000 to the Dom Pedro Segundo Railway of Brazil, with payment terms of cash on delivery. Rather than await payment until after the locomotives' lengthy sea voyage to Brazil, Baldwin negotiated an immediate payment of $187,200 from Drexel, with the banker taking a fee for the service and the engines' bills of lading as collateral. See BLW Contract Book, 1: 306–10, BLW Papers, Pennsylvania State Archives.

43. For the bankers' role in railway promotion and finance, see Carosso, *The Morgans,* chaps. 7 and 10.

44. Oberholtzer's *Jay Cooke,* 2:131, describes how Baldwin partner George Burnham accompanied Cooke and others on an 1869 inspection tour over the first stages of the Northern Pacific. In the year to December 1871 the Northern Pacific ordered ninety-four engines from Baldwin, and such large orders continued for decades (BLW Order Books, BLW-NMAH).

45. Carosso, *The Morgans,* pp. 241 and 245; BLW Order Book for 1877, p. 67, BLW-NMAH.

46. Holton, *Reading Railroad,* 1:240–44. This addition of the Reading's business dates after the period of Baldwin's growth to dominance yet is discussed here to show how such informal alliances with investment bankers greatly aided the builder's business. Geographical proximity also cemented the Baldwin-Reading alliance.

47. Ward, *J. Edgar Thomson,* pp. 161–62. Ward notes that this informal group shifted in membership but possessed assets sufficient to make them "equal to other major regional investment groups: the Joy and Forbes combinations in Boston and the Vanderbilt and Gould cabals in New York" (p. 162).

48. Ibid., pp. 192–94. Ward describes further investments of Thomson et al. in the Northern Pacific, the Texas and Pacific, and the Lake Superior and Mississippi— all carriers with financially troubled histories, particularly after the 1873 depression. Nonetheless, the Baldwin company benefited from its ties to these carriers—in 1873 Baird sat on the boards of the Kansas Pacific and the Texas and Pacific, while another partner, George Burnham, was a director of the LS&M—selling engines to them all. In the early

1870s the Philadelphia interests also financed the construction of the Denver and Rio Grande, cementing yet another important locomotive customer for Baldwin (Grodinsky, *Transcontinental Railway Strategy*, p. 89).

49. Holley knew the Baldwin partners from his earlier work editing the *Railroad Advocate*. In the mid-1860s he began to construct an American Bessemer plant, casting his first high-quality ingot in April 1865. The ingot or billet was to make a tire (or replaceable rim) for a locomotive driving wheel. Holley marked it with Baldwin's name, and a delegation from Baldwin was there to observe this success—suggesting that the firm had underwritten some of Holley's developmental costs. See McHugh, *Alexander Holley*, pp. 190–93.

50. For the Pennsylvania Steel Company, see Morison, *Men, Machines, and Modern Times*, p. 163. Standard Steel was founded in 1870 as the William Butcher Steel Works of Lewistown, Pa., to make crucible steel—an older process than Bessemer's—for locomotive tires. When Butcher failed ca. 1873, Baldwin's partners took a controlling interest, reorganizing the company as the Standard Steel Works (Inc.) in 1875. See *Baldwin History*, p. 190. Baldwin's takeover of Standard amounted to vertical integration; however, its original motive was probably to acquire a reliably high-quality source of steel, more than to lower its costs. Ca. 1900 the two firms truly became integrated. Note that another company originally named the William Butcher Steel Works (of Philadelphia) became the Midvale Steel Company.

51. Beginning in the 1870s Baldwin had cost-accounting controls that improved its ability to bid accurately on new work, and by 1900 the firm had an estimating department to work up bids for potential sales.

52. Baldwin's Order Books noted contract terms; smaller carriers frequently purchased engines with a 25–50 percent payment on delivery, remitting the remainder within one to two years with interest charges. During the Gilded Age the builder even leased engines to bankrupt carriers such as the Reading.

53. OUT 9/63: M. W. Baldwin & Co. to Enoch Lewis (general superintendent, Pennsylvania Railroad, Altoona), Nov. 10, 1863, BLW-HSP.

54. M. Baird & Co., *Illustrated Catalogue of Locomotives* (1871). Baldwin's forty standard models compares favorably with Grant's twenty classes in that year; data from Grant Locomotive Works, *Description of Locomotives Manufactured by the Grant Locomotive Works*.

55. *Baldwin History*, pp. 67–68. To save on construction costs, during the 1870s many new lines, particularly in the West, were built to a narrower gauge than the standard 4 feet 8½ inches. In his superb history of these carriers George Hilton provides a table of U.S. locomotive builders' output for 3-foot-gauge lines, the most common narrow gauge. Baldwin's 45 percent share of this business was more than its overall share of U.S. locomotive output for the decade (32 percent) (*American Narrow Gauge Railroads*, p. 140). For Baldwin's share of the total market, see data in White, *Short History*, p. 21.

56. *Baldwin History*, p. 69.

57. Baldwin's production for industrial customers slowly rose from 7 percent of total output in 1870 to 10 percent in 1900. Data from BLW Order Books, BLW-NMAH. In comparison, Rogers, the industry leader in 1860, largely ignored the industrial engine market, making no engines for industrial customers in 1860, 1870, or 1880 (Moshein and Rothfus, "Rogers Locomotives").

58. M. Baird & Co. Scrapbook, Division of Transportation, NMAH.

59. Baldwin sent its own commission agents around the world to drum up business while also relying on commission merchants abroad. The former group had the training

required to advise potential customers on technical matters, while the merchants, tied to local commercial elites, could arrange the financial, freight, and rate-of-exchange aspects of a sale. This bifurcated arrangement of semi-independent salesmen served Baldwin into the 1920s, when the firm began replacing them with salaried sales personnel.

60. Baldwin, Vail, and Hufty Scrapbook, p. 231, Division of Transportation, NMAH.

61. Baldwin secured higher capital and labor productivity than British builders, which more than offset higher U.S. wages. Its efficiency derived from a larger plant, more intensive use of power tooling, and such American methods as the use of jigs and templates to make parts interchangeable. By the 1880s these factors allowed Baldwin to turn out a mainline engine with 6.25 man-years of labor, while the large Crewe Works of England's London and North Western Railway required 11.5 man-years (Crewe was not an export builder, however). Data from Richards, "Economic and Labor Factors," p. 104.

62. "American Locomotives on the Burma Railways," *Board of Trade Journal,* p. 449. British contract locomotive builders suffered from the fact that most English railways built their own locomotives; hence the contract shops were smaller than American builders. Cases of entrepreneurial failure also held Britain back. For example, most foreign engines were to custom designs that American firms like Baldwin were happy to build. But British builders often clung to their national design traditions in locomotives, telling "their customers exactly what they proposed to supply, rather than consult their wishes" (*American Engineering Competition,* p. 69). Finally, British export builders, such as Kerr Stuart, often had very primitive managerial systems compared with American firms; L. T. C. Rolt, *Landscape with Machines* (London: Longman, 1971), p. 103. For an econometric view of this question, see Floud, "Adolescence of American Engineering Competition," pp. 57–71.

63. "British and American Exports of Locomotives," *Railroad Gazette* 33 (Feb. 1, 1901): 79. Britain's decline stemmed partly from the great engineers' (i.e., machinists') strike in Britain late in the 1890s, while the American advance was aided by the severe domestic depression, which forced firms like Baldwin to look abroad for sales. When prosperity returned after 1900, U.S. locomotive builders were flooded with domestic orders, temporarily lessening their capability of, and interest in, meeting foreign needs.

64. Between 1875 and 1886 Baldwin accounted for 75 percent of all U.S. locomotive exports. Comparison of data from *Railroad Gazette* 19 (Mar. 25, 1887): 198 and from "Development of the American Locomotive," *Journal of the Franklin Institute,* p. 269.

65. Data from White, *Short History,* p. 21.

66. In 1880 Cambria Iron had 4,200 employees on the payroll, Lackawanna Iron and Coal had 3,000, and Carnegie's Edgar Thomson Works employed 1,500, compared with 2,648 at Baldwin.

67. While no data on Baldwin's actual profits in this period survive, the company was sufficiently profitable during the mid-1870s to pay off the ownership interest of a retiring partner, Matthew Baird.

68. Blood testimony in *Report upon Labor and Capital,* 3:163; *Railroad Gazette* 10 (Mar. 15, 1878): 137.

69. Schenectady was shut down in late 1877 (*American Machinist,* Dec. 1877, p. 9), as was Pittsburgh in the fall of 1879 (*American Machinist,* Feb. 22, 1880, p. 10). Baldwin's substantial market share after 1870 allowed it to avoid total suspension, even in the worst depressions, for the remainder of the nineteenth century.

70. *Railroad Gazette* 13 (Sept. 2, 1881): 480.

71. Since Baldwin, like most industrial companies of the period, primarily relied on profits for investment capital, one might argue that the company did not have to worry

about debt servicing. The Baldwin partners were quite careful, however, to ensure that capital improvements were warranted by the volume of business (see Vauclain and May, *Steaming Up,* p. 115). Decisions to invest their capital in new plant and equipment had to be made in relation to real market forecasts, rates, and returns—particularly since the partners also invested in other ventures.

72. For example, mass-producing blast furnaces mechanized throughout the Gilded Age, increasing capital per man by 79 percent between 1869 and 1899, while Baldwin's capital-labor ratio grew by only 16 percent (1870 vs. 1900, each case measured in current dollars). Data on steel manufacturers in Chandler, *Visible Hand,* p. 266.

73. *Railway Age,* Dec. 26, 1902, p. 717.

74. See esp. Chandler, *Visible Hand;* McCraw, *Prophets of Regulation;* and Lamoreaux, *Great Merger Movement.*

75. Lamoreaux, *Great Merger Movement,* p. 27. This behavior contrasted with that of mass-production manufacturers, which often pushed through recessions with high output, trying to generate enough cash to meet their high fixed charges.

76. Philip Scranton has described similar collusive policies among twentieth-century machine tool builders ("Diversity," p. 65).

77. "Meeting of Locomotive Builders," *American Engineer,* p. 12. Although Baldwin sent no representative to the meeting, this source indicates that some builders were represented "by letter."

78. A source from 1869 asserts that this meeting secured an agreement on prices that quickly fell apart. See "Locomotive Builders Meeting," *Van Nostrand's Eclectic Engineering Magazine,* p. 65.

79. IN 1/62: Rogers Locomotive and Machine Works (per M. K. Jesup) to M. W. Baldwin & Co., Jan. 17, 1862, BLW-HSP.

80. "Locomotive Builders Meeting," *Van Nostrand's Eclectic Engineering Magazine,* p. 65.

81. These are the only known members, but the LBA must have included other builders.

82. "Locomotive Builders' Association," *Railroad Gazette* 4 (Oct. 19 and 26, 1872): 451 and 465.

83. Commission merchants serving as market intermediaries long predated the Industrial Revolution. Glenn Porter and Harold Livesay have described the merchants' role in early industrialization, focusing in part on the railway equipment industry. Until the 1860s Matthias Baldwin used the commission houses for some raw materials purchasing, but he seems to have avoided these intermediaries when selling engines. As Porter and Livesay note, by the 1870s most industrial firms were bypassing the merchants, so it was relatively easy to make the LBA resolution stick (*Merchants and Manufacturers,* chaps. 5–7).

84. *Railroad Gazette* 21 (Nov. 1, 1889): 714.

85. List of members found in ALMA document dated June 18, 1897, folder 27, box 1, Pittsburgh Locomotive Works Collection, Archives Center, NMAH. For more on the ALMA, see M. Baird & Co. Scrapbook, p. 164, Division of Transportation, NMAH.

86. The document cited in the previous note gave Baldwin a quota of 30.65 percent; second-ranking Schenectady had 15.5 percent, with the rest ranging from 11.75 percent (Brooks) to zero (Rhode Island). Baldwin built 39 percent of total industry output for the decade 1890–99, suggesting that the firm may have paid the over-quota penalty on some domestic sales while also producing for the unregulated export market.

87. See lists in Burnham, Williams & Co. Scrapbook, p. 243, Division of Transportation, NMAH, and in folder 27, box 1, Pittsburgh Locomotive Works Collection, Archives Center, NMAH.

88. The ALMA firms countered the temptation to undercut each other secretly by requiring all members to inform the association secretary of all bids proffered and accepted, using coded telegrams.

89. Surviving cost data for a 4-6-0 engine built by Baldwin in 1897 give the builder's cost as $9,105.65, while the sales price was $9,550—hardly an exorbitant markup, although the ALMA fixed price for this engine was $9,350. See note in S. M. Vauclain's Production Notebook for 1897, box 9, SMV-DEG.

90. "The Locomotive Building Interest," *Iron Age,* Apr. 1, 1880, p. 9.

91. Charles T. Parry to "Messrs. J. Brown and others," June 20, 1872, uncataloged, DEG.

92. The lack of surviving data on Baldwin's profits for the period 1867–1900 makes it more difficult to evaluate the company. There are ledgers for the years from 1839 to 1866, but the firm changed its accounting practices over the life of these books, hindering proper interpretation of the data. Also, while a balance was struck at year's end, a given year's profit often included payments made on past-due accounts from previous years. So the data must be used charily. In comparing profits from 1858 and 1866, readers are reminded of the considerable wartime inflation that buoys the 1866 total.

93. Profitability was determined by subtracting annual labor charges and 3 percent of the fixed capital figure (to account for depreciation) from the value-added total. That figure was divided by the capital total to determine rate of return. Despite the precision of the equation, the manufacturing census data are often so wildly inaccurate that the resultant numbers are hardly worth the trouble to determine them. But this is the only source available, so I include the general findings in the text with this strong cautionary note attached.

94. Fishlow, "Productivity and Technological Change," p. 604.

95. In 1880 the national motive power fleet numbered 17,900 engines; twenty years later, 37,600 locomotives were in service. Data from *Historical Statistics,* pt. 2, pp. 729, 731.

96. White, *Short History,* p. 21. Baldwin's ability to improve market share during the price fixing of the 1890s underscores how competition could run concurrently with collusion. Evidently Baldwin beat out its rivals with earlier delivery dates and export sales.

97. At its opening, Alco included Brooks, Cooke, Dickson, Manchester, Pittsburgh, Rhode Island, Richmond, and Schenectady, with Montreal and Rogers added in the next few years. Ibid., p. 23.

98. Victor Clark notes that "in the spring of 1892 the five principal firms manufacturing steam locomotives . . . took steps to form a combination, but the trust movement made no further headway . . . at that time" (*History of Manufactures,* 2:347). This implies that Baldwin took part in these negotiations, but it would later decline to join Alco. I would speculate, however, that Baldwin did not participate in the 1892 discussions and, lacking the industry leader, the merger was put off.

99. Measured in constant dollars, Baldwin's capital-labor ratio grew from $1,074 in 1880 to $1,518 in 1888. Note that Naomi Lamoreaux's explanation for the horizontal-combination movement by manufacturers in the 1890s mirrors the locomotive builders' experience, namely that rising capital costs coupled with narrowing profit margins impelled collusive attempts and then outright consolidations. Additionally Lamoreaux

underscores the role played by dynamic technical change in fostering price competition. This also occurred in locomotives of the 1880s and 1890s (*Great Merger Movement,* pp. 187–89).

100. At the time, Hoadley had already consolidated three stationary steam engine firms, including the well-known Corliss works. See Phenis, *Yankee Thrift,* pp. 78–79.

101. For Hoadley, see ibid., p. 80. For Fiske, see *National Cyclopedia of American Biography,* 14:524–25. For Leiter, see *Railroad Gazette* 33 (May 10, 1901): 320, and his entry in *DAB,* vol. 11, pt. 1, suppl. 1, pp. 491–92. At the height of Leiter's wheat corner in the spring of 1898, he owned 18 million bushels outright with 22 million more in futures, but forecasts of a record-breaking crop that summer broke the corner, and his losses totaled as much as $10 million.

102. From 1900 to 1909 Alco (and its predecessors) accounted for 46 percent of U.S. locomotive production (vs. 39 percent for Baldwin). In the next decade Alco took 37 percent to Baldwin's 46 percent (data from White, *Short History,* p. 21). By 1920 Alco had shut down many of its plants, consolidating operations at Montreal and Schenectady.

103. Testimony of John H. Converse in *Report of the Industrial Commission,* 14:234.

104. See Chandler, *Visible Hand,* pp. 337–39, and Lamoreaux, *Great Merger Movement,* pp. 188–90, for discussions of the determinants of success or failure in mergers.

105. In the 1920s Baldwin and Alco faced a new competitor in the Lima Locomotive Works of Lima, Ohio. Its rise from small logging and industrial engines during the nineteenth century to mainline engines in successful competition with the two majors may have been aided by Baldwin and Alco's renewed collusion ca. World War I.

106. In 1908 the combined output of U.S. and Canadian locomotive builders fell by 68 percent compared with 1907. Data from *Railroad Gazette* 43 (Dec. 27, 1907): 767 and from *Railway Age Gazette* 45 (Dec. 25, 1908): 1617. Note that Alco suffered a loss of $760,000 in its first fiscal year after the panic (1908–9). No comparable data are available for Baldwin, since it was a partnership, but Alco's losses lend indirect support to the argument that one great locomotive combine would have run into difficulties during recessions.

107. This competition in an oligarchic industry likely arose from Alco's need to service its $50 million capitalization.

108. Bruce, *Steam Locomotive in America,* p. 73. Orders for new engines peaked in 1905 at sixty-three hundred units. Because of this high demand, output lagged a year or so behind orders. Then the Panic of 1907 obscured the fundamental reversal in domestic demand for engines.

109. The machine tool and shipbuilding industries fell into bad slumps in the interwar years.

3: THE CHARACTER OF INNOVATION IN LOCOMOTIVE DESIGN

1. Chandler, *Visible Hand,* pp. 282, 143. For the subject of standardization in nineteenth-century railroad technology, see also Usselman, "Running the Machine."

2. The Master Car Builders Association discussed rolling-stock issues.

3. *Seventh Annual Report of the American Railway Master Mechanics Association* (1874), p. 173; hereafter volumes in this series are cited in short form, viz. *Seventh Master Mechanics Report* (1874).

4. *Thirty-second Master Mechanics Report* (1899), pp. 35–36.

5. The carriers increased their volume of traffic partly by running more trains, but two problems arose with this strategy. First, it required more inputs—such as engines, cars, and train crews—which limited the productivity improvements. The railways also

found that additional trains quickly clogged their lines. They needed better trains, which required better locomotives.

6. Historians of technology have traditionally noted the benefits of technological standards, pointing for instance to American railways' adoption of a standard track gauge, which created a national railway system. Such standards were beneficial in a number of respects, but they also proved constraining. Indeed, one can see much of the railways' 160-year technical history as an effort to lift or sidestep the productivity constraint imposed by the standard Stephenson track gauge. While allowing national shipments, it forever fixed one crucial dimension of railway operations. As a nineteenth-century authority on engineering standards, Charles Fitch, noted, "The uniform system . . . stands in the way of its own advancement" ("Report on the Manufactures of Interchangeable Mechanism," in *Tenth Census of the United States*, 2:663; hereafter cited as "Fitch Report").

7. Johnson, "Problem of Motive Power," p. 6.

8. Historians interested in the social construction of technology are well advised to look at such design negotiations in capital goods, since the mediation between buyer and seller often leaves a more explicit record of divergent motivations than is the case when design and production issues are reconciled in house, as is generally the case with manufactured products. For the social derivation of technology, see Bijker et al., *Social Construction of Technological Systems*.

9. *Baldwin History*, p. 27. Locomotive types are defined by their wheel arrangements under the Whyte system. Thus a 4-4-0 has a four-wheel leading truck, four drivers, and no trailing wheels.

10. White, *American Locomotives*, pp. 152–53. The additional pair of drivers directly provided the improved power and traction, but it was unfeasible without equalizers, which assured that each driver would ride over rough track without derailing or losing traction.

11. Ibid., p. 153. While Campbell's general plan for the 4-4-0 was patented, he evidently had less success than Harrison in collecting royalties or preventing unauthorized use of the design.

12. In the late 1830s Baldwin refused to sell a right for use of his patented half crank, and he most likely believed that his 1842 patent on the flexible beam was too important to allow its use by other U.S. builders. For the former, see Ferguson, *Early Engineering Reminiscences*, p. 188. For the flexible beam, see White, *American Locomotives*, p 398

13. By 1856 Winans had so alienated the other builders by his high-handed assertion of dubious claims and threats of legal actions that seven builders, led by railroading expert Zerah Colburn, joined to oppose one of his infringement suits. See IN 1/56: Zerah Colburn to M. W. Baldwin & Co., Feb. 2, 1856, and IN 10/57: Danforth, Cooke & Co. to M. W. Baldwin & Co., Nov. 13, 1857, BLW-HSP.

14. IN 9/49: J. Spaulding to M. W. Baldwin & Co., Oct. 25, 1849, BLW-HSP. This collection contains numerous examples of such letters throughout the 1840s.

15. Licht, *Working for the Railroad*, pp. 40–41.

16. Baldwin, Vail, and Hufty, *Locomotive Engine Catalog* (1840).

17. OUT 3/44: M. W. Baldwin & Co. to C. F. Hagedorn (Bavarian consul to the U.S.), Apr. 13, 1844, BLW-HSP.

18. In 1841 Norris marketed four classes of 4-2-0s. See catalog of William Norris & Co. (1841) in Trade Catalogue Collection, HAG.

19. OUT 10/45: Baldwin and Whitney to John B. Jervis, Oct. 22, 1845, BLW-HSP.

20. Steven Usselman notes that railway managers had one overriding goal when considering technical innovations: "to concentrate on the single objective of meeting

fixed costs by generating traffic and moving it through the system as cheaply as possible" ("Running the Machine," pp. vii–viii). The railways' desire for ever more powerful engines to carry an increasing traffic burden accords with this analysis.

21. IN 1/45: J. Edgar Thomson to M. W. Baldwin, Apr. 10, 1845, BLW-HSP.

22. IN 1/45: W. H. Clement (Cincinnati) to Baldwin and Whitney, Feb. 11, 1845, BLW-HSP.

23. OUT 3/44: Baldwin and Whitney to James Murray, Apr. 10, 1844, BLW-HSP.

24. Entire specification reprinted in White, *American Locomotives,* pp. 461–62.

25. Numerous letters between Baldwin and Latrobe discussing the first specification are in OUT 8/47 and IN 10/47. For the second specification, see the same authors in OUT 1/48 and IN 1/48, BLW-HSP. Ultimately Baldwin built only three of these engines.

26. OUT 1/48: M. W. Baldwin to Henry R. Campbell, Mar. 30, 1848, BLW-HSP.

27. The costs of drawings and patterns for a single new engine design are given in "Petition of M. W. Baldwin for Extension of Patent," June 16, 1856, Patent Extension Files, National Archives.

28. OUT 1/48: M. W. Baldwin to J. Edgar Thomson (Pennsylvania Railroad, Harrisburg), Mar. 15, 1848, and to Governor Paine (president, Vermont Central), May 9, 1848, BLW-HSP. There is no indication that either road bought the design.

29. OUT 1/48: M. W. Baldwin to Governor Paine, May 9, 1848, BLW-HSP; *Baldwin History,* pp. 45–46. This mile-a-minute design appears on page 15 as class A, no. 1.

30. OUT 6/48: M. W. Baldwin to Henry R. Campbell, July 25, 1848, BLW-HSP.

31. "Western Locomotives," *Colburn's Railroad Advocate,* Jan. 12, 1856, p. 3.

32. *Baldwin History,* p. 41. Baldwin purchased patent rights needed for this design from Harrison and Campbell.

33. Comparison of data in White, *American Locomotives,* p. 76, vs. Burnham, Williams & Co., *Catalogue of Locomotives* (1908), p. 44.

34. See Bruce, *Steam Locomotive in America,* for the subject of locomotive types.

35. Forcing was achieved by sending exhaust steam from the cylinders up the smokestack, creating a forced draft that sucked air through the fire. The added air promoted combustion, which in turn created more steam. Such forced draft was rare outside of locomotive practice. One way to measure boiler forcing is the ratio of heating surface (the steam-producing areas of a boiler) to grate surface (the heat-producing area). The normal nonrail ratio was about 35 to 1, while locomotive boilers reached a 60-to-1 ratio or more. See Graham, *Audels Engineers and Mechanics Guide,* 5:1976.

36. White, *American Locomotives,* p. 105.

37. Some readers may object here, noting that certain railroads made great efforts to create locomotive fleets composed of standard designs. This was certainly true and is discussed below. But even on these roads standard designs were quickly perceived as a bar to future development.

38. Clearances were ruling limits over the road like the load-bearing capacity of bridges and the dimensional limits at stations, roundhouses, and tunnels beyond which engines and cars could not pass. Weights of track often varied between main and branch lines, with the latter's lightweight rails incapable of carrying large road engines safely.

39. BLW, "Wheel Spacing and Loads of Heavy Passenger Locomotives," tracing #15735, BLW-STAN. Weights are engine only (without tender) in 2,000-pound tons.

40. In such situations railroads often resorted to "doubleheading," or coupling two engines (or more) to a train, a common practice today with diesel locomotives. But unlike diesels, doubleheaded steam engines needed an engineer and fireman in each locomotive.

This added labor expense made the practice undesirable and promoted the development of more powerful engines.

41. White, *American Locomotives,* pp. 427–29.

42. By this term I mean to connote a pride in technical creation, no different from the creative drive of an artist. Technical virtuosity could have positive results, as in the creation of new locomotive designs capable of burning coal — an accomplishment of such master mechanics as John Wootten — or it could result in less beneficial changes, merely reflecting the power to tinker. Arnold Pacey discusses technical virtuosity in *Culture of Technology.*

43. As Usselman notes of the master mechanics, "Their position as the unchallenged experts on most technical affairs enabled them to shape the course of innovation more than anyone else at the railroad" ("Running the Machine," p. 180).

44. For a fictional but highly interesting account of a master mechanic's work, see Hill, *Jim Skeevers' Object Lessons,* esp. pp. 115–23.

45. *Second Master Mechanics Report* (1869), p. 38.

46. Among the recommendations and standards the association urged master mechanics to adopt were the Franklin Institute or U.S. Standard screw threads, standard sizes for driving-wheel centers and tires, a standard cross-section for tires, standard axle dimensions, boiler steel specifications, and pipe thread standards. See *Thirty-third Master Mechanics Report* (1900), pp. 95–107 and 371–81.

47. Alexander L. Holley, "Rail Patterns," *Railroad Gazette* 13 (Mar. 11, 1881): 139–40. Holley had bitter scorn for this practice, noting that "some of these tinkers . . . fondly believe that having 'Stiggins' rail pattern' talked over in the mills and railroad offices gives them a certain immortality. It does." An even more telling example is the case of standard couplers. A common design for couplers was essential if freight cars were to interchange from carrier to carrier. In 1888 the Master Car Builders established such a standard with the Janney coupler. But the MCB would only mandate an acceptable profile for the coupler jaws; the parts themselves did not have to be standard. By 1899 nearly eighty different coupler designs were in interchange service across the nation, some designed by railway officers. The officers responsible for buying this bewildering variety created a maintenance headache, since carriers had to stock replacement parts for dozens of different brands to fix cars in interchange service. As John White notes, "The adoption of the Janney coupler . . . did not usher in an age of standardization," since "each make was unique insofar as individual parts were concerned" (*American Railroad Passenger Car,* p. 570).

48. Reading between the lines of Joseph Bryan's remarks, quoted at this chapter's opening, one can see an argument against the mere tinkering that often resulted from the master mechanics' desire to demonstrate their technical virtuosity.

49. "Building Locomotives," *Machinery,* p. 206.

50. This latitude in producing blueprint jobs is evident in an 1882 report by a Pennsylvania Railroad inspector overseeing engines under construction at Baldwin. See S. M. Vauclain, "Baldwin Locomotive Works — Comments on the System and Shop Practices," box 1, SMV-DEG.

51. "The Best Type of Locomotive," *Railroad Gazette* 30 (Jan. 21, 1898): 40.

52. Burnham, Williams & Co. to Theodore Voorhees (first vice president, Philadelphia and Reading), Jan. 2, 1900, file: Locomotives, Locomotive Purchases for 1900, box 978, Reading Company Collection, HAG.

53. Bell, *Early Motive Power,* pp. 88–95, 124.

54. Colvin, *Sixty Years,* pp. 100–101. This incident probably occurred before 1900.

55. Klein, *Union Pacific,* p. 498, and "Statement of Engines," file: Locomotives—Inventories and Statements (1889–99), box 977, Reading Company Collection, HAG.

56. Warner, *Motive Power Development,* chap. 2.

57. The Associated Lines controlled by E. H. Harriman ca. 1900 included the Southern Pacific, Union Pacific, and three smaller carriers. One of the Harriman-standard engines appears on page 201.

58. Details drawn from Warner, *Motive Power Development.* Another source notes that in 1898 the Pennsylvania had six general classes of engines, but within a class there were up to thirty different designs. This variety came despite thirty years of effort to standardize its locomotives. See "Interchangeability in Locomotive Parts," *Railroad Gazette* 30 (June 10, 1898): 412.

59. *Historical Statistics,* pt. 2, p. 728.

60. BLW, "Wheel Spacing and Loads of Heavy Freight Locomotives," tracing #15734, BLW-STAN.

61. Data for 1870 from White, *American Locomotives,* p. 76. The 1915 figure is that listed for heavy Consolidations, Mikados (2-8-2s), and Pacifics in BLW, *Catalogue of Locomotives* (1915).

62. The impossibility of constructing a division between custom and standard engines does not arise from a lack of data. The problem is that the company so integrated its custom and standard work in the production process that it never had a need to differentiate the two in its records. The division between custom and standard engines became increasingly imperceptible after the 1880s, for reasons described below.

63. M. Baird & Co., *Illustrated Catalogue of Locomotives* (1871), p. 47.

64. Warner, *Motive Power Development,* p. 31, and Burnham, Parry, Williams & Co., *Illustrated Catalogue of Locomotives* (1881), p. 45.

65. Warner, *Motive Power Development,* p. 31; William Austin, "Trip to Reading, Nov. 13, 1879," WLA(B)-HAG; Burnham, Parry, Williams & Co., *Illustrated Catalogue of Locomotives* (1881), pp. 46–47.

66. In 1891 Baldwin reported that it had 136 different complete designs on file for Consolidations alone. See Burnham, Williams & Co. to B. B. Adams (editor, *Railroad Gazette*), May 28, 1891, BLW Order Book, 9:158, BLW-NMAH.

67. "Heavy Consolidation Locomotives—Philadelphia & Reading," *Railroad Gazette* 21 (Jan. 4, 1889): 2–3. The Northern Pacific models were adapted slightly to burn bituminous coal.

68. Comparison of models listed in OUT 8/46: M. W. Baldwin to Ferrier Wicksted, Oct. 22, 1846, BLW-HSP; and in Baldwin's 1871 and 1881 catalogs. The 1871 and 1881 totals lack a number of special service engines to standard designs that Baldwin marketed.

69. *Baldwin History,* pp. 73–75; Burnham, Parry, Williams & Co., Street Car Sales Brochure, folder: Test Department Specifications, box 6, WLA-HAG.

70. Burnham, Parry, Williams & Co., *Illustrated Catalogue of Locomotives* (1881), pp. 129–42. Included in this total are such models as Forney 0-4-2T and 0-4-4T types and double-ender tank engines. Not included are standard American and other mainline types also widely used in suburban service.

71. "American Locomotives for Foreign Railways," *Engineering Magazine,* p. 859.

72. BLW Order Book, 8:84, BLW-NMAH; William Austin, "Trip to Brooklyn and New York, Aug. 8, 1888," WLA(B)-HAG. U.S. practice for such an engine was 4-foot 8½-inch gauge, steel firebox and flues, Baldwin reversing gear, and no buffers.

73. BLW Order Book, 9:158, BLW-NMAH.

74. Comparison of BLW Order Book for 1860 vs. data from Order Book, 9:158, BLW-NMAH (other sources give Baldwin's 1860 output as eighty-three engines).

75. Data from Junior Law Book, 1:4, BLW-DEG, and *American Machinist* 22 (May 4, 1899): 373. Quotation from Arthur L. Church, "Memorandum with Respect to William Henszey, Mar. 15, 1906," folder: Correspondence, 1910–11, box 1, SMV-HSP.

76. Baldwin's president, Alba B. Johnson, wrote of his opposition to standard designs in "Problem of Motive Power," pp. 11–12. For a description of the USRA designs, see Huddleston, "Uncle Sam's Locomotives," pp. 30–38.

77. Bruce, *Steam Locomotive in America,* p. 85.

78. Ibid., p. 31.

79. *Twenty-fifth Master Mechanics Report* (1892), pp. 57, 68, 71.

80. "Compound Locomotives Built in the United States," *Railroad Gazette* 32 (Apr. 20, 1900): 257.

81. *Annual Reports of the Patent Office,* 1877–1900.

82. Test Department Specifications, Oct. 1, 1885, folder: BLW Test Department Specifications, 1885–1905, box 6, WLA-HAG; *Railroad Gazette* 23 (Aug. 14, 1891): 566; see also Custer, *No Royal Road,* chaps. 23–25.

83. Goss, "Tests of the Locomotive," pp. 826–54.

84. "An Experimental Locomotive to Be Installed," *Railway Master Mechanic* 14 (Feb. 1903): 56.

85. "The Use of Locomotive Laboratory Experiments," *Railroad Gazette* 30 (June 24, 1898): 458.

86. *Railway and Locomotive Engineering* 13 (Nov. 1900): 477–78.

87. See "Report of Test Department on African Coal from Cape Government Railways," Feb. 23, 1898, in BLW Ledger, "Railroads of South America," p. 188, BLW-STAN.

88. A correspondent to *Railway Age Gazette* noted in 1911 that "scientific designing intelligently applied to operating conditions . . . offers a means for greatly reducing the cost per ton-mile; a number of roads have already taken advantage of this possibility." Letter signed "Operation," *Railway Age Gazette* 51 (1911): 1165.

89. Baldwin's increasing influence over design can be measured indirectly by the growing number of draftsmen at the firm. In 1899 the company employed 125 draftsmen and made 901 engines. In 1913 output was up 129 percent, while the drafting force had increased by 260 percent, to 450 men. Force total for 1913 from "Baldwin Locomotive Works," *Railway Magazine,* p. 461.

90. BLW, *Record of Recent Construction #1,* p. 9.

91. Similar diversity in locomotive design occurred in nineteenth-century Britain. See Kirby, "Product Proliferation," pp. 287–305. User initiative in innovation has survived to the modern era in many capital goods, even though the equipment itself is far more standardized than were steam locomotives. The Douglas DC-3 airliner resulted from a set of performance specifications written by TWA that closely mirrored locomotive specifications. The chairman of Pan American Airways, Juan Trippe, pushed Boeing to develop its 747. And America's most modern locomotive, Amtrak's Genesis design, resulted from close collaboration between the customer and the builder, General Electric.

92. Note also that American railway accounting procedures "overstated operating costs and understated capital consumption" (Chandler, *Visible Hand,* p. 115). Such biases worked against standard locomotives and favored custom designs.

93. As I argued above, in American railroading of the 1880s the 119 rail patterns decried by Alexander Holley are as revealing of the actual conditions in that era as the adoption of one standard gauge.

4: MANAGEMENT AT BALDWIN, 1850–1909

1. Data from the 1850 U.S. Manufacturing Census, with class variety from BLW Register of Engines for 1850, BLW-NMAH. Data are for the fiscal year ending June 1, 1850.

2. Market shares computed from data in White, *Short History*, p. 21.

3. JoAnne Yates argues that "individuals, not systems, were still primary" until the science of management developed in the 1880s (*Control through Communication*, p. 3). See also Litterer, "Systematic Management," pp. 461–76; Nelson, *Managers and Workers*, chap. 3; and Chandler, *Visible Hand*, pp. 272–81. These and other authors overlook earlier examples of systematized managerial practice largely because they seek the origins of recognizably modern forms of management, particularly white-collar bureaucracies.

4. Chandler, *Visible Hand*, p. 272.

5. In describing Baldwin's early development of sophisticated controls, I also attempt where possible to describe the practices of other capital equipment firms, since I believe that companies throughout the sector had particular reasons to innovate in management. But I concede at the outset that one study of one firm will not of itself revise the entire literature on the systematic management movement.

6. "Center firms" is the appellation used by Thomas McCraw to describe those companies Alfred Chandler has made the focus of his distinguished career. McCraw precisely defines their attributes as including "major economies of scale . . . vertically integrated structures, with highly organized, functional divisions" combining "mass production with mass distribution" (*Prophets of Regulation*, p. 98). While Baldwin did secure economies of scale, they were hardly of the magnitude found in mechanized mass production, nor did the company achieve mass distribution.

7. The fact that Baldwin was not a center firm accords with Chandler's general analysis of their attributes; however, he mistakenly includes Baldwin in this category (*Visible Hand*, pp. 359, 511). I would argue that few if any capital equipment companies ever became center firms, since these builders were so directed by their markets in both product design and the scale of production. For these reasons Baldwin's management necessarily focused on issues of design and production rather than sales.

8. One might reasonably question the meaning of the center-firm concept itself, given that it does not apply to Baldwin: an enterprise with a 140-year history, ranking among the top U.S. industrial employers for seventy years, which garnered more than a 30 percent market share during the same period, and which used a range of novel managerial systems and sophisticated tooling and skills to make a highly complex product that lay at the core of industrial society. Certainly such a company should not be described as a peripheral firm—McCraw's alternative (*Prophets of Regulation*).

9. Lubar, "Managing the Industrial Revolution," pp. 2–3.

10. Philip Scranton has described how Philadelphia textile mill owners in the first half of the nineteenth century believed "that close management by [the] interested parties was the only way to run a mill" (*Proprietary Capitalism*, p. 107).

11. Such advancement is described in Roe, "Industrial Background of 1855," pp. 1446–47, and in Scranton, *Proprietary Capitalism*, pp. 338–47.

12. To my knowledge, no historian has looked closely at industrial management before the 1880s, and the producer-ethos concept is my own. Discussions supporting the prevalence of such beliefs among nineteenth-century owner-managers appear in

Rogers, *Work Ethic in Industrial America,* and in Ernst, "Yellow Dog Contract," pp. 251–74, esp. 258. While I only wish to advance the case of a producer ethos at Baldwin, Daniel Ernst has described a general mindset among nineteenth-century proprietary capitalists that illustrates much of what I mean by this term. As he notes, "These strong-minded businessmen identified themselves closely with their enterprises. . . . By following (and stimulating) the precepts of the work ethic, they had conserved a business patrimony or created one." The Baldwin partners fit this description quite well. For more on Ernst's views, see his "The Closed Shop, the Proprietary Capitalist, and the Law, 1897–1915," in *Masters to Managers,* ed. Jacoby, p. 135.

13. Frederick W. Taylor wrote in 1895 that "the clerk in the factory is the particular horror of the old-style manufacturer. He realizes the expense each time he looks at him, and fails to see any adequate return" ("Piece-Rate System," p. 862). Although clerks were not managers, they came under the category of management personnel—in the companies that would have them.

14. Roper, "Largest Locomotive Works in the World," p. 4. Roper's characterizations of bosses as "drones" and "lookers-on" are suggestive of the producer ethos.

15. Private shorthand notes of William Austin, June–July 1910, box 5, WLA-HAG.

16. Chandler, *Visible Hand,* p. 8.

17. The transformation ca. 1900 of large industrial companies from partnerships to corporations is described in Navin and Sears, "Rise of a Market," pp. 105–38. Note, however, that partnerships remain common today in large accounting, legal, and real estate development firms. Like Baldwin, such companies rely on this legal form to foster and reward individuals' service to the firm.

18. The fact that Baldwin ultimately gave up its partnership in favor of incorporation in 1909 accords with the generally acknowledged advantages of the corporate form: access to capital markets, unlimited life, and limited liability. Herein I simply wish to see nineteenth-century business practice on its own terms rather than as a primitive antecedent to modern developments.

19. Quoted in Smith, *Elements,* pp. 23–24. Baldwin became a publicly held company in 1911, and sixteen years later it became the target of a successful hostile takeover by the Fisher brothers of Detroit, who installed their own candidate as president shortly thereafter.

20. At Baldwin's major competitors, the four locomotive-building firms at Paterson, N.J., the proprietors also generally "started their careers as workers, apprenticed to learn a skill, and then opened small shops or factories of their own" (Gutman, *Work, Culture, and Society,* p. 221). The same was true of the proprietors of the Mason and Taunton companies, which also built locomotives (Lozier, "Taunton and Mason," pp. 197–201, 398).

21. Alba Johnson, quoted in Smith, *Elements,* p. 26. Since my research failed to uncover an actual partnership agreement, Johnson's extended remarks in this source are particularly helpful.

22. See wills of Edward H. Williams (1899 #2,061), William Henszey (1909 #718), and William C. Stroud (1891 #1,311), Office of the Registrar of Wills, Philadelphia City Hall.

23. Under Pennsylvania law, limited liability in partnerships extended only to limited partners (those contributing capital but inactive in daily management); general partners remained fully and personally liable for a firm's debts.

24. Carosso, *The Morgans,* p. 633.

25. R. G. Dun report on Burnham, Parry, Williams & Co., Jan. 29, 1874, Pa. vol. 136, p. 321C, HAG. Saul Engelbourg argues that standards of business morality improved

during the nineteenth century as common-law precedents, statutes, and administrative rulings mandated accepted rules of behavior. He notes, "Dishonesty lessens when the rules are clarified and enforced" ("Power and Morality: American Business Ethics," in *U.S. Economic and Business History,* ed. Blicksilver, p. 297). I do not argue against this analysis, but I would note the difference between legality and morality. In the absence of legal safeguards, ethical behavior or integrity became even more important as a safeguard to the participants in business transactions.

26. R. G. Dun report on M. W. Baldwin & Co., Sept. 1863, Pa. vol. 219, p. 332, HAG. Baird became Baldwin's boiler shop foreman in 1838 (White, *American Locomotives,* p. 449). He briefly left Baldwin's employment in 1850 to enter a marble monument company partnership with his brother, in which his share upon its dissolution in 1852 was $30,000. Evidently these funds served as his initial investment in the Baldwin partnership (R. G. Dun report on John Baird, Jan. 31, 1853, Pa. vol. 133, p. 121, HAG). Baird's percentage of ownership is given in BLW Ledger, 1854–58, pp. 67 and 77, BLW-HSP.

27. Five of these twelve were still partners ca. 1907 when one, Alba Johnson, wrote, "Not a single man who is at present a partner in the Baldwin Locomotive Works has brought a cent into that concern" (quoted in Smith, *Elements,* p. 24). Other evidence suggesting the lack of any capital contributions is noted below.

28. Carosso, *The Morgans,* p. 169.

29. The quotation, describing the Morgan firm, is from ibid.

30. Johnson, quoted in Smith, *Elements,* p. 26.

31. Williams came to Baldwin in 1870 from a career on the Pennsylvania, having risen to become its general superintendent (1866–70). Wilson, *Pennsylvania Railroad Company,* 1:191.

32. See Baird's entry in *DAB,* 1:511, and Williams's obituary in Philadelphia *Public Ledger,* Dec. 22, 1899, p. 2.

33. Protégés responded with lifelong gratitude: Austin named a daughter Mabel Henszey Austin, and Vauclain had a son called Parry.

34. Burnham's career is described in a full-page obituary in Philadelphia *Public Ledger,* Dec. 15, 1912, magazine section, p. 1. Pettit's career is summarized in *Baird v. Pettit.* Some readers may object to my including a draftsman in "management," which only underscores the difficulty of definitional boundaries in this age before professional managers. I consider draftsmen to be management personnel because their drawings directed shop floor workers. In general my categorization of workers and managers follows the division established by the company. The salaries of all employees not engaged in actual production were charged to overhead, and such personnel are defined as management.

35. Alba Johnson, quoted in "Fifty Years at Baldwin's," p. 1.

36. IN 5/50: M. W. Baldwin to George Burnham, Aug. 10, 1850, and OUT 7/57: M. W. Baldwin & Co. to Holley and Colburn (*American Engineer*), July 25, 1857, BLW-HSP.

37. Quotation from an English visitor to the factory ca. 1890 (*American Engineering Competition,* p. 75).

38. "Obituary of Charles T. Parry," *Twentieth Master Mechanics Report* (1887), pp. 200–207.

39. Ibid., p. 200.

40. The new draftsmen and Baldwin's displeasure are described in "Fifty Years at Baldwin's." Victor Clark notes that the 1860s saw a general rise of mechanical drawing in

industry as companies used such plans to increase their control over work and workers (*History of Manufactures*, 2:81).

41. While this is large according to orthodox business management historiography, in locomotive building alone four firms—Rogers, Baldwin, Norris, and Mason—employed over four hundred men each in 1860. So it is likely that similarly extensive managerial hierarchies existed elsewhere.

42. My description of Baldwin's 1860-era management structure as extensive may seem to contradict the point made earlier that the company sought to limit growth of nonproducing personnel. While the number of managers did increase at this particular time, Baldwin's overall emphasis was on improved systems rather than more overseers.

43. "Obituary of Charles T. Parry," p. 200.

44. "Fitch Report," p. 664.

45. To continue the cylinder-casting example, after Armory practice was instituted ca. 1865, a molder who cast an oversized cylinder would find his work rejected by his own foreman, and if the error was his fault, he would not receive the piece rate for that casting.

46. "Fitch Report," p. 664.

47. In his examination of interchangeability in consumer products, David Hounshell notes an increase in managerial supervision as American System firms hired inspectors to ensure that parts were made to gauge dimensions (*American System to Mass Production*, p. 119). In capital equipment firms like Baldwin, however, the annual output of finished products was far smaller, making it possible to maintain interchange standards with few or no inspectors.

48. Most historians who have looked at interchangeability in industry have focused on its substantial marketing and/or production benefits, the primary factors behind its adoption at Baldwin as well. But at many metalworking firms systematic management reforms and the installation of Armory practice both date roughly to the 1880s, and they should be seen as related developments. The leading authority on interchangeable production in the Gilded Age, Charles Fitch, clearly valued its promotion of "administrative conditions of order and simplicity," essential themes of systematic management.

49. The superbly detailed records of the Baldwin company's early years now at the Historical Society of Pennsylvania unfortunately terminate with the founder's death in 1866, and most other archival sources do not pick up the trail until the early 1870s. Although its legal name became M. Baird & Co. after 1866, the company was generally known as the Baldwin Locomotive Works, in honor of the founder.

50. "Baldwin Locomotive Works," *American Artisan*, p. 483. By 1881 early Bell telephones had replaced the shop telegraph.

51. It is unclear if this decrease occurred during the 1867 or the 1870 reorganization, but in either case it resulted from admitting new partners (note that Baird's initial one-third interest in the 1854 firm had been increased during Baldwin's lifetime to a 50 percent share).

52. Baird's shares are revealed in M. W. Baldwin's will, #456 of 1866, Office of the Registrar of Wills, Philadelphia City Hall, and in R. G. Dun report on Burnham, Parry, Williams & Co., Mar. 26, 1873, Pa. vol. 136, p. 682, HAG. Henszey's, Longstreth's, and Williams's shares of 12½ percent each are listed in a Dun report on M. Baird & Co., Jan. 15, 1870, Pa. vol. 136, p. 656, HAG. It is unlikely that any of Baird's five partners of 1870 contributed any capital upon their admission (with the possible exception of Williams). The last-mentioned source above lists Burnham and Parry as men of "small

means" (p. 656), and Alba Johnson asserts that Burnham and Henszey had not "brought a cent" into the firm (quoted in Smith, *Elements*, p. 24).

53. R. G. Dun reports on M. Baird & Co., June 19, 1868, and on Burnham, Parry, Williams & Co., Dec. 16, 1873, Pa. vol. 136, pp. 656 and 682, HAG. Assuming these values are accurate, the dollar value of Baird's interest increased from $500,000 in 1868 to over $1.1 million in 1873, despite contemporary deflation and his declining percentage of ownership.

54. Vast quantities of Baldwin's engineering records survive, and while they are not entirely complete, enough remains to show clearly the evolution of drawing room practice. The earliest records, such as the Order Book (1839–57) and Miscellaneous Record Books (1846–57), are chronological lists of orders for engines and parts, having information on who ordered a given component, its dimensions, and the date it was completed (BLW-DEG). As such, these records combined sales, engineering, and production management functions.

55. See "Driving and Truck Wheel Sizes" (1850–55) and "Sizes of Driving Wheel Centers" (1852–68), BLW-DEG.

56. Specification Books (1854–1938), BLW-DEG.

57. Vansant, *Royal Road to Wealth*, pp. 79 (quotation), 80. These drawings were also called general plans or, more commonly, erecting drawings.

58. The plans in the Card Books, called card drawings, were mounted on heavy cardboard stock for durability.

59. The sources of these illustrations are Junior Law Book, 1:50 and 2:30, folder 1, box 10, BLW-DEG.

60. The Card and List Systems also imposed order on purchasing, mandating standard sizes for ordering materials as well as for finishing them in Baldwin's own shops.

61. Engine Order Book, 1908, Scrapbook, p. 1288, Division of Transportation, NMAH.

62. Entirely custom engines often required a longer schedule, especially for drafting work, while small switchers or industrial engines were often built in five or six weeks. But the eight-week schedule was the norm for standard mainline locomotives.

63. See Litterer, "Systematic Management," pp. 369–91, and Chandler, *Visible Hand*, pp. 272–73. Metcalfe was a U.S. Army ordnance officer in charge of production at the Frankford Arsenal, a high-volume manufacturer of ammunition located in Philadelphia. Note that Baldwin's original system focused purely on production management, while Metcalfe's included aspects of cost accounting.

64. "Visit to the Norris Locomotive Works," *United States Magazine*, pp. 161–62.

65. Chandler, *Visible Hand*, p. 272. Capital equipment companies have largely been overlooked by business historians, an omission this study alone can hardly fill. But the incentives behind Baldwin's development of sophisticated production management controls would also have been felt by most capital equipment builders of the period.

66. See "Instructions to Foremen" and "Time-Keeping and Pay Regulations," Apr. 16, 1888, M. Baird & Co. Scrapbook, Division of Transportation, NMAH. The former document lists thirty-one instructions, indicating that foremen of the 1880s had lost much of their earlier autonomy.

67. Historians who have written on the inside-contract system agree on both its predominance during the nineteenth century and its relative neglect in the historiography of American industry. For a review essay, see Englander, "Inside Contract System," pp. 429–46, esp. 429–31. Historians tend to disagree about contracting's characteristics, largely because contractors' ill-defined position allowed the system to vary greatly

between firms and between industries. Montgomery ("Workers' Control of Machine Production") places inside contracting in the context of the "helper system" of the iron and steel industries, while Nelson (*Managers and Workers,* chap. 3.) roots it in a need for decentralization in American System manufacturing.

68. Richards, "Compensation of Skilled Labor," p. 111, and Charles T. Parry to "Messrs. J. Brown and others," June 20, 1872, uncataloged, DEG.

69. A surviving contract for locomotive tender water tanks suggests that Baldwin experimented with contracting as early as 1854, shortly after Baird became a partner (OUT 12/53: contract dated Feb. 1, 1854, BLW-HSP). But both Richards ("Compensation of Skilled Labor") and longtime Baldwin officer Samuel Vauclain (Vauclain and May, *Steaming Up,* p. 181) date the origin of contracting at Baldwin to 1872. This was a relatively late date, for contracting had been widespread among both builders and manufacturers since before the 1850s.

70. Like many historians, I have been frustrated by the scarcity of original sources on contracting, so my description is mostly based on surviving material from 1890 to 1915.

71. Contracted work listed in shorthand notes of William Austin, June–Sept. 1910, box 5, WLA-HAG. By 1914 contractors had undertaken twenty-one separate tasks in Baldwin's boiler shop alone. See list in S. M. Vauclain's Production Notebook for 1914, box 3, SMV-DEG. As production volume increased, contracting evidently became more cost-efficient for simpler tasks.

72. Testimony of Alba Johnson in *Report of the Commission on Industrial Relations* (hereafter cited as CIR *Report*), 3:2822–23. This is the best available source on contracting at Baldwin, but it calls for careful analysis since its references to contracting often commingle that system with straight piecework. The 25 percent of the force on hourly rates included draftsmen, watchmen, some laborers (not all), and power-plant staffs.

73. "Agreement with Charles B. Allen," BLW Contract Book, 1:113–14, BLW Collection, Pennsylvania State Archives. This source contains only one inside contract.

74. Baldwin debited the wage costs on Allen's contract from his account for finished springs, so the firm had a clear statement of his profits in its own books.

75. Cotton, "Baldwin's 'Contract System,'" p. 37.

76. In public remarks ca. 1900 the Baldwin partners tended to obfuscate regarding the ability of all workers to become contractors (see Johnson testimony in CIR *Report,* 3:2822–23). Although Dan Clawson asserts that Baldwin had no "fixed, permanent contractors" ca. 1900 (*Bureaucracy and the Labor Process,* p. 92), and that senior employees competed for such positions, both common sense and other sources suggest that this is incorrect (see Cotton, "Contract System," p. 37, and Johnson testimony in CIR *Report,* 3:2832). Baldwin had every reason to pick as contractors men with a proven record, ability, and experience—particularly because by 1900 it did not accept competitive bids from the men. The company might occasionally select other men to keep contractors from becoming complacent, but this was hardly open competition for contracting positions.

77. Comparison of BLW, "Rules for Time-Keeping and Payment of Wages," Feb. 10, 1905, p. 11, box 5, WLA-HAG, vs. BLW, "Shop Rules," July 1910, p. 5, box 3, SMV-DEG.

78. Author's telephone conversation of Aug. 17, 1989, with John Kirkland, Baldwin's chief of inspection ca. 1942.

79. Data from BLW Order Books for 1870 and 1880, BLW-NMAH.

80. Converse, "Some Features," p. 6.

81. Englander, "Inside Contract System," p. 443. This notion that the owners of nineteenth-century industrial companies utilized contracting because they lacked technical knowledge seems wide of the mark in many cases. Such firms as Baldwin, Colt,

Pratt and Whitney, Whitin, Wheeler and Wilson, and Brown and Sharpe—all of which used contractors—owed their origins precisely to the technological inventiveness or ability of their founders.

82. Many historians believe that contractors engaged in a continuous process of innovation to lower their own costs and improve their profits. This was no doubt true in many firms including Baldwin; however, the innovative role of contractors is easily overstated. Englander asserts that contractors were generally able "to direct technological change within the firm" ("Inside Contract System," p. 438). Such power would have surprised the Baldwin partners, who took out sixteen patents for new production machinery between 1882 and 1895 and spent an average of $164,000 a year on new tooling from 1886 to 1904. It seems likely that contractor innovations at any company were of low cost and relatively minor—although numerous, perhaps—since the contractor generally had to pay for innovations himself and had only a year or less to benefit from them before contracts were renegotiated. For Baldwin's annual tooling expenses, see S. M. Vauclain, Memo Book, box 7, SMV-DEG. For further discussion of the relatively minor role of contractors in innovation, see Albert D. Pentz, "Machine Shop Practice," *Engineering Magazine* 6 (Jan. 1894): 552–53.

83. Of fourteen Baldwin contracts listed in president William Austin's 1910 notes, six involved more than two hundred workers each. Employment on the eight other contracts ranged from eight to eighty men. See shorthand notes of William Austin, box 5, WLA-HAG. Quotation from Phillips, "Reconciliation of Capital and Labor," p. 922.

84. Englander, "Inside Contract System," pp. 442 and 437 (quotation), and Clawson, *Bureaucracy and the Labor Process,* p. 112.

85. The Colt Armory was another contract shop that paid its contractors' labor directly and therefore could ascertain contractors' profits (Roe, *English and American Tool Builders,* p. 178). The same was true of Winchester Repeating Arms (Chandler, *Visible Hand,* p. 271) and the Taunton Locomotive Works (Lozier, "Taunton and Mason," pp. 469–70).

86. Baldwin's Timekeeping Rules of 1878 required all workers to report their time spent *on each job;* such detailed record-keeping can only mean that the company had a system of unit cost accounting at this time. This challenges the point made by Chandler that firms using contracting "knew relatively little about the precise costs of labor" (*Visible Hand,* p. 272). Baldwin's 1878 rules are in "Rules and Regulations," *American Machinist,* Feb. 1878, p. 10. (Although this article does not identify the rules as Baldwin's, a comparison with the firm's 1891 rules shows that it was indeed the source.)

87. Quotation from Cotton, "Contract System," p. 38. See also Levasseur, *American Workman,* pp. 171–72; Johnson testimony in CIR *Report,* 3:2832; and Henszey, "Organization and Methods," p. 402.

88. Johnson testimony in CIR *Report,* 3:2822.

89. Uncited newspaper clipping ca. 1905 in Engine Order Book, 1908, Scrapbook, p. 1295, Division of Transportation, NMAH.

90. Contracting was particularly advantageous when sales declined and the volume of work slowed, since the system automatically trimmed managerial costs, with contractors reverting to the ranks of skilled workers as workmen with lesser skills were laid off.

91. White, *American Locomotives,* p. 14; White, "Holmes Hinkley," p. 52; Emerson Company, "Report on Brook's Works," p. 25, file 4, box 18, Harrington Emerson Papers, Pennsylvania State University Library; *Machinists Monthly Journal* 3 (June 1891): 131 (Richmond); Fowler, "American Engineering Practice," p. 170 (Schenectady).

92. Englander, "Inside Contract System," p. 429 (Whitin); *American Machinist,*

May 25, 1893, p. 1 (Bement); *American Machinist,* Apr. 11, 1895, p. 283 (GE); Thompson and Murfin, *The I.W.W.,* p. 42 (Pressed Steel); *Railroad Gazette* 15 (Mar. 16, 1883): 170 (Pullman); CIR *Report,* 3:2885 (Cramps); Montgomery, *Fall of the House of Labor,* p. 211 (Hoe).

93. Nelson, *Managers and Workers,* p. 37.

94. The new firm was composed of these three men and their previous partners in M. Baird & Co.: Henszey and Longstreth. A month later these principals admitted Edward Williams's confidential secretary, John H. Converse, into the partnership—again without requiring an initial capital contribution.

95. R. G. Dun reports on Burnham, Parry, Williams & Co., May 15, 1873, and May 20, 1875, Pa. vol. 219, p. 332, HAG. Evidently the new firm paid Baird $10,000 in cash immediately upon his retirement. The figure of $1.125 million represented one-third of the company's total capital value in 1873, underscoring how the Baldwin principals benefited from the partnership form. Because Baird could sell out to his former partners, the liquidity of his ownership interest was promoted, enhancing its value. Thomas Navin and Marion Sears overlook the ability of a retiring partner to sell out to a successor firm, and therefore they underestimate both the liquidity and the value of such an ownership interest ("Rise of a Market," p. 108).

96. R. G. Dun reports on Burnham, Parry, Williams & Co., Dec. 16, 1873, and Sept. 16, 1882, Pa. vol. 136, pp. 389 and 682, HAG. Note that if the new partners had run into financial difficulties after the panic, Baird would have provided assistance because their continued vitality was to his benefit—at least until his interest was fully paid off. As a Dun credit reporter noted shortly after the panic, "The fortune of Matthew Baird stands at their back, ready for any Emergency" (report of Sept. 26, 1873, Pa. vol. 219, p. 332, HAG).

97. Surviving pay regulations from M. Baird & Co. (1867–72) required the company's timekeeper to keep wage accounts only on a per-employee basis. By 1878 the rules had changed, and timekeepers also assembled labor cost data by the "name of job and machine or engine." This change can only mean that the firm sought the new data for a unit cost accounting system.

98. Chandler believes that cost sheets in metalworking industries "were only just being perfected" in 1880 (*Visible Hand,* p. 268).

99. Garner, *Evolution of Cost Accounting,* p. 29.

100. A complete cost sheet for a Souther-built engine of 1851 is in "Cost of Building Locomotives in America," *Engineer,* pp. 5–6. A textile machine builder, the Boston Manufacturing Company, was developing unit cost data ca. 1820 (Gibb, *Saco-Lowell Shops,* pp. 41–42).

101. The episodes of locomotive price fixing described in chap. 2 provided an additional disincentive to developing cost-accounting data. Why go to all that trouble over costs when simple collusion could set bids, hold up prices, and ensure adequate profits? On the other hand, the inside-contract system readily complemented cost-accounting controls and establishing bids for engine sales. A main function of the contractors was to take over from the firm the risk of fluctuating labor costs; the fixed contract prices simplified accounting and bidding.

102. For a description of these controls, see Henszey, "Organization and Methods," pp. 401–9. Walter Chrysler's autobiography provides an interesting comparison of cost accounting in locomotive building and in volume manufacturing. After leaving his post as works manager at American Locomotive's Schenectady plant ca. 1910 for a similar job at Buick Motor Company, Chrysler found Buick's accounting practices to

be very rudimentary compared with the detailed and sophisticated cost controls Alco found necessary to bid successfully for locomotive contracts. While Buick admittedly was a young firm, this comparison again suggests that the Chandlerian emphasis on manufacturers as managerial pathfinders is overly narrow. Walter P. Chrysler and Boyden Sparkes, *Life of an American Workman* (New York: Dodd, Mead, 1950), p. 133.

103. Henszey, "Organization and Methods," p. 401.

104. Salary list from S. M. Vauclain's Production Notebook for 1909, box 7, SMV-DEG.

105. Nelson, *Managers and Workers,* p. 22.

106. *Railroad Gazette* 27 (Aug. 9, 1895): 530. General Electric and the Schenectady Locomotive Works had a similar agreement.

107. Among the many studies that consider aspects of this period in American business history are Chandler, *Visible Hand;* Lamoreaux, *Great Merger Movement;* Nelson, *Managers and Workers;* and Zunz, *Making America Corporate.*

108. Austin began at the company as an eighteen-year-old draftsman in 1870, while Johnson started out as a clerk at age nineteen. Vauclain was a machinist's apprentice at the Pennsylvania Railroad before coming to Baldwin in 1884 as foreman of its tender shop. As before, none of these new partners made any initial capital contribution upon admission to the partnership (Johnson, quoted in Smith, *Elements,* p. 24).

109. Johnson, quoted in ibid., p. 23.

110. Vauclain and May, *Steaming Up,* p. 164.

111. During World War I Baldwin created a munitions subsidiary, the Eddystone Ammunition Corporation, which proved quite lucrative, and in 1919 it reorganized its sales department, replacing some commission merchants with its own salaried personnel (*Baldwin History,* pp. 132, 141).

112. Baldwin took twenty-two years (1906–28) to effect the complete transfer of operations from Broad Street to Eddystone.

113. Nelson's *Managers and Workers,* p. 116, has a list of firms with extensive welfare programs ca. 1905–15. Almost all were mass producers of machinery, iron and steel, textiles, or other products.

114. The American Locomotive Company experimented with Scientific Management policies ca. 1906–8, but labor's opposition quickly "led to the repudiation of time study and the incentive wage" there (Nelson, *Frederick W. Taylor,* p. 153).

115. The augmentation of staff under Scientific Management and the Emerson quotation are in Chandler, *Visible Hand,* p. 277.

116. This and subsequent Johnson quotations are from CIR *Report,* 3:2831. Johnson was hardly alone among industrial managers in opposing Scientific Management. Frederick Taylor was barely seated after presenting his "Piece-Rate System" paper to the American Society of Mechanical Engineers before other engineers were objecting to his ideas. Even at the birthplace of Taylorism, Midvale Steel, the company's superintendent had disbanded most of Taylor's systems by 1914, noting that they incurred "too much overhead" expense. See testimony of William Barba in CIR *Report,* 3:2855.

5: THE BALDWIN WORKFORCE, 1860–1900

1. Because this period saw little overt labor strife, it is more difficult to retrieve workers' views of the firm. Conflict leaves a much bigger mark than consensus.

2. Bingham, "Labor Situation," pp. 40–41.

3. The sparring analogy is not an abstract notion of my own creation. According to David Montgomery, the workforces of metal trades employers like Baldwin at this

time were exerting "a form of control of productive processes which became increasingly collective, deliberate, and aggressive," and such worker power was exerted in "a struggle, a chronic battle in industrial life" ("Workers' Control of Machine Production," p. 487).

4. These questions are seldom encountered in labor historiography—largely because its laudable goal over the last twenty years has been to recover labor's agency in creating the contours of work and opposing managerial encroachments. Many fine studies have resulted, but the goal itself has often foreclosed examination of periods of labor peace and the reasons for it, consideration of nonunion workplaces, or a focus within one firm.

5. For a study of the relationship between labor management and competitiveness, although with a focus on mass producers, see Lazonick, *Competitive Advantage.*

6. As Lazonick notes, cooperative labor-management relations "encourage not only the greater utilization of an existing [production] technology but also greater efforts in effort-saving technological change" (ibid., p. 8).

7. Charles T. Parry to "Messrs. J. Brown and others," June 20, 1872, uncataloged, DEG.

8. Just as Baldwin secured competitive advantages on its shop floor, competitive pressures could also have an effect there. In 1860 and in 1910–11 Baldwin's producer culture failed to avert conflict because particular conditions in the industry caused management to limit wage costs. As these cases show, the "conflict model" of labor history does have some applicability to Baldwin, but those two strikes were exceptional incidents, divided by fifty years of cooperation that this model is inadequate to explain.

9. It could not have escaped the partners' notice that strikes, rather than labor harmony, frequently followed Taylor, Gantt, Barth, Emerson, and other Scientific Management consultants wherever they went.

10. An 1880 study found that locomotive builders employed less capital per employee than did Armory-practice mass producers of sewing machines and firearms (although builders' capital burden was still high). See "Fitch Report," p. 49.

11. Fincher, "Early History," p. 521.

12. Note also that industrial employers' production needs and training methods, like apprenticeship programs, largely defined the character of these trades from their origins; they were not self-defining crafts like many preindustrial trades.

13. OUT 2/60: M. W. Baldwin & Co. to John Stuart, Mar. 20, 1860, and to J. Kinsley Smedley, Mar. 12, 1860, BLW-HSP.

14. The Civil War saw some shortages in particular trades and skills, but these were temporary, and the overall supply was sufficient to see total employment reach nine hundred men in 1864.

15. I found this informality somewhat surprising, particularly for such a large company. But Walter Licht's study of the personnel practices of Philadelphia's employers confirms that few nineteenth-century firms resorted to hiring halls, employment agencies, or help-wanted ads when seeking skilled workers (*Getting Work,* chap. 2).

16. The residents of the Preston boardinghouse and their occupations are given in the manuscript schedules of the 1870 U.S. Population Census, 15th ward, 43d district, p. 145. Unmarried workers like Skelly commonly lived in such boardinghouses.

17. For the importance of personal ties in securing work during the nineteenth century, see Licht, "Case Studies," p. 171.

18. Calkins, *Memorial,* p. 54.

19. OUT 10/49: M. W. Baldwin to Alfred Vail (Speedwell Iron Works), Nov. 21, 1849, BLW-HSP.

20. Terms from OUT 3/44: Baldwin & Whitney to L. O. Reynolds, Aug. 5, 1844,

BLW-HSP. Boys seeking an early release from their indentures often had to pay Baldwin for the loss of their services, since in the company's view they were denying Baldwin the benefit of the training they had received. Rather than pay, most dissatisfied boys simply ran away.

21. See OUT 10/46: M. W. Baldwin to J. H. Cleveland, Sept. 9, 24, and 28, 1846, BLW-HSP.

22. Matthias Baldwin had felt that the retained earnings of apprentices were sufficient to ensure the completion of their terms, but Baird believed that a fully contractual relationship was better. IN 10/53: W. Flisher Mitchell to M. W. Baldwin, Nov. 23, 1853, and OUT 7/53: George Burnham to John Clayton, Sept. 19, 1853, BLW-HSP.

23. Labor historians commonly believe that apprenticeship had withered by 1860 with the rise of factory production (see Rorabaugh, *Craft Apprentice*). While this is true in manufacturing, the system retained great vitality among machinery builders well into the Gilded Age and beyond.

24. I traced apprentices' residences in the manuscript schedules of the 1870 U.S. Population Census. Twenty-six boys (72 percent) lived with their parent(s) and eight resided in boardinghouses, while two lived in rental housing as married heads of households.

25. OUT 10/49: M. W. Baldwin (per George Burnham) to Elisha Hollingsworth (Harlan and Hollingsworth, Wilmington), Nov. 24, 1849, BLW-HSP.

26. OUT 2/60: M. W. Baldwin & Co. to R. J. Hollingsworth, Mar. 13, 1860, BLW-HSP.

27. The main sources on Baldwin's apprenticeships are two books of indentures and another ledger recording days worked (BLW-HSP). This last volume is incomplete, particularly through the mid-1850s, and all three volumes have some duplication of names. But if used carefully and in combination with other sources, these books give an accurate portrait of the program between 1854 and 1868.

28. Of 310 apprentices, 279 have indentures that give a place of residence (although the form asked where boys were "from," an imprecise term open to various interpretations). Sixty-seven percent (188) gave Philadelphia, 7 percent (20) were from counties adjoining the city, 8 percent (21) came from elsewhere in the state, an equal number hailed from other states, and 10 percent (29) were immigrants. Data compiled from Apprentice Books, BLW-HSP.

29. Among the master mechanics who sent their sons or other relatives to Baldwin were M. O. Davidson (Havana Railroad), Franklin Roop (North Pennsylvania Railroad), and William Hardman (Georgia Railroad and Banking Company).

30. Because of his extensive activities in the Presbyterian Church and in charitable relief, Matthias Baldwin frequently received requests to place boys from ministers, Sunday school teachers, and the like. See IN 1/60: Jno. Playton to M. W. Baldwin & Co., Feb. 9, 1860, BLW-HSP.

31. City data from the *U.S. Population Census for 1860*, p. 438; apprentice data (for boys entering the program between 1860 and 1868) derived by linking names and ages (86 total) to entries in the manuscript schedules of the 1860 and 1870 censuses. Two-thirds of the immigrant apprentices were Irish.

32. Fathers' ethnicity derived by linking names (83 total) to entries in the manuscript schedules of the 1860 and 1870 censuses.

33. To derive the data in this paragraph I linked fathers' names—since they generally co-signed indentures—with their occupations as listed in the manuscript schedules of the 1860 U.S. Population Census and Philadelphia city directories for 1860, 1865, and 1870.

Linkages were attempted only for the 1860–68 group of apprentices (238 total). Since the sources are all Philadelphia-based, non-Philadelphian fathers were not counted, even when their occupations were known. Because most were Baldwin business associates, the "middle-class" percentage would have been higher had they been included. Ninety-three linkages were made: sixteen middle-class fathers (generally white-collar), fifty-six skilled tradesmen, and twenty-one unskilled parents.

34. Of the fifty-six fathers in skilled trades, thirty-seven were metalworkers. In the latter group, all seventeen machinist fathers sent their sons into the same trade.

35. Calvert, *Mechanical Engineer in America,* p. 12; see also pt. 1 for the concept of engineers' shop culture.

36. The four apprentices known to have become master mechanics were James Boon (Union Pacific), John Cook (Georgia Railroad and Banking Company), N. W. Sample (Denver and Rio Grande), and John Wootten (Philadelphia and Reading).

37. I constructed a tally, based on the length of the boys' actual service, to determine the ratio of apprentices at work to the total force. In the years for which the total workforce size is available (see appendix A), apprentices accounted for 9–10 percent of total employment (14 percent of the skilled force). By contrast, in 1855 the ratio of apprentices to total workers at the Norris Locomotive Works, located across the street from Baldwin, was 25 percent ("Visit to the Norris Locomotive Works," *United States Magazine,* pp. 165–66). In 1868 the ratio at the Whitin Machine Works, a textile machinery builder in Massachusetts, was 21 percent (Navin, *Whitin Machine Works,* p. 66).

38. The earnings of a skilled Baldwin journeyman are found in the wage account book (1856–57) of Joseph Shear, collection of Mrs. William Vauclain (photocopy held by author).

39. Apprentices also received a certificate upon completion of their terms—if you will, a diploma. Interestingly, the major illustration on these certificates showed Oliver Evans's amphibious boat, the *Orukter Amphibolos*—not a Baldwin locomotive. This suggests that the company saw its apprentices as descendants of Philadelphia's distinguished heritage of mechanicians, described in chap. 1. For the certificate, see Vauclain, "System of Apprenticeship," p. 323.

40. The war accounted directly for 5 percent of the voluntary departures and probably was a major factor for many in the "left," "gone," and "runaway" categories. One timeless issue in apprenticeship, runaways, can be largely traced to homesickness. Sixty-one percent of runaways were from families outside the Philadelphia area.

41. Without personnel records it is impossible to know how many apprentice graduates stayed at Baldwin and for how long. But the graduates were traced in *Gopsill's Philadelphia City Directory for 1880* to determine if they stayed in the trade of their indenture. Of the forty-five men who were linked, thirty-five (78 percent) remained in their original trades, with most probably still at Baldwin, since the firm paid higher wages than other Philadelphia employers.

42. Calkins, *Memorial,* p. 220.

43. During the Gilded Age Baldwin took on boys whom it continued to call apprentices, but "most . . . learn how to make only a few of the parts: to run a planer or to do the lathe work; [and] only a small number ever acquire the entire trade." *Pennsylvania Industrial Statistics Report,* vol. 21 (1893), p. D-16.

44. Ibid., p. D-80. An 1896 study by the *American Machinist* of 116 "machinery building establishments" found that 73 percent took apprentices (including seven of the eight locomotive builders polled, Baldwin being the only exception). This high

proportion is further evidence of how the practice of capital equipment builders diverged from better-known developments in mass-production manufacturing. See "Status of Apprenticeship," *American Machinist* 19 (1896): 1184–1203.

45. Burnham, Parry, Williams & Co., *Illustrated Catalogue of Locomotives* (1881), p. 51.

46. The contract was not announced until Tuesday, December 18 (see Philadelphia *Evening Bulletin* of that date, p. 8), but the men were there on Monday, according to "Russian Contract for the Baldwin Locomotive Works," *Railroad Gazette* 10 (Apr. 26, 1878): 206–7 (quotation from p. 206).

47. "System in the Baldwin Locomotive Works," *American Machinist,* Aug. 25, 1883, p. 5.

48. The speaker is Baldwin's president in the 1920s, describing the firm's general labor management policy—unchanged from the 1870s. Vauclain and May, *Steaming Up,* p. 183.

49. During the 1884–85 recession the workday was cut from ten to eight hours while the company tried to improve its export business with cuts in locomotive prices.

50. Testimony of Charles Harrah in *Report of the Industrial Commission,* 14:350.

51. Vauclain and May, *Steaming Up,* p. 178.

52. Wage account book of Joseph Shear, collection of Mrs. William Vauclain. The firm set rates for each component; for example, the finishing work for a set of pumps paid $3.00.

53. "Fitch Report," p. 664.

54. "The Grant Locomotive Works," *American Artisan,* Mar. 15, 1871, p. 163; Navin, *Whitin Machine Works,* p. 145; Gibb, *Saco-Lowell Shops,* p. 53.

55. Hughes, "Trade Unions," p. 47. There was some truth to the notion that piecework could result in substandard work.

56. Although most historians see Taylorism as a fundamental break from the past, in fact large employers in metalworking had long used piece rates, a truth that lessens the "revolutionary" impact of Taylor's ideas. While David Montgomery notes this pre-1890, pre-Taylor adoption of piece rates (*Fall of the House of Labor,* p. 130), other labor historians such as Dan Clawson imply that time rather than piece payment was normal until the Taylor era (*Bureaucracy and the Labor Process,* p. 169).

57. Roland, "Six Examples," p. 834 (Sellers); Fowler, "American Engineering Practice," pp. 15 (Pond) and 208 (Brown and Sharpe).

58. Once workers' skills had become defined or circumscribed by the few components they made under piecework, they were well on their way to becoming industrial operatives rather than craftsmen. Such men possessed real (if narrow) skills, but their expertise came to be defined by the organization of work, as established by employers, rather than by craft training.

59. Actually this is not quite true. Many workers struck because their foremen refused to put them on piecework, which paid better than day work.

60. IN 7/56: Samuel Barry to M. W. Baldwin, Sept. 17, 1856, BLW-HSP. Subsequent quotations from testimony of John Tobin in CIR *Report,* 3:2840, and Arnold, "Production Up to the Power Limit," p. 923.

61. Quotation from Taylor, *Principles of Scientific Management,* p. 17. See also Montgomery, *Fall of the House of Labor,* p. 151, and Clawson, *Bureaucracy and the Labor Process,* p. 169.

62. Roland, "Six Examples," p. 833.

63. This is described in an extensive article about Baldwin, "Production Up to the Power Limit," by Horace L. Arnold, and is confirmed in *American Engineering Competition,* pp. 75–76, wherein Baldwin's piece-rate policy and high wages are favorably compared with British practices.

64. The following example demonstrates this savings. Assume that Baldwin produces one hundred locomotives, selling for $10,000 each. Per-unit labor charges are $3,000, materials cost is $3,500, and fixed overhead expenses are $2,500. At this production level (100 units), the company's unit profit is $1,000. With a 50 percent improvement in productivity (150 units) and unchanged piece rates, the labor and materials costs per engine will remain the same ($3,000 and $3,500 respectively), but the fixed costs or overhead will drop to $1,670 for each locomotive. With unchanged piece rates, the company's profit grows from $1,000 to $1,830 per engine, and workers' take home pay also increases with the higher production level (and/or more men are employed). For more on this point, see Roland, "Six Examples," p. 835.

65. William Lazonick describes this surplus above direct labor costs as "value created." As he notes, "The sharing of value created [between workers and companies] need not represent a zero sum situation. By providing incentives for workers to supply more effort and for managers to invest in effort-saving technological change, the distribution of value created can result in . . . value gains," or increased output and productivity (*Competitive Advantage,* p. 8). Lazonick's focus is on mass producers, but the locomotive builder's combined burdens in labor and capital gave it a particular incentive to follow this course. With its adoption of cost accounting in the 1870s, Baldwin could chart the potential gains of such a policy in detail.

66. *American Engineering Competition,* p. 76.

67. It is commonly argued that workers imposed stints in direct response to such declines in output. While this was no doubt true, particularly among manufacturers, builders like Baldwin faced such precipitous falls in sales that they overwhelmed any attempt by workers to even out the flow of work. Few would soldier when one-third of the workforce had been laid off (particularly since those who tried it would soon follow). Furthermore, the company itself created a form of stinting during recessions with its policy of shortening the hours of work to keep more men employed.

68. Data from sheet in S. M. Vauclain's Production Notebook for 1893, box 7, SMV-DEG. See also Partington, *Railroad Purchasing,* p. 222 (orders data), and "Strike at Baldwin's," Philadelphia *Evening Bulletin,* Aug. 17, 1893, p. 2 (piece-rate cut).

69. Arnold, "Production Up to the Power Limit," p. 919.

70. Weeks, "Report on the Statistics of Wages in Manufacturing Industries," pp. 189, 191–94, 522, 528, 532.

71. Surviving data on wages paid to Baldwin's workers are scarce and scattered. The 1863 figure—the only such data known to the author for this early period—is from OUT 3/63: table titled "Rates of Wages Paid by M. W. Baldwin & Co., Apr. 4, 1863," BLW-HSP. The 1893 figure is from S. M. Vauclain's Production Notebook for 1893, box 7, SMV-DEG. Both sources give the average wage for the workforce as a whole in a given week. Those weeks may have been exceptional (the 1863 figure may have been high, given wartime labor scarcity). In each case I weighted the given wage ($8.74 in 1863, $13.06 in 1893) by the Burgess cost-of-living index (*Historical Statistics,* pt. 1, p. 212) to derive a comparative measure of real purchasing power ($10.93 in 1863, $17.41 in 1893). Finally, since the data are averages for the force as a whole, the changing distribution of skilled and unskilled workers (in 1863 vs. 1893) necessarily affects the figures. Inferential evidence suggests that the percentage of high-skill, high-pay workers was larger in 1863 than in 1893.

72. Arnold, "Production Up to the Power Limit," p. 922.

73. Fraser, *America at Work,* p. 45, and ibid., pp. 923, 919.

74. Weeks, "Report on the Statistics of Wages in Manufacturing Industries," p. 194.

Comparative data on accident rates in the steel industry vs. the "machine building industry" (i.e., capital equipment firms) are found in Cheney and Hanna, "Accidents," p. 9. The rate of severe injuries in locomotive building was half that in the steel industry, but the injury rate among workers in locomotive plants was double that of machine builders generally.

75. These are the actual causes of death of three Baldwin men as listed in *Pennsylvania Industrial Statistics Report,* vol. 20 (1892), p. 110F. Eleven-year death statistics from CIR *Report,* 3:2867.

76. CIR *Report,* 3:2924.

77. Newly hired employees accounted for most of the dead and injured (ibid., p. 2866). Baldwin was not unconcerned about the combined influences of inexperience and haste on the accident rate. Ca. 1900 the following notice was posted in the plant: "Owing to the . . . number of men who are continually being taken on . . . every precaution should be taken to prevent accidents . . . [in] your effort to get out the most work in the least possible time" (cited in Tolman, *Social Engineering,* pp. 114–15). But posting a notice was different from ensuring that its instructions were followed.

78. *O'Dowd v. Burnham* and *Campbell v. Baldwin Locomotive Works.* O'Dowd and Campbell sued Baldwin for damages after their injuries. In such cases the firm invariably mounted vigorous defenses and appeals, suggesting a general strategy to discourage all claims by maintaining a reputation for legal intransigence—a common stance among industrial employers. Baldwin generally won these cases, often under the fellow-servant rule (applied in *O'Dowd*), although it lost in *Campbell.* Injured men who did not sue often received low-paying sinecures such as watchman's jobs.

79. OUT 10/61: M. W. Baldwin & Co. to C. G. Steer, M.D., Nov. 23, 1861, BLW-HSP.

80. *Pennsylvania Industrial Statistics Report,* vol. 15 (1887), p. 32B.

81. The association incorporated in 1870 (*Laws of the General Assembly of the State of Pennsylvania* [1870], pp. 1295–96); among its nine incorporators were the fathers of two Baldwin apprentices, suggesting the seniority and special status of the incorporators.

82. *Pennsylvania Industrial Statistics Report,* vol. 15 (1887), pp. 38B–39B.

83. A search for the incorporators in the manuscript schedules of the 1870 U.S. Population Census uncovered three out of the nine: an erector, a blacksmith, and an iron molder. This suggests that skilled men had stronger ties across trades than they had with the semiskilled and laboring men of their own departments. The limit probably rose as Baldwin's force grew, but in its first year only 30 percent of the force could have joined the association.

84. A good primary source on developments in New York is "Appeal to the Machinists and Blacksmiths of America," *Machinists' and Blacksmiths' International Journal,* p. 693. See also Montgomery, *Fall of the House of Labor,* pp. 193–94. The strike was no problem to Singer, since it had fifty thousand finished machines laid up in inventory, and it could sit the workers out (*New York Times,* June 6, 1872, p. 4). The builders of custom capital equipment lacked this latitude when labor unrest threatened production.

85. Philadelphia *Public Ledger,* July 27, 1872, p. 1 (Paterson), and June 12 and 15, 1872, both p. 1 (cabinetmakers).

86. Charles T. Parry to "Messrs. J. Brown and others," June 20, 1872, uncataloged, DEG.

87. Vauclain and May, *Steaming Up,* p. 181, and Richards, "Cooperative Contract System," p. 3.

88. Clawson, *Bureaucracy and the Labor Process,* p. 94.

89. Converse, "Some Features," p. 6.

90. Annual wages extrapolated from totals for June–August 1910 in shorthand notes of William Austin, box 5, WLA-HAG. In a few cases this source lists much higher annual incomes for contractors, but the circumstances surrounding these notes by Baldwin's president suggest that such high pay was exceptional. Recollect that in years of lean sales, contractors' incomes, like pieceworkers', declined in proportion to output.

91. Foremen's wages from sheet in S. M. Vauclain's Production Notebook for 1909; average workman's wage from sheet in his Production Notebook for 1912, box 7, SMV-DEG.

92. BLW, "Time-Keeping and Pay Regulations," Apr. 24, 1882, M. Baird & Co. Scrapbook, Division of Transportation, NMAH. See also Johnson testimony in CIR *Report*, 3:2827.

93. Fowler, "American Engineering Practice," p. 173. This source suggests that Nelson is in error in asserting that contractors alone reaped the rewards of the system (*Managers and Workers*, p. 37).

94. Hourly rates are described in CIR *Report*, 3:2822; for piece rates under contracting, see "Strike at Baldwin's," Philadelphia *Evening Bulletin*, Aug. 17, 1893, p. 2.

95. Contractors also desired to use piece rates because it would greatly simplify their effort to make a profit on contracts, given that such rates allowed them to predict and contain their total labor cost. This dynamic and Baldwin contractors' use of piece rates challenge Clawson's implication (*Bureaucracy and the Labor Process*, p. 169) that contract workforces received time wages.

96. "Strike at Baldwin's," Philadelphia *Evening Bulletin*, Aug. 17, 1893, p. 2.

97. "Laying Off Hands," Philadelphia *Evening Bulletin*, Aug. 18, 1893, p. 2.

98. CIR *Report*, 3:2840. For the general subject of internal promotion in industry, see Sundstrom, "Internal Labor Markets," pp. 424–45.

99. Arnold, "Production Up to the Power Limit," p. 923.

100. Converse, "Some Features," p. 9.

101. The only plantwide ethnicity data on Baldwin date to 1910. The force then totaled 16,211 men and included these nationalities: American (49 percent), Irish (14 percent), Polish (9 percent), Russian (7 percent), German (6 percent), Austrian (5 percent), Italian (4 percent), and English (2 percent). The remaining 4 percent encompassed thirty-one nationalities (CIR *Report*, 3:2819–20). Philadelphia's population as a whole was far less "ethnic" than that of most other industrial cities. Yet compared with other very large industrial employers, Baldwin was exceptional in having a force half composed of native-born white Americans. Contemporary steelworkers were 38 percent native-born whites, coal miners 32 percent, and male workers in cotton textiles only 28 percent (data for 1907–8, in "Report on Conditions of Employment in the Iron and Steel Industry," 3:83). Baldwin's comparatively high proportion of native-born white males resulted from its Philadelphia location and its high skill needs. This shared ethnicity heightened the native-born white proprietors' ability to identify with the skilled portion of their workforce — in contrast to the unskilled, where foreign-born laborers predominated.

102. Many Baldwin workers chose this job description, rather than trade names, when responding to the enumerator of the 1870 census.

103. This period of conflict is described by Clawson, *Bureaucracy and the Labor Process*, Montgomery, *Fall of the House of Labor*, and McGuffie, *Working in Metal*, among others.

104. Charles T. Parry to "Messrs. J. Brown and others," June 20, 1872, uncataloged, DEG, and Converse, "Some Features," p. 8.

105. If Baldwin gained a competitive advantage from its labor policies, what was happening at other locomotive builders? Were they at a disadvantage in labor relations, forced to cap wages to remain competitive because their workers were not as productive? Too many factors intervene to know for sure. However, at least some of Baldwin's competitors encountered periods of labor unrest in the Gilded Age, with strikes at Cooke, Rogers, Grant, and Richmond.

6: BUILDING LOCOMOTIVES, 1850–1900

1. The chief factors affecting the timing of changes in production were the great fluctuations in locomotive demand and the fact that locomotives saw their greatest increases in size and weight during and after the 1890s. But the mix of production factors—as economists describe land (or raw materials), labor, and capital—was not equally susceptible to change at any moment, as Edward Ames and Nathan Rosenberg emphasize in "Enfield Arsenal."

2. This last debate and its partisans are summarized in Dubofksy, "Technological Change and American Worker Movements." See also Laurie and Schmitz, "Manufacture and Productivity," p. 78; Rosenberg, "Technological Change in the Machine Tool Industry"; Smith, *Harpers Ferry Armory;* Hounshell, *American System to Mass Production;* and Hoke, *Ingenious Yankees.*

3. The selection of an appropriate time-data series for a statistical profile of Baldwin's production is hampered by a lack of sources and by the volatility of demand. The choice of a decennial portrait as given in table 6.1 was largely dictated by the availability of data on the firm in the manuscript schedules of the U.S. Manufacturing Census. Note also that at each year in the table, Baldwin's sales in the contemporary business cycle for locomotives were either well into a period of expansion or at the peak of that cycle. It may therefore be assumed that the plant was operating near or at capacity at each decennial, aiding temporal comparisons of productive efficiency.

4. Map of the Baldwin factory ca. 1850 in "Miscellaneous Records Book 1846–75," p. 63, folder 7, box 9, BLW-DEG.

5. Mean capital and employment data for Philadelphia industry from Laurie and Schmitz, "Manufacture and Productivity," p. 66.

6. The figure of $549 is derived by dividing the mean capital of $7,078 by the average employment of 12.9 workers for all Philadelphia industry.

7. Kenneth Sokoloff has demonstrated that, like Baldwin, most early factory producers "had quite modest investments in machinery and tools per unit of labor," undercutting the notion that such firms were devoted to machine-intensive production technologies ("Investment in Fixed and Working Capital," pp. 545–56, quotation from p. 556).

8. See IN 1/46: Hollingsworth, Harvey & Co. (Wilmington) to Baldwin & Whitney, Jan. 26, 1846, and IN 7/44: H. C. Seymour (Erie Railroad) to Baldwin & Whitney, July 10, 1844, BLW-HSP; and "Miscellaneous Records Book," p. 4, folder 8, box 9, BLW-DEG.

9. OUT 2/50: M. W. Baldwin (per George Burnham) to William Ash, June 15, 1850; OUT 7/50: memo of files ordered from Marriott and Atkinson, July 18, 1850; and M. W. Baldwin (per George Burnham) to Johnson Cammell & Co., Sept. 24, 1850, BLW-HSP.

10. OUT 8/47: George Burnham to M. W. Baldwin (Cape Island, N.J.), Aug. 14, 1847, BLW-HSP.

11. That engines were finished one at a time is evident in a series of letters describing

the assembly process written by George Burnham to M. W. Baldwin during the summers of 1847 and 1849. See OUT 3/47, 8/47, and 5/49, BLW-HSP.

12. IN 9/50: William Hamm to George W. Hufty, Sept. 6, 1850, BLW-HSP.

13. OUT 6/48: George Burnham to M. W. Baldwin (Cape Island, N.J.), Aug. 4, 1848, BLW-HSP. This letter suggests that Baldwin began piecework ca. 1848, but probably on a very limited basis.

14. Logic says most industrialists would have emulated Baldwin in first preferring to boost productivity through such low-cost organizational changes as piecework, leaving costly investments in capital-intensive production technologies for later. Even in American System manufacturing it is counterintuitive to believe that industrialists would proceed "from technology to organization" (Chandler, *Visible Hand,* p. 272). Indeed, as Robert Gordon has shown, they did not. According to Gordon, interchangeable manufacture at government and private armories originated and continued for decades with a primary reliance on improved hand skills and organizational techniques. Production machinery was secondary ("Mechanical Ideal into Mechanical Reality," esp. p. 759).

15. Note that the cost of labor (i.e., wages) did go up with the adoption of piece rates even if the quantity of labor (man-hours) stayed relatively constant. But the workforce's productivity increased even more. For example, in 1850 Baldwin required 10.8 man-years to build one engine; in 1852 it took 8.2 man-years.

16. OUT 1/51: M. W. Baldwin to Henry R. Campbell, Mar. 27, 1851, BLW-HSP. In this letter Baldwin complains of a profit margin of only 1 percent on revenues of $340,000 (over an undisclosed period).

17. The exact construction date is unknown; however, the letter cited in the next note suggests it was built in 1853.

18. For the 1853 expansion, see OUT 3/53: M. W. Baldwin to B. Cridland, Mar. 5, 1853. For 1857, see IN 1/57: W. R. Stockton (Philadelphia) to Matthew Baird, Feb. 6, 1857, and same to same, Mar. 16, 1857, BLW-HSP. For the size of the factory in 1860, see *Maps of the City of Philadelphia, Surveyed by Ernest Hexamer & William Lochner* (Philadelphia, 1860), HSP.

19. IN 1/50: Bancroft and Sellers to M. W. Baldwin, Feb. 21, 1850 (planer); IN 1/60: Bement and Dougherty to C. T. Parry, Mar. 27, 1860 (lathe); and IN 4/56: Hoff, Fontaine & Co. to M. W. Baldwin & Co., Apr. 7, 1856 ("chipping machine," or slotter), BLW-HSP. This is hardly an exhaustive list of all tooling acquired during the decade.

20. Roe, *English and American Tool Builders,* p. 256. Sellers started in 1848, and Bement began ca. 1852 (pp. 247, 254). Baldwin secured a Bement slide lathe for the Trinidad Railroad in 1856 (OUT 3/56: bill of lading for Trinidad Railroad, Oct. 16, 1856). For similar aid, see OUT 7/50: George Burnham to Bancroft and Sellers, Sept. 21, 1850, and OUT 5/49: George Burnham to M. W. Baldwin, Aug. 20, 1849, BLW-HSP.

21. An 1871 description of Bement's work on hand includes tools for seventeen railroads, five locomotive builders, and other firms making railway cars, steam engines, bridges, and ships (*Railroad Gazette* 3 [1871]: 321). Sellers' 1874 catalog, *A Treatise on Machine Tools,* shows that firm's reliance on locomotive firms and railways (Trade Catalog Collection, HAG).

22. Percentage based on value of output given in Robertson, "Changing Production," p. 489. According to Robertson, the Philadelphia firms produced 30 percent of total output in 1860, while all six New England states had 40 percent. Nathan Rosenberg's leading study ("Machine Tool Industry") slights the roles of the railway and locomotive sectors — and the Philadelphia tool builders — in the development of the machine tool industry

while concentrating on its relations to American System mass-producing manufacturers. For a discussion of how railway needs reordered the machine tool industry, see Sellers, "Progress of the Mechanical Arts."

23. Although Baldwin adopted certain machine tools during the decade, its capital-labor ratio declined between 1850 and 1860. The firm's investments only provided a foundation for future growth, with far more mechanization occurring in the 1860s. This accords with Louis Hunter's observation that the machine tool industry did not make significant contributions to stationary engine building until after the Civil War (*Industrial Power in the United States*, vol. 2, *Steam Power*, pp. 237–42).

24. "Japanese Visit M. W. Baldwin & Co.'s Locomotive Works," Philadelphia *Inquirer*, supplement, p. 1.

25. "Obituary of Charles T. Parry," p. 200.

26. This relatively early adoption of mechanical drawings marked another difference between the methods used in building capital equipment and those employed in manufacturing American System consumer products like guns and watches. As Robert Gordon and Patrick Malone note, such drawings appeared in manufacturing only after 1870 (*Texture of Industry*, pp. 374–77).

27. A surviving 1855 truck-frame drawing includes dimensions allowing use of this basic design in engines made to three different gauges of track (drawing #1879-B-1, WLA-HAG). Other engineering records provide further evidence of this standardizing process (see "Sizes of Driving Wheel Centers" [1852–68], BLW-DEG).

28. See "Baldwin Locomotive Works," *Engineering*, p. 141; William Austin, "Standard Reference Book" (ca. 1874), box 1, WLA-HAG; and Junior Law Books, BLW-DEG.

29. List titled "BLW STD. Boilers 6/26/89," drawing #1879-F-27, WLA-HAG.

30. "Baldwin Locomotive Works," *Engineering*, p. 140.

31. Watson, "Changes of One Lifetime," p. 888. Although Watson was discussing general machinists' practice of the 1850s, when he had been apprenticed to a marine engine builder, his comments describe Baldwin's problems exactly.

32. The leading accounts on interchangeable manufacturing include Smith, *Harpers Ferry Armory;* Hounshell, *American System to Mass Production;* and Hoke, *Ingenious Yankees.* Other valuable sources on this subject are "Fitch Report," pp. 611–704; Mayr and Post, *Yankee Enterprise;* Howard, "Interchangeable Parts"; Gordon, "Mechanical Ideal into Mechanical Reality"; and Fries, "British Response."

33. In a sense Baldwin's motivation to ease repairs mirrored the U.S. Armories' original goal in seeking interchangeable production. But the two cases quickly diverge and have relatively little in common. In wanting interchangeable parts to ease battlefield repair of weapons, the U.S. Army was its own customer and could therefore focus on the goal with little concern for costs. Baldwin lacked this luxury. Also, the technical parameters surrounding interchangeability in military small arms differed markedly from those in locomotives. For more on the Armories, see Smith, *Harpers Ferry Armory,* and Gordon, "Mechanical Ideal into Mechanical Reality." In Hounshell's leading account of American System firms (*American System to Mass Production*), the repair-parts motivation is seldom discussed as a factor in reaching for interchangeable work.

34. An 1839 start date is found in Church, "Extra Work Department," p. 29.

35. For the 1865 date, see ibid., p. 28; quotation from M. Baird & Co., *Illustrated Catalogue of Locomotives* (1871), p. 134. For most historians of technology, interchangeability "consists of making the parts of a mechanism so uniform in size that each mating part will go together and function properly without fitting" (Col. E. C. Peck, quoted in Howard,

"Interchangeable Parts," p. 636). This absolute concept arises from American System mechanisms (particularly guns) and from some of their historians. It is an overly rigorous definition that, contrary to Howard's assertion (p. 637), was not universally accepted in the nineteenth century (see Hoke, *Ingenious Yankees*, pp. 26–31). Interchangeability in locomotive components often met this definition, but not necessarily.

36. OUT 6/48: George Burnham to M. W. Baldwin (Cape Island, N.J.), Aug. 18, 1848, BLW-HSP.

37. IN 9/46: Isaac Dripps (Camden and Amboy Railroad) to M. W. Baldwin, Sept. 19, 1846, BLW-HSP.

38. OUT 9/56: M. W. Baldwin & Co. to Jas. S. Cox, Sept. 16, 1856, BLW-HSP. Tires required a relatively high level of precision to be interchangeable. In 1867 the shrink fit allowance—the variation between the diameters of the tire and the wheel center— for Krupp's Cast Steel tire was one-seventy-second of an inch per foot of diameter; for Butcher's American steel tire it was one-eightieth of an inch. These two allowances indicate that Baldwin worked to quite close tolerances, since it could determine the difference between one-seventy-second and one-eightieth—a figure of 0.0013. Even if machinists' actual practice varied to the extent of rounding off the last digit, this was still work to the thousandth of an inch, quite exacting precision in the metalworking world of 1867. For shrinkage allowances, see "Dimensions of Detail Parts," folder 5, box 27, BLW-DEG.

39. OUT 6/48: M. W. Baldwin (per George Burnham) to Hendricks and Brother (New York City), Oct. 30, 1848, BLW-HSP. Because flue and tire gauges were such simple tools, they probably suggested themselves, as it were. Yet they were essential in allowing Baldwin to purchase such finished parts from outside vendors.

40. OUT 8/46: M. W. Baldwin to R. R. Montgomery, Sept. 18, 1846, BLW-HSP.

41. While the average size of batch orders only grew from 2.6 in 1850 to 2.7 in 1860, the number of engines ordered in quantities of at least two grew from 16 to 54 over the same period. Author's analysis of 1850 and 1860 outputs in BLW Order Books, BLW-NMAH.

42. OUT 9/56: M. W. Baldwin & Co. to Jacob Perkins, Nov. 10, 1856, BLW-HSP.

43. OUT 12/53: M. W. Baldwin & Co. to E. Miller (chief engineer), Feb. 27, 1854, BLW-HSP. As the first case of general interchangeability stipulated by contract, the North Pennsylvania order whets interest, although further details are not known. The contract does not stipulate which parts should interchange, and there are no special remarks with this order in Baldwin's construction lists, the Register of Engines. I would speculate that only the main, moving parts were brought to interchange standards, and that the work was laboriously done by cut-and-try methods.

44. IN 9/54: Charles T. Parry (Corning, N.Y.) to George W. Hufty, Sept. 4, 1854, BLW-HSP.

45. "Obituary of Charles T. Parry," pp. 201–2. Baldwin's old 4-4-0 is shown as class C, no. 1, on page 15; its new design appears on page 60.

46. "Baldwin Locomotive Works," *American Artisan*, p. 483. This tie to Brown and Sharpe is the one known instance where Baldwin's interchange efforts had a direct link to New England Armory practitioners.

47. "Fitch Report," p. 48, and Burnham, Parry, Williams & Co., *Illustrated Catalogue of Locomotives* (1881), p. 57.

48. The Law Books begin ca. 1872; precisely how the use of gauges was mandated to draftsmen or workers before then is unclear. A law on gauging is shown on page 111.

49. "Fitch Report," p. 48.

50. Ibid., and "Baldwin Locomotive Works," *American Artisan,* p. 483.

51. Baldwin ordered quantities of costly semifinished materials like boiler flues, steel tires, forged axles, and sheet copper from specialist firms, with the expense reaching over 50 percent of the final selling price of a locomotive ("Fitch Report," p. 46). Curbing wastage of such expensive materials was another benefit of the new interchange standards.

52. "Workmen," *Engineer,* p. 51.

53. Robert Gordon has provided an insightful breakdown of the constituent elements encompassed in the term *skill,* including planning, judgment, dexterity, and resourcefulness ("Mechanical Ideal into Mechanical Reality," pp. 768–70). Interchangeable production required more dexterity and judgment on the part of Baldwin's machinists and less planning and resourcefulness compared with their work before interchange standards.

54. Capital measured in constant (1850) dollars. This increase shows how Baldwin used its relative prosperity of the 1860s to reinvest profits in plant and equipment.

55. Baldwin's entry in the manuscript schedules of the 1870 U.S. Manufacturing Census lists this sort of general-purpose tooling in the plant.

56. This accords with Daniel Nelson's observation of the American System factory as a "near-ideal work environment" whose workers "were on the whole a prosperous and contented group" ("The American System and the American Worker," in *Yankee Enterprise,* ed. Mayr and Post, p. 172).

57. Such tests have been a foundation of the modern historiography of interchangeability ever since Edwin Battison demolished Eli Whitney's claims to be the father of interchangeable production by tests of Whitney guns, described in his "Eli Whitney and the Milling Machine." Subsequent tests have been done by Smith, Howard, Hounshell, Hoke, and Gordon. Similar tests of interchangeability are impossible with Baldwin locomotives because too few survive, and the ones that do exist have undergone so much modification and wear.

58. In specifications and locomotive contracts of the early interchange era, Baldwin did not explicitly claim interchangeable parts, using instead such equivocal language as "the engines to be substantially duplicates." See M. Baird & Co. to John E. Wootten, Mar. 19, 1873, file: John E. Wootten, Incoming Correspondence, Locomotives and Rolling Stock, 1873, box B-175, Reading Company Collection, HAG.

59. In 1872 Thurston visited Baldwin and wrote that "all work is . . . made to gage and the several parts are 'assembled,' to make the complete machine, without the expense attending the old process of 'cutting and carving' in fitting up." See "Letter from Professor R. H. Thurston," *Scientific American,* p. 40.

60. The presence of two fitters, men who literally fitted parts together, is indicated in *Pennsylvania Industrial Statistics Report,* vol. 4 (1877), p. 546.

61. Under an absolute definition of interchangeability, such fitting or adjustments would disqualify these parts from being considered truly interchangeable. To a railroad such a distinction was academic. Minor fitting was far better than having to send a broken axle back to Baldwin for fabrication of an exact duplicate, or having to forge and turn a new axle in its own shops while the engine sat idle.

62. White, *American Locomotives,* p. 79.

63. J. T. Redmon (master mechanic) to Burnham, Parry, Williams & Co., Feb. 2, 1878. Letter printed by Baldwin as an advertising circular, folder: BLW Test Department Specifications, 1885–1905, box 6, WLA-HAG. It is unclear whether Baldwin made all the required parts in that three-day period or drew some of them from finished inventory

stocks. In either case interchangeability standards were essential to providing such fast service.

64. M. B. Prichard to Burnham, Parry, Williams & Co., Mar. 13, 1877. Letter printed by Baldwin as an advertising circular, folder: BLW Test Department Specifications, 1885–1905, box 6, WLA-HAG.

65. Once railroads could purchase replacement parts from Baldwin, they were less inclined to make their own (although many continued to use their existing repair shops and staffs for this purpose). Baldwin's sales of replacement parts grew from roughly 8 percent of total output ca. 1855 to 15 or 20 percent by 1900. While providing healthy profits, this market also required Baldwin to maintain extensive records and many gauges and templates for engine designs long after they had become obsolete in new construction.

66. The question of whether interchangeability affected costs and prices of products in American System industries has received a variety of answers. Smith (*Harpers Ferry Armory*) and Hounshell (*American System to Mass Production*) argue that prices (or costs) necessarily increased, at least in the short run; Hoke (*Ingenious Yankees*) and Fries ("British Response") believe that interchangeable production lowered costs and selling prices. As Fitch notes, locomotive price changes over time arising from evolving production methods are concealed by overall deflation, fluctuating materials costs, and (I would add) the dizzying technical changes in the products themselves ("Fitch Report," pp. 46–47). Table 6.2 shows, however, that Baldwin's costs of production declined after the company achieved interchangeability.

67. The factors described in the previous note thwart locomotive price comparisons over time. But comparisons of different sales of the same design at the same time, differing only in units produced, indicate unchanged selling prices despite the productive efficiencies that derived from larger orders.

68. Data from author's analysis of Baldwin's sales listed in Order Books and Registers of Engines, BLW-NMAH. In 1900 the average batch order size in the domestic railroad segment of Baldwin's sales (i.e., not including export and industrial customers) was 9.1 engines. A batch is defined as one sale to one railroad of one design in quantities of two or more.

69. A survey of the catalogs of various locomotive builders gives the following dates of origin for interchangeable production: Grant 1864, Baldwin 1865, and Rhode Island 1871. A Mason catalog from 1879 has no mention of interchangeability, and an informal survey of builders' ads in the trade press also suggests that Baldwin and Grant led the industry in this accomplishment.

70. Even if interchangeability did not lower engine prices, it could still result in an increasing volume of business for Baldwin by heightening throughput—a real advantage, since locomotive builders often competed on terms of early delivery dates rather than price alone.

71. The subject of interchangeability in locomotives cannot be left without some remarks on how Baldwin's record relates to the larger historiography of this subject. In general this study supports the arguments of Donald Hoke, and therefore challenges much of David Hounshell's account of interchangeable production. In brief: Baldwin achieved interchangeable production by developing its own form of Armory practice, but with little or no discernible link to the U.S. Armories. Interchangeability in production depended on such preparatory developments as piecework, product redesign, and standard parts. In combination these developments lowered production costs. There was no need for absolute precision in all parts—the Armories' goal—for a mechanism to have

interchangeable parts. The degree of precision required was tempered by the adjustability of the mechanism. Interchangeability was a technical ideal, but in competitive private sector markets, the ideal was modified to suit economic ends.

72. Freedley, *Philadelphia and Its Manufactures,* p. 315.

73. Sinclair, "At the Turn of a Screw."

74. Information on tooling from "Local Affairs," Philadelphia *Ledger and Transcript,* p. 3. Baldwin's 1873 capitalization is listed in the R. G. Dun credit reports as ranging from $3.075 million to $3.5 million. See entries for Burnham, Parry, Williams & Co. in R. G. Dun reports, HAG.

75. "Local Affairs," Philadelphia *Ledger and Transcript,* p. 3.

76. Laurie and Schmitz used the raw data from the U.S. Manufacturing Census for Philadelphia to come to this conclusion ("Manufacture and Productivity," quotations from pp. 78 and 88), which is repeated in Montgomery, *Fall of the House of Labor,* p. 55.

77. It is impossible because the manufacturing census data are grossly inaccurate for most firms in the industry. In Baldwin's case other sources allow me to correct for these errors (e.g., the 1870 census gives Baldwin $12.5 million in capital, a figure Laurie and Schmitz must have used, when actual capitalization was roughly $2 million). Judith McGaw found similar problems in the manufacturing census data on the paper industry, calling them "usually inaccurate . . . often extremely inaccurate" (*Most Wonderful Machine,* p. 407). At a minimum, this characterization calls the validity of Laurie and Schmitz's findings into question.

78. As this sequence suggests, Laurie and Schmitz overlook the possibility that industrial managers of this era consciously chose complementary long-term strategies of improving productivity (through capital investment) and enlarging market share (with highly competitive pricing). Such policies would secure growth and higher unit profits over the long run, even as they depressed return on investment in the short term. Philip Scranton kindly provided this insight, which Baldwin's experience supports.

79. Although Baldwin's financial records for this period do not survive, it is fair to assume that the company was profitable, since it made installment payments on Baird's interest in the firm throughout the depression.

80. The presence of productive efficiencies accruing to firms with economies of scale is often testable using the Stigler survivor test. This posits that in an industry composed of firms of various sizes, those companies operating at the optimal size for best efficiency will survive the longest. Under this criterion Baldwin would appear to have been more efficient than its smaller competitors, although the boom-and-bust cycles and episodes of price fixing in the locomotive industry cloud the survivor test.

81. For the 1880 building, see "Notes of Travel," *Railroad Gazette* 12 (Oct. 15, 1880): 546; for 1884, "Fire at the Baldwin Locomotive Works," *Iron Age* 34 (Aug. 7, 1884): 30; for 1890, "Baldwin Locomotive Works," *Railroad Gazette* 23 (Apr. 10, 1891): 246. Baldwin also built a new six-story structure in 1889 at Fifteenth and Spring Garden Streets, originally used for pattern storage and then converted to a machine and erecting shop for electric trucks; see "Enlarging Locomotive and Car Works," *Railroad Gazette* 21 (July 5, 1889): 443.

82. U.S. patents 257,061 (1882), 305,709 (1884), 343,842 (1886), and 410,051 (1889).

83. "Fitch Report," pp. 58–59. Fitch was ostensibly speaking of the whole industry here, but it is clear that his report was largely based on developments at Baldwin.

84. This was the sort of "technological convergence" described in American System industries by Nathan Rosenberg in "Technological Change in the Machine Tool Industry."

85. Note, however, that Baldwin's special-purpose plate planers, automated drilling machines, and the like were adjustable, allowing them to be used to make parts in various sizes—unlike much of the special-purpose tooling in American System manufacturing.

86. [James W. See], *Extracts from Chordal's Letters*, p. 127.

87. L. B. Paxon to John E. Wootten, Nov. 12, 1881, file 33, box B-228, Reading Company Collection, HAG. Its original system was an Edison incandescent plant.

88. Vauclain and May, *Steaming Up*, p. 117. Naomi Lamoreaux notes that between 1880 and 1900 the capital invested in American manufacturing increased by 75 percent (*Great Merger Movement*, p. 29). It is interesting to speculate whether this capital growth depended on the development of electric lighting to allow the capital to be utilized with sufficient intensity. For a general study of electrification, see Devine, "Shafts to Wires."

89. Richmond, "Operating Machine Tools by Electricity," p. 670.

90. Flather, "Modern Power Problem," pp. 637–38.

91. "An Electric Traveling Crane," *Railroad Gazette* 21 (Feb. 8, 1889): 95.

92. S. M. Vauclain, Memorandum Book of Tooling Costs, box 7, SMV-DEG, and Outerbridge, "Labor Saving Machinery," p. 651. The cranes were first used on December 19, 1890; hence the 1890 productivity data in tables 6.3 and 6.4 are essentially pre-electrification.

93. "Extension of Electric Driving at the Baldwin Locomotive Works," *American Machinist*, pp. 137–38.

94. "Report of the Committee on Power Transmission by Shafting vs. Electricity," *Thirty-third Master Mechanics Report* (1900), p. 328.

95. S. M. Vauclain, Memorandum Book of Tooling Costs, box 7, SMV-DEG. Baldwin generally used a direct motor drive for its larger tools (like wheel lathes), while smaller tools were brought together for a group drive with belts and countershafts from one motor. Total horsepower from ibid., p. 327.

96. These general findings are echoed in Hunter and Bryant, *Industrial Power in the United States*, vol. 3, *Transmission of Power*, chap. 4.

97. Vauclain quoted in Crocker, "Electric Distribution of Power in Workshops," p. 8.

98. Of $3.1 million spent on tooling between 1888 and 1904, $469,000 (or 15 percent) went into materials-handling devices. Figures computed from S. M. Vauclain, Memorandum Book of Tooling Costs, box 7, SMV-DEG.

99. In 1886 Baldwin had 4 acres of blacksmith shops and made ten engines a week. By 1893 the smiths occupied only 2 acres, yet output had more than doubled. See *Twenty-sixth Master Mechanics Report* (1893), p. 60.

100. Annual expenses given in sheet of overhead and direct labor costs (1881–1910) found in Assorted Train File, box 1, SMV-DEG.

101. "Nine Hundred Men Work Undisturbed While Their Workshop Is Rebuilt," Philadelphia *Inquirer*, July 24, 1900, p. 2.

102. Computed from data in White, *Short History*, p. 21, with the decade set at 1900–1909.

103. Nelson, *Managers and Workers*, p. 22.

104. Revenues given in clipping from Philadelphia *Record*, Sept. 30, 1907, in Engine Order Book, 1908, Scrapbook, p. 1288, Division of Transportation, NMAH.

105. National data from *Historical Statistics*, pt. 1, p. 225.

106. Times based on days elapsed from initial ordering of raw materials to final testing of finished engines. Such times varied greatly depending on the general volume of work in the plant, but the sixty-day figure was the norm.

7: TRIUMPH AND ECLIPSE, 1900–1915

1. The dinner is described in BLW, *Record of Recent Construction #33.*

2. Data in S. M. Vauclain's Production Notebook for 1902, box 9, SMV-DEG.

3. Market share tabulated from data in White, *Short History,* p. 21. For Alco, see p. 23.

4. Pusateri, *History of American Business,* p. 172. The seven controlling parties were Vanderbilt, Morgan, Gould, Harriman, Hill, the Rock Island, and the Pennsylvania.

5. Percentage increase derived from a weighted output index (passenger miles and freight ton-miles) constructed by Martin, *Enterprise Denied,* p. 374. With traffic growing by 107 percent while mileage increased by only 34 percent, it is clear that the major carriers made great improvements in productivity—thanks in part to powerful new locomotives.

6. *Historical Statistics,* pt. 2, p. 728.

7. *Baldwin History,* p. 103.

8. For railway administration of technology, see Usselman, "Running the Machine."

9. The CMOs often purchased engines in a variety of classes that shared common detail parts. But they also frequently sought designs adapted to particular 75- to 100-mile divisions of their roads, increasing the variety in their locomotive fleets to improve operating efficiency. There is no doubt that the variety of engines increased on most roads and in Baldwin's production during this period. For the heightened pace of innovation, see Bruce, *Steam Locomotive in America,* pp. 75–82.

10. The railways' desire for better engines ca. 1900 is evident in the percentage of the national motive power fleet that was composed of modern designs. In 1901, 35 percent of the national fleet was less than a decade old. By 1907 that figure reached 60 percent. See BLW, "Motive Power Situation," p. 39.

11. "Improvements at the Baldwin Locomotive Works," *Railroad Gazette,* p. 829.

12. S. M. Vauclain, Memorandum Book of Tooling Costs, box 7, SMV-DEG.

13. "Acceptance Tests for Locomotives," *Railroad Gazette* 32 (Oct. 19, 1900): 682.

14. Wherever possible Baldwin gave priority to domestic customers over foreign ones, and certain U. S. carriers such as the Pennsylvania and the Santa Fe were particularly favored.

15. Market shares computed from data in White, *Short History,* p. 23. For Alco's expansion, see articles in *Railroad Gazette,* June 13, Oct. 3 and 31, and Dec. 26, 1902.

16. In 1890 Baldwin required 4.75 man-years of labor to build one locomotive; by 1906 that figure had risen to 6.54 man-years, despite millions invested in new facilities and tooling.

17. A list of sixty-two boiler shop contracts from 1914 shows the extent to which Baldwin used contractors to oversee relatively simple tasks, such as supervising chippers and caulkers and even laborers. See sheet titled "Boiler Shop, Best Averages" in S. M. Vauclain's Production Notebook for 1914, box 3, SMV-DEG.

18. Henszey, "Organization and Methods," p. 402.

19. See articles of those titles in (respectively) *Railway Age Gazette* 49 (July 8, 1910): 65 and *Railroad Gazette* 32 (Oct. 19, 1900): 681. The quotation following in the text is from the latter, p. 682.

20. For different views on these developments, see Chandler, *Visible Hand;* Noble, *America by Design;* and Beniger, *Control Revolution.*

21. Johnson quoted in Smith, *Elements,* p. 23.

22. This attitude continued at Baldwin into the 1920s, well after incorporation. In 1921 president Samuel Vauclain wrote an article entitled "Is Your Business a Debating

Club?" Its subheading summarized his argument: "Don't let the organization germ bite too deeply—for the surest way of all to kill a good idea is to commit it to a committee!" *Collier's,* Apr. 2, 1921, pp. 10, 11, 18.

23. This development is evident in a comparison of BLW, "Rules for Timekeeping and Payment of Wages," Feb. 10, 1905, 11, box 5, WLA-HAG, vs. BLW, "Shop Rules," July 1910, 5, box 3, SMV-DEG. The new employment department took over responsibility for hiring from the foremen, who retained their power to fire workers.

24. Locomotive sales in 1906 reached $31.2 million (data from sheet in S. M. Vauclain's Production Notebook for 1907, box 7, SMV-DEG). Because Baldwin was a partnership at this time, exact data on sales and profits are difficult to find. But its average annual sales (1902–11) were $27.2 million, with net profits averaging $2.6 million (BLW *Annual Report* for 1911, p. 5).

25. Johnson, "Market for Locomotives," p. 547.

26. As described in chap. 5, the average real wage per employee increased by 60 percent from 1863 to 1893. From 1898 to 1906 average wages per man increased slightly in current dollars, from $12.72 a week to $12.91; but when inflation is factored in, this period saw a 15 percent decline in real wages. Wages for 1898 given in S. M. Vauclain's Production Notebook for 1898, data for 1906 from his Production Notebook for 1912, boxes 8 and 9, SMV-DEG. The Burgess cost-of-living index was used to convert these data to real wages (index in *Historical Statistics,* pt. 1, p. 212). Although Gilded Age deflation and subsequent inflation had a major impact on workers' real pay, I believe that the long-term changes in their income resulted from deliberate company policies that in turn arose from gains and losses in labor productivity (charted by cost accounting)—not simply from external monetary forces.

27. Surviving data on bids in the Vauclain papers (DEG) suggest vigorous and close bidding between Baldwin and Alco in this period—particularly for orders from carriers that had not established a close design relationship with a single builder. Such pricing behavior was far more common in mass-producing industries, where unit costs declined substantially with overall growth in sales.

28. The issue of escalating locomotive prices became so sensitive that in 1907 Baldwin opened its order and price data for public review. A BLW officer, Lawford H. Fry, wrote a detailed article for *Railroad Gazette* to disprove price gouging by the locomotive builders ("Cost of Locomotives", quotation following in the text is from p. 803).

29. *Baldwin History,* p. 92.

30. Smith, *Elements,* p. 28. Baldwin's special relationship with the Pennsylvania continued after 1900. Between 1900 and 1909 the carrier contracted to place almost all its outside orders for engines with Baldwin. These agreements gave the builder a healthy source of profits—the sales were priced at cost plus 10 percent—while the Pennsylvania received guaranteed precedence over other carriers in Baldwin's production schedules. See agreements in BLW Contract Book, 3:159–60 and 288–93, BLW Collection, Pennsylvania State Archives.

31. "Baldwin Locomotive Works and the Metric System," *Iron Age,* pp. 1885–86.

32. Baldwin pursued this policy so vigorously that it could hire untrained eighteen-year-old boys, place them at lathes, slotters, or planers, and get adequate productivity from them in less than a month. See folder: Employees' Records, 1919, box 2, SMV-HSP.

33. Converse, "Some Features," p. 4.

34. Henszey, "Organization and Methods," pp. 402–3.

35. The "new apprenticeship" is discussed in Nelson, *Managers and Workers,* chap. 5; McGuffie, *Working in Metal,* chap. 2; and Noble, *America by Design,* chap. 10.

36. Vauclain, "System of Apprenticeship," pp. 323, 330.

37. As Monte Calvert puts it, men of the shop-culture engineering elite wanted to preserve or renew apprenticeships for their workers because "as long as both made their way through the same apprenticeship system, this personal contact led to understanding and respect between the men of the office and shop" (*Mechanical Engineer in America*, p. 73).

38. This description of the program's structure is drawn from Vauclain, "System of Apprenticeship." First-class boys had elementary school educations, the second class were high school graduates, and college men composed the third group.

39. These eleven categories of work in one trade testify to the division of labor Baldwin had instituted.

40. Diaries of Vernon Gotwals Sr. (1886–1972), owned by his son, who kindly provided them to the author for transcription of relevant entries.

41. Remarks of Samuel Vauclain following "Report on the Apprenticeship System," *Forty-first Master Mechanics Report* (1908), p. 194.

42. "A Simple Apprenticeship System," *American Machinist* 33 (Nov. 10, 1910): 876.

43. For example, Vernon Gotwals had a forty-eight-year career at Baldwin, ending up as chief inspector in the drafting room. Another apprentice, François de St. Phalle, became Baldwin's vice president in charge of foreign sales in the 1920s, while a third, Elmer Hemberger, was an inside contractor in the 1930s. Fifty of the two hundred graduates remaining at the firm in 1908 "occupied places of responsibility as heads of department, foremen, assistant foremen, contractors and leading workmen" (all three classes of apprentices were represented among these fifty graduates). Sample, "Apprenticeship System," p. 177.

44. Note, however, that the program never sought to provide an exploitable pool of labor per se. For one thing, it was purposely limited to roughly 3 percent of the labor force.

45. See bibliography for articles by Converse, Fry, Henszey, Johnson, Sample, and Vauclain.

46. David Montgomery describes mounting strife in labor-management relations in the late-1890s, followed by Progressive mediation efforts—such as the Murray Hill agreement, negotiated in 1900 under the auspices of the National Civic Federation, which established a nine-hour workday for machinists—and then renewed labor-management conflict (*Fall of the House of Labor*, chap. 6).

47. Bingham, "Labor Situation," pp. 40–41. All quotations in this paragraph are from this source.

48. Data from Philadelphia *Record*, Sept. 30, 1907, clipping in Engine Order Book, 1908, Scrapbook, Division of Transportation, NMAH. Baldwin's non-locomotive revenues included sales of repair parts, electric trucks for trolleys and interurbans, and contracts for engine repairs.

49. Johnson, "Chester's Opportunity."

50. Albro Martin's *Enterprise Denied* describes the regulatory stranglehold placed on American railroads after the Hepburn Act. His thesis that regulation was the proximate cause of railroading's long decline has been rejected by some historians, who also point to the post-1920 rise of such new railway competitors as automobiles and motor trucks. The Hepburn Act's effect on railway equipment builders such as Baldwin became evident very quickly (although other factors also played a role in the builders' decline). Because the locomotive industry was a leading indicator of railroading's financial health, Baldwin's change in fortunes after 1907 lends indirect support to Martin's argument.

51. Carosso, *The Morgans*, p. 534.

52. Martin, *Enterprise Denied*, p. 6.

53. In April 1908 Baldwin built fifty-two locomotives and lost $27,000 on the batch. See folder: Defense of the Drawing Room, box 9, WLA-HAG.

54. Will of William Henszey, #718 of 1909, Office of the Registrar of Wills, Philadelphia City Hall. For Henszey's percentage of ownership, see sheet in S. M. Vauclain's Production Notebook for 1909, box 7, SMV-DEG. For its value, see Vauclain and May, *Steaming Up*, p. 168.

55. Evidently the partners were divided ca. 1900 on the desirability of incorporating. By reforming the partnership in 1901 and 1907, Baldwin chose to continue proprietary management. But no new partners were admitted in either year, ensuring the ultimate death of the partnership. I speculate that a minority of the partners wanted incorporation before 1909; they lacked the votes to achieve that goal but could veto any new additions to the firm. According to another source, the partners remained divided on whether to incorporate right through 1909 (*Hardin v. Robinson*).

56. Vauclain and May, *Steaming Up*, p. 291.

57. The strike is described in Foner, *Labor Movement in the United States,* vol. 5, *AFL in the Progressive Era,* chap. 6.

58. Report of George Dearnley to E. L. Walker, Feb. 23, 1910, folder: Correspondence, 1910–11, box 1, SMV-HSP.

59. William Austin to William Penn Evans, May 2, 1910, folder: Technical Correspondence and Reports, 1908–10, box 5, WLA-HAG.

60. The Duluth story ran with a photo of Philadelphia's finest firing into the plant (Duluth *News Tribune,* Feb. 27, 1910, p. 1). *Bulletin* story quoted in Foner, *AFL in the Progressive Era,* p. 147.

61. Foner, *AFL in the Progressive Era,* pp. 151–54. The PRT strike ended with a compromise settlement in the third week of April (pp. 159–62).

62. Baldwin's workforce for the week ending March 12 totaled 12,059 men, a high point for the year to date. Average wages were off during the week, however, indicating that some men had briefly joined the sympathy strike (data from S. M. Vauclain's Production Notebook for 1910, box 7, SMV-DEG). This source shows Foner to be mistaken in reporting that six thousand Baldwin men went out in the first week of the strike or at any point in 1910, for that matter.

63. Quotation from W. J. Bender's report to Mr. Kenny, folder: Correspondence, 1910–11, box 1, SMV-HSP.

64. Report of H. Ireland to Mr. Walker, Mar. 10, 1910, folder: Correspondence, 1910–11, box 1, SMV-HSP.

65. Minutes of the Central Labor Union, Mar. 13, 1910, p. 152, Urban Archives, Temple University.

66. Alco had seen labor disturbances in 1908, protesting experiments with time study and incentive wages by Frederick Taylor's follower, Harrington Emerson. The company soon dropped these planks of Scientific Management, and it accepted unions in its plants beginning in 1909. The IAM knew that Alco's management would be less inclined to move against its locals there if Alco's main competitor was also organized. For Scientific Management and organizing at Alco, see *Machinists Monthly Journal* 22 (Jan. 1910): 42, 58–59, and Nelson, *Frederick W. Taylor,* pp. 129–130, 153.

67. Petition details from "First Floor 16th St. Shop" to Samuel Vauclain, Mar. 10, 1910, folder: Correspondence, 1910–11, box 1, SMV-HSP. Force totals from S. M. Vauclain's Production Notebook for 1910, box 7, SMV-DEG.

68. The 1914 Congressional Commission on Industrial Relations (cited here as CIR *Report*) elicited substantial testimony on Baldwin's 1910 labor unrest from company officers and union leaders. The latter had a litany of complaints against the firm, many clearly ex post facto justifications rather than actual matters raised in 1910. The issues noted here are found in the actual petitions and complaints of 1910 as well as in the CIR *Report*.

69. When the *New York Times* announced the issue in a news item, it made no mention of any ongoing labor problems ("Baldwin Co. Bond Issue," *New York Times*, Apr. 13, 1910, p. 13).

70. "Baldwin to Add 4000 to Force," Philadelphia *Public Ledger*, Mar. 13, 1910, p. 1.

71. The evidence of strong immigrant participation comes from the March 14 report of a Vauclain spy who attended a meeting of Baldwin men at the Labor Lyceum and reported that "the majority" of speakers addressed the crowd in their native languages. See handwritten report of Mar. 14, 1910, folder: Correspondence, 1910–11, box 1, SMV-HSP.

72. *Machinists Monthly Journal* 22 (Sept. 1910): 854.

73. Moore, "Strike at Baldwin's," p. 90. Moore says there also was a small Industrial Workers of the World local at Baldwin. I am indebted to Andrea Kluge for providing this source.

74. *Machinists Monthly Journal* 22 (Dec. 1910): 1168.

75. See rules 4, 8, 18, and 22 in Piecework Section, BLW "Shop Rules," July 1910, box 3, SMV-DEG. For the half-holiday concession, see Ken Fones-Wolf citing the Philadelphia *North American* in "Mass Strikes and Corporate Strategies," p. 452.

76. I have had these notes translated from their Pittman shorthand into clear text. They reveal a great concern on Austin's part that contractors were out of control. As one manager told Austin, "The shop is run more by the contractors than the foreman." Conversation with Bernard Converse, Aug. 14, 1910, p. 20, William Austin notes, box 5, WLA-HAG.

77. In 1907 the average weekly wage per man was $13.05; for 1910 it reached $13.94 (current dollars). S. M. Vauclain's Production Notebook for 1910, box 7, SMV-DEG.

78. Durland, "Industrial Deadlock," p. 3.

79. A pamphlet describing the association and its benefits is in case 48, file 332-19, Boston Chamber of Commerce Collection, Baker Library, Harvard Business School. In 1914, 67 percent of Baldwin's 5,530 employees were members of the association, suggesting that it offered valuable benefits to workers ("Status of Employees of the Baldwin Locomotive Works," *Manufacturers' Record,* p. 65).

80. *Machinists Monthly Journal* 23 (July 1911): 683.

81. *New York Times,* June 9, 1911, p. 3.

82. In the week ending June 10 the payroll totaled 13,035 men. A week later—the first full week of the strike—6,927 were paid. Getting half the force out was no small accomplishment for the unions, but it would prove insufficient. Figures from S. M. Vauclain's Production Notebook for 1911, box 7, SMV-DEG.

83. By July 1, 7,686 men were at work, and three weeks later the total passed 10,000. Strike support was evenly divided between the Broad Street and Eddystone plants.

84. Typescript report of M. Sullivan, "Metal Trades in Philadelphia," June 1914, 5, box 19, Commission on Industrial Relations Papers, record group #174, National Archives.

85. In his autobiography written twenty years later, Vauclain opens his description of the strike by saying, "When your company has recently been reorganized, there is serious

illness in your family, [and] you are called out on the road to settle a selling difficulty, you don't like it when the men you believe are getting the best of it walk out on you" (Vauclain and May, *Steaming Up,* p. 175).

86. See letters to Samuel Vauclain, June 1911, p. 173, M. Baird & Co. Scrapbook, Division of Transportation, NMAH.

87. CIR *Report,* 3:2833 (time-and-a-half concession), 2823 (apprenticeship program's end).

88. "The BLW Loyal Legion," *Railway Age Gazette* 52 (Feb. 23, 1912): 351. This is the only known source measuring seniority in Baldwin's workforce. It is imprecise, for many senior men no doubt struck in 1911 and never came back, while membership included management and draftsmen as well as production workers.

89. The 1909 board included only former partners or their heirs. With reincorporation the board acquired new blood, including seven outside directors (a majority) of whom five represented the banks or brokerage houses involved in the public offering. Alba Johnson served as the firm's new president (1911–19), replacing William Austin.

90. In 1909 Burnham, Williams & Co. (the partnership) gave way to Baldwin Locomotive Works (a privately held corporation), which was restructured in 1911 to become The Baldwin Locomotive Works (publicly held). Three brokerage houses combined in the 1911 transaction: Drexel & Co.; Montgomery, Clothier and Tyler; and White, Weld & Co. The lead broker, Drexel, in turn called on its allied houses, J. P. Morgan & Co. (New York) and Morgan, Grenfell & Co. (London), to help underwrite and distribute the Baldwin issue (Carosso, *The Morgans,* p. 604).

91. Chandler and Salsbury, *Pierre S. du Pont,* p. 294. In fact Baldwin's 1911 capitalization was rather conservatively structured. The common carried no guaranteed dividend, the preferred paid 7 percent, and in the previous decade Baldwin earned an average of 14 percent annually on the amount of preferred that it would issue in 1911. See prospectus titled "The Baldwin Locomotive Works" in Papers of P. S. du Pont, group 10, series A, file 562: "Baldwin Loco. Works," Longwood Manuscripts, HAG.

92. Du Pont bought eighty shares of Baldwin preferred in 1911 for his own and family accounts and purchased another sixty shares in 1914. Papers of P. S. du Pont, group 10, series A, file 562: "Baldwin Loco. Works," Longwood Manuscripts, HAG.

93. Between 1910 and 1915 Baldwin's net profits averaged 9.8 percent of sales, with annual profits ranging from 2.6 percent (1914) to 12.8 percent (1915).

94. *Baldwin History,* p. 120 (erecting shop); BLW *Annual Report* for 1911, p. 7 (East Chicago plant). Alco also contemplated a new western plant at this time, believing, like Baldwin, that it would allow substantial economies on freight expenses for shipping engines to customers west of the Mississippi. Neither plant was ultimately built.

95. Memo in S. M. Vauclain's Production Notebook for 1911, box 7, SMV-DEG.

96. Martin, *Enterprise Denied,* pp. 310, 135, 294, 308, 374, 136. Martin's thesis is that a generally hostile regulatory climate coupled with the specific refusal of the ICC to raise rates crippled railroading's profitability after 1906. But profits per se are not Martin's concern. Instead he analyzes how the decline in profitability—due to inflation, the wage demands of organized labor, and the freeze in rates—affected the railroads' ability to make the capital investments necessary to carry an increasing burden of traffic. This analysis required Martin to posit a relationship between the volume of traffic and the level of investment in the pre-Hepburn period (he chose the average for the years 1905–7). Using this relationship as his base, he then carried the data forward to derive the figure of $5.6 billion that the railroads "needed" for investment between 1907 and 1914 but

did not receive because of regulation. Baldwin's rapidly plummeting fortunes after 1907 support Martin's general thesis; however, there are data (described below) that challenge the validity of his 1905–7 base.

97. Ibid., p. 132.

98. Johnson, "Chester's Opportunity," pp. 5–7. This included projected annual capacities of 3,000 engines at Eddystone and 1,560 at Chicago as well as continued operations at Broad Street.

99. Bruce, *Steam Locomotive in America,* p. 73. Baldwin's output peak (for U.S. customers) came a year after the high point in orders because production lagged behind demand in this bull market.

100. After 1907 annual orders by Class 1 companies (for all builders) surpassed three thousand units only three times: 1910, 1912, and 1913 (data from Park, "U.S. Steam Locomotive Builders," p. 53). Appendix A shows Baldwin's declining output after 1907. Because they include export sales—particularly strong during World War I—those data only suggest the mounting debility of U.S. carriers.

101. The thirty-year service life is given in Bruce, *Steam Locomotive in America,* p. 69. In 1901, 35 percent of the U.S. motive power fleet was less than a decade old; that figure reached 61 percent in 1907. See BLW, "Motive Power Situation," p. 39.

102. It ended in 1907; thereafter the national motive power fleet steadily aged. The trend after 1907 supports Albro Martin's thesis of a regulatory stranglehold on the investment capability of American railroads. But the abnormally large purchases of locomotives between 1905 and 1907 suggest that Martin's counterfactual analysis, based on the rate of investment in those years, overstates the railways' forgone investment needs after 1907.

103. Raffe Emerson to Harrington Emerson, Aug. 21, 1904, file 1, box 6, Harrington Emerson Papers, Pennsylvania Historical Collections, Pennsylvania State University Library. I am indebted to James Quigel for this source.

104. Johnson, "Efficiency of Modern Railway Equipment," p. 424.

105. In 1903 the average tractive effort (a measure of hauling capacity) of engines on Class 1 railroads was 21,781 pounds; ten years later it stood at 30,258 pounds. By 1925 average tractive effort exceeded 40,000 pounds. *Historical Statistics,* pt. 2, p. 728.

106. Baldwin's new president, Alba Johnson, was eminently aware of the potential harm of the Hepburn Act and other regulatory legislation. But he believed this was an artificial problem (which it was) that would pass once reasonable men could act (but they didn't). For example, he was confident "that much of this legislation was clearly unconstitutional, and must surely be set aside by the courts" ("Market for Locomotives," p. 548).

107. Between 1915 and 1918 the vast majority of Baldwin's output went to Europe in the form of engines for the warring Allies' home needs, small locomotives for hauling men and ammunition in the trenches, and large Consolidations (2-8-0s) for the American military railway service in Europe. Baldwin and its wartime subsidiaries also made 6.5 million artillery shells and a number of heavy gun mounts. All this work pushed sales from $13.6 million in 1914 to $123.2 million in 1918. See BLW *Annual Report* for 1918.

108. Alco began to diversify in 1905, entering the automobile and motor truck markets. Although the venture proved a failure—it was discontinued in 1914—it does suggest the myopia of Baldwin's managers, who would not look beyond the dwindling locomotive market until 1929. In that least auspicious of years Baldwin acquired control of three companies: Midvale (a specialty steel producer), the Southwark Foundry and Machine Company, and George D. Whitcomb (makers of gasoline locomotives for

industry). The company was finally impelled to make these acquisitions because the Fisher brothers (of Detroit's Fisher Auto Body) had themselves acquired a controlling interest in Baldwin in 1927, and Edward Fisher was "anxious to amplify . . . operations . . . by the addition of other lines of manufacture." Samuel Vauclain quoted in "Record of conversation between S. M. Vauclain, J. P. Sykes and J. L. Vauclain," Jan. 15, 1929, p. 4, file: Domestic BLW, box 8, SMV-DEG.

109. Alco would enjoy more success than Baldwin in the diesel engine market that rose after the late 1930s. But both firms were ultimately liquidated, Alco in 1969 and Baldwin in 1972; newcomers General Motors and General Electric have taken over the locomotive industry in recent decades.

110. The 1930s depression hammered capital equipment builders, already weakened by lackluster sales in the 1920s, and it was the ultimate cause of Baldwin's bankruptcy. But two more immediate issues pushed the firm into insolvency. Traditionally, capital goods companies banked large cash surpluses to tide them over in such market downturns. Vauclain drew down Baldwin's surplus in the 1920s, however, by raising dividends, building Eddystone, and financing acquisitions late in the decade. With its working capital depleted by 1930, the firm took advantage of its corporate status to float short-term bond issues in 1930 and 1933 (the latter refinancing the former). Inability to service its 1910 and 1933 debt resulted in voluntary reorganization in 1935. In this instance ostensible advantages of the corporate form contributed to Baldwin's fall.

111. Baldwin built its last locomotive, a diesel switcher, in 1956. Thereafter the company made railway rolling stock, construction equipment, and other heavy engineering products until it was liquidated in 1972 by its corporate parent, Armour & Co. (McKelvey, "Whistle Blows No More," p. 25).

112. Daniel Nelson compiled a list of the seventy largest (by employment) industrial factories of 1900, nineteen of which were capital equipment builders (*Managers and Workers*, pp. 7–8). The list suggests the structural changes that occurred in the American economy during the twentieth century, with thirteen of these nineteen no longer in business. With one exception—Newport News Shipbuilding, a defense contractor— the only capital goods plants to survive are those of electrical equipment manufacturers. These firms had a crucial advantage over their fellow capital equipment builders: they retained proprietary control over their own product technologies.

113. Burnham, Williams & Co., *Catalogue of Locomotives* (1908), and BLW, *Catalogue of Locomotives* (1915). These totals do not include compressed-air, electric, or many other types of Baldwin-standard engines.

114. Assertions of price fixing by Baldwin and Alco are found in Walter Isaacson and Evan Thomas, *The Wise Men* (New York: Simon & Schuster, 1986), pp. 113–14.

115. For the steam-to-diesel transition, see Churella, "Corporate Response"; Arns, "Embrace"; and Marx, "Technological Change."

116. Undoubtedly, GM's success in diesel locomotives resulted directly from integrating the Chandlerian triad of manufacturing technologies, managerial structures, and marketing capacities to capitalize upon technological innovations by GM's research and development chief, Charles Kettering. Its initial decision to enter locomotive building, however, was somewhat irrational given railroading's long-term decline. As Churella notes, in essence GM's top officers stumbled accidentally into the locomotive industry, acquiring Electro-Motive for reasons that had very little to do with railroading ("Corporate Response," pp. 125 and 137–38).

117. Churella estimates that GM spent at least $15 million during the 1930s to enter the diesel locomotive market ("Corporate Response," p. 176). While the huge automaker

had comparatively vast resources to draw upon, such a sum would not have taxed Baldwin's finances overly, even in the depression, had it not squandered profits on excessive dividends and the unjustifiable Eddystone complex in the previous decade. On the other hand, for those believing that diesel power would only supplement steam, $15 million was a major gamble for any firm in the locomotive industry since average output between 1921 and 1940 stood at only 962 units a year (domestic and foreign production by all builders, *Historical Statistics,* pt. 2, pp. 696 and 728). Total locomotive sales (steam and diesel) remained lackluster until 1947, when American railroads finally committed themselves to replacing their steam power with diesels. After the transition, sales again fell off.

118. Sloan, *My Years with General Motors,* p. 351.

119. Electro-Motive and other modern capital equipment makers prefer to make standard products in part because they use patents to derive competitive advantages within their industries. Indeed, in modern capital goods it seems likely that patents are far more important in fostering standard designs than is the ability to offer cost savings through standardization. In other words, such standardization serves makers' interests primarily, not customers'. After 1850 patenting strategy played an insignificant role in the steam locomotive industry.

120. While Electro-Motive's policy of standard designs greatly improved its control over its markets, the carriers retain a substantial, if general, influence over locomotive innovation in the diesel age. For example, the railways' demand for higher-horsepower engines has continued down to the present, impelling the diesel builders to produce generations of successively more powerful standard products.

121. John Kirkland treats Binkerd's blind adherence to steam simply as one man's private opinion rather than corporate policy (*Diesel Builders,* p. 25). But two Baldwin publications reprinted Binkerd's 1935 speech professing faith in the future of steam power (BLW, "Motive Power Situation," pp. 61–76, and "Muzzle Not the Ox That Treadeth Out the Corn," *Baldwin Locomotives,* Apr.–July 1935, pp. 11–20). Furthermore, Baldwin's top management appointed a fervent steam advocate, Ralph Johnson, as its chief engineer in 1939. He remained in that position through 1948, indicating the company's lukewarm interest in diesel power.

122. Arns, "Embrace," p. 45, and Reck, *On Time,* pp. 109–10.

123. This material is drawn from Arns ("Embrace"), who notes that Baldwin and Alco could hardly complain about the WPB's decision, made in the interest of wartime efficiency. Those firms simply were not ready to enter the mainline freight diesel market in 1942, and in any event their own salaried executives on the WPB had authorized this division of effort and markets. But the WPB decision did give GM a substantial competitive advantage after the war by providing four years in which to extend and perfect its technical support services for diesels.

124. Immediately after the war major coal haulers like the Pennsylvania, the Norfolk and Western, and the Chesapeake and Ohio attempted to hold off internal combustion by developing novel steam designs, which Baldwin built for them. Although shortsighted, the builder's jaundiced view of the diesel had a measure of support in the market.

125. Baldwin also continued its steam business into the postwar period, making its last conventional steam locomotive for a class-one American carrier in 1949 and its last export steam engines in 1955. Beyond focusing on a single technology, Electro-Motive followed other general principles of volume manufacturing. It adhered rigidly to standard models, whereas Baldwin willingly customized its diesels to suit individual railways' preferences. Following the example of the auto industry, Electro-Motive created a comprehensive

parts and service support network that promoted customer loyalty while also allowing the railways to shed much of their extensive infrastructure devoted to steam maintenance, with its expensive union labor force.

CONCLUSION: BALDWIN, THE CAPITAL EQUIPMENT SECTOR, AND THE NINETEENTH-CENTURY ECONOMY

1. As Louis Galambos and Joseph Pratt note of the mid-nineteenth century, "The business system . . . was chaotic: information was more limited than it is today . . . the risks of investment much higher; price competition was often fierce, and shady practices were common" (*Rise of the Corporate Commonwealth,* pp. 24–25).

2. In its growth and long-term success as a large-scale enterprise, Baldwin's collaborative business strategy stands as a midpoint alternative in Chandler's stark dichotomy that firms chose either the external workings of free markets or internal managerial controls to allocate resources. In turn, Baldwin's concern for minimizing risks and optimizing its capacities suggests that the disembodied Chandlerian logic of efficiency calculations has only limited applicability in explaining the behavior of firms—even very large ones. While Baldwin did ultimately fail, it took a radical innovation, the diesel locomotive, originating largely outside its collaborative network to overcome its risk-minimizing measures. Nonetheless, Baldwin's fall—which originated with its inadequate response to the 1905 market climax rather than with the diesel revolution—does demonstrate the potential hazard when firms simply respond to customers' demands at the expense of attempting to guide their own destinies.

3. As business historians further explore historic choices in organizational formats, they likely will find that social and cultural factors (as well as economic and technical issues) played a key role in the rise of managerial bureaucracies and ideologies ca. 1900, just as earlier social views and cultural beliefs influenced Baldwin's producer culture. Tracing the social construction of business management depends, however, on acknowledging that managers and management are different historical phenomena, with the former guiding developments in the latter (for more on this point, see Zunz, *Making America Corporate,* intro.).

4. Tracing the causes of decline in the U.S. capital equipment sector requires much further study; but the shortcomings of American managerial capitalism appear to have played a key role. Some support for this notion is evident in modern America's practice of importing many capital goods from other advanced industrial nations, such as Japanese machine tools, German printing presses, and Canadian railway cars. We still need such products—and high-wage workers still build them, albeit foreign workers—but corporate America has evidently lost the will to make many of them (see Michael L. Dertouzos et al., *Made in America* [Cambridge: MIT Press, 1989], pp. 201–16 and 232–47). For GM's efforts to sell off Electro-Motive, see Stephen Franklin, "GM to Keep Locomotive Operations," Chicago *Tribune,* October 27, 1993, sec. 3, p. 1.

5. Paul T. Kidd, *Agile Manufacturing Forging New Frontiers* (Wokingham, England: Addison Wesley, 1994), chap. 2. The fact that volume manufacturers are mirroring practices that originated among nineteenth-century capital equipment builders reinforces the point made in the introduction that the two formats are descriptive terms that fall along a continuum of firm choices and strategies rather than represent a strict dichotomy.

6. Recently the Union Pacific teamed up with General Electric Transportation Systems and two other firms to develop a locomotive that will use both diesel fuel and liquefied natural gas (Leo O'Connor, "Building Natural Gas Locomotives," *Mechanical Engineering* 116 [Apr. 1994]: 82–84). Also note that even today, very large and expensive

machine tools, such as those built by the Wisconsin toolmaker Giddings and Lewis, "are always somewhat customized to suit . . . customer needs" (Frederick Mason, "Machining Center Speeds Big Part Throughput," *American Machinist* 137 [Jan. 1993]: 60).

7. GE Fanuc, a capital equipment firm making automated production equipment for assembly lines, is supplementing its standard designs with a new line of custom products made to suit customer specifications. A modern volume manufacturer, Dell Computer, makes large quantities of customized personal computers to order.

BIBLIOGRAPHY

UNPUBLISHED MATERIAL

Baldwin Locomotive Works. "The Motive Power Situation of American Railroads."
   Typescript report, Sept. 10, 1937. Baker Library, Harvard Business School.
Brown, John K. "The Baldwin Locomotive Works, 1831–1915: A Case Study in the Capital
   Equipment Sector." Ph.D. diss., Univ. of Virginia, 1992.
Churella, Albert J. "Corporate Response to Technological Change: Dieselization and the
   American Railway Locomotive Industry during the Twentieth Century." Ph.D. diss.,
   Ohio State Univ., 1994.
Curtis, Julia B. "The Organized Few: Labor in Philadelphia, 1857–1873." Ph.D. diss.,
   Bryn Mawr College, 1970.
Fowler, Henry. "American Engineering Practice." Typescript report by the chief me-
   chanical engineer of the Midland Railway (England), Nov. 1, 1905. Division of
   Transportation, NMAH.
Goldberg, Judith L. "Strikes, Organizing, and Change: The Knights of Labor in
   Philadelphia, 1869–1890." Ph.D. diss., New York Univ., 1985.
Graves, Carl Russell. "Scientific Management and the Santa Fe Shopmen of Topeka,
   Kansas, 1900–1925." Ph.D. diss., Harvard Univ., 1980.
Heinrich, Thomas R. "Ships for the Seven Seas: Philadelphia Shipbuilding in the Age of
   Industrial Capitalism, 1860–1900." Ph.D. diss., Univ. of Pennsylvania, 1993.
Johnson, Alba B. "Chester's Opportunity." Speech delivered to the Civic Advancement
   Association of Chester, Pa., Jan. 25, 1913. HSP.
Lozier, John William. "Taunton and Mason: Cotton Machinery and Locomotive Manu-
   facture in Taunton, Massachusetts, 1811–1861." Ph.D. diss., Ohio State Univ., 1978.
Lubar, Steven. "Managing the Industrial Revolution." Paper given at the American
   Historical Association Convention, 1985.
Park, Donald, K. "An Historical Analysis of Innovation and Activity: American Steam
   Locomotive Building, 1900–1952." Ph.D. diss., Columbia Univ., 1973.
Tripathi, R. S. "A Study of Labor Skills Required for Specific Production Jobs Used in
   the Manufacture of Steam Locomotives at the Baldwin Locomotive Works." M.B.A.
   thesis, Wharton School, Univ. of Pennsylvania, 1949.
Usselman, Steven W. "Running the Machine: The Management of Technological
   Innovation on American Railroads, 1860–1910." Ph.D. diss., Univ. of Delaware, 1985.

Whitney, James S. "Apprenticeship and a Boy's Prospect of a Livelihood." Paper given to the Philadelphia branch of the American Social Science Association, Mar. 21, 1872. HAG.

PUBLIC DOCUMENTS

Annual Report of the Secretary of Internal Affairs of the Commonwealth of Pennsylvania, Industrial Statistics. Vols. 1–32 (1872–1905).

Annual Reports of the U.S. Patent Office. 1830–1900.

Cheney, Lucian W., and Hugh S. Hanna. "Accidents and Accident Prevention in Machine Building." Bureau of Labor Statistics *Bulletin,* no. 216. Washington, D.C.: U.S. Government Printing Office, 1917.

Fitch, Charles H. "Report on the Manufactures of Interchangeable Mechanism." In U.S. Department of the Interior, Census Office. *Tenth Census of the United States* (1880). Vol. 2, *Report on the Manufactures of the United States,* pp. 611–704. Washington, D.C., 1883.

Laws of the General Assembly of the State of Pennsylvania. Harrisburg, 1870.

"Letter of the Secretary of the Treasury." *House Executive Document #21.* 25th Cong., 3d sess. Washington, D.C., 1838. [Woodbury Report.]

Licht, Walter. "Case Studies of Philadelphia Firms." In *The Organization of Work, Schooling and Family Life in Philadelphia, 1838–1920,* ed. Michael B. Katz. Report for the U.S. Department of Education, May 1983.

Report of the Board to Recommend a Standard Gauge for Nuts, Bolts, and Screw-Threads for the United States Navy, May 1868. Washington, D.C., 1880.

"Report on Conditions of Employment in the Iron and Steel Industry in the United States." *Senate Executive Document #110.* 62d Cong., 1st sess. Washington, D.C.: U.S. Government Printing Office, 1913.

U.S. Congress. *Report of the Commission on Industrial Relations.* Vols. 3 and 14. Washington, D.C.: U.S. Government Printing Office, 1915.

———. *Report of the Committee of the Senate upon the Relations between Labor and Capital.* Vol. 3. Washington, D.C., 1885.

———. *Report of the Industrial Commission.* Vol. 14. Washington, D.C.: U.S. Government Printing Office, 1901.

U.S. Department of Commerce, Bureau of the Census. *Historical Statistics of the United States.* Vols. 1 and 2. Washington, D.C.: U.S. Government Printing Office, 1975.

U.S. Department of the Interior, Census Office. *Manufactures of the United States in 1860.* Washington, D.C., 1865.

Weeks, Joseph D. "Report on the Statistics of Wages in Manufacturing Industries." In U.S. Department of the Interior, Census Office, *Tenth Census of the United States* (1880), 20:167–533. Washington, D.C., 1886.

CASE LAW

Baird v. Pettit. 70 Pennsylvania Reports (1872): 477.

Baldwin et al. v. Payne et al. (6 Howard) 47 U.S. Reports (1848): 301.

Baumgardner et al. v. Burnham et al. 93 Pennsylvania Reports (1880): 88.

Campbell v. Baldwin Locomotive Works. 202 Federal Reporter (1913): 505.

Finnerty v. Burnham. 205 Pennsylvania Reports (1903): 305.

Hardin v. Robinson. 178 New York Supreme Court Reports (1918): 724.

Jones v. Burnham. 217 Pennsylvania Reports (1907): 286.

Lach v. Burnham. 134 Federal Reporter (1904): 688.

MacLean v. Burnham. 4 Sadler, Pennsylvania Supreme Court (1885–89): 505.

O'Dowd v. Burnham. 19 Pennsylvania Superior Court Reports (1902): 467.

Politowski v. Burnham. 214 Pennsylvania Reports (1906): 165.

BOOKS

American Engineering Competition. New York: Harper & Brothers, 1901.

M. Baird & Co. *Illustrated Catalogue of Locomotives.* Philadelphia, 1871.

Baldwin Locomotive Works. *Catalogue of Locomotives.* New York: Rand McNally, 1915.

———. *History of the Baldwin Locomotive Works.* Philadelphia: Bingham, n.d. (ca. 1924).

———. *The Story of Eddystone.* Philadelphia, 1928.

Baldwin, Vail, and Hufty. *Locomotive Engine Catalog.* Philadelphia, 1840. Rpt., Oakland, Ca.: Grahame H. Hardy, 1946.

Bell, J. Snowden. *Early Motive Power of the Baltimore and Ohio Railroad.* New York: Angus Sinclair, 1912.

Beniger, James R. *The Control Revolution.* Cambridge: Harvard Univ. Press, 1986.

Berg, Walter Gilman. *American Railway Shop Systems.* New York: Railroad Gazette, 1904.

Bijker, Wiebe E., et al., eds. *The Social Construction of Technological Systems.* Cambridge: MIT Press, 1987.

Blicksilver, Jack, ed. *Views on U.S. Economic and Business History.* Atlanta: Georgia State Univ. Press, 1985.

Blodget, Lorin. *Census of Manufactures of Philadelphia.* Philadelphia, 1883.

Brandes, Stuart. *American Welfare Capitalism, 1880–1940.* Chicago: Univ. of Chicago Press, 1970.

Braverman, Harry. *Labor and Monopoly Capitalism.* New York: Monthly Review Press, 1974.

Bruce, Alfred W. *The Steam Locomotive in America.* New York: W. W. Norton, 1952.

Bruchey, Stuart W., ed. *Small Business in American Life.* New York: Columbia Univ. Press, 1980.

Burawoy, Michael. *Manufacturing Consent: Changes in the Labor Process under Monopoly Capitalism.* Chicago: Univ. of Chicago Press, 1979.

Burnham, Parry, Williams & Co. *Illustrated Catalogue of Locomotives.* Philadelphia, 1881.

Burnham, Williams & Co. *Catalogue of Locomotives.* Philadelphia: Edgell Press, 1908.

———. *Exhibit of Locomotives by the Baldwin Locomotive Works.* Philadelphia, 1893.

Burns, Arthur F. *Production Trends in the United States since 1870.* New York: National Bureau of Economic Research, 1934.

Calkins, Rev. Wolcott. *Memorial of Matthias W. Baldwin.* Philadelphia, 1867.

Calvert, Monte. *The Mechanical Engineer in America, 1830–1910: Professional Cultures in Conflict.* Baltimore: Johns Hopkins Press, 1967.

Carlson, W. Bernard. *Innovation as a Social Process: Elihu Thomson and the Rise of General Electric, 1870–1900.* Cambridge: Cambridge Univ. Press, 1991.

Carosso, Vincent P. *The Morgans: Private International Bankers, 1854–1913.* Cambridge: Harvard Univ. Press, 1987.

Chandler, Alfred D., Jr. *The Railroads: The Nation's First Big Business.* New York: Harcourt, Brace & World, 1965.

———. *The Visible Hand: The Managerial Revolution in American Business.* Cambridge: Harvard Univ. Press, 1977.

Chandler, Alfred D., Jr., and Stephen Salsbury. *Pierre S. du Pont and the Making of the Modern Corporation.* New York: Harper & Row, 1971.

Clark, Victor S. *History of Manufactures in the United States*. 3 vols. New York: McGraw-Hill, 1929.

Clarke, T. C. *The American Railway*. New York, 1889.

Clawson, Dan. *Bureaucracy and the Labor Process: The Transformation of U.S. Industry, 1860–1920*. New York: Monthly Review Press, 1980.

Cochran, Thomas C. *Frontiers of Change: Early Industrialism in America*. New York: Oxford Univ. Press, 1983.

Colburn, Zerah. *The Locomotive Engine*. Philadelphia, 1853.

Colvin, Fred H. *Sixty Years with Men and Machines*. New York: McGraw-Hill, 1947.

Commons, John R., et al. *History of Labor in the United States*. Vol. 2. New York: Macmillan, 1918.

Copley, Frank B. *Frederick W. Taylor, Father of Scientific Management*. New York: Harper & Brothers, 1923.

Custer, Edgar A. *No Royal Road*. New York: H. C. Kinsey, 1937.

Cutcliffe, Stephen H., and Robert C. Post, eds. *In Context: History and the History of Technology, Essays in Honor of Melvin Kranzberg*. Bethlehem, Pa.: Lehigh Univ. Press, 1989.

Day, Charles. *Industrial Plants*. New York: Engineering Magazine, 1911.

Diary of the Japanese Visit to Philadelphia in 1872. Philadelphia, 1872.

Dolzall, Gary W., and Stephen F. Dolzall. *Diesels from Eddystone: The Story of Baldwin Diesel Locomotives*. Milwaukee: Kalmbach Books, 1984.

Douglas, George H. *All Aboard! The Railroad in American Life*. New York: Paragon, 1992.

Ferguson, Eugene S. *Engineering and the Mind's Eye*. Cambridge: MIT Press, 1992.

———, ed. *Early Engineering Reminiscences (1815–40) of George Eschol Sellers*. Washington, D.C.: Smithsonian Institution, 1965.

Ferrell, Mallory Hope, ed. *Centennial Limited Edition of the 1871 Grant Locomotive Works Catalogue*. Boulder, Colo.: Pruett, 1971.

Fishlow, Albert. *American Railroads and the Transformation of the Ante-Bellum Economy*. Cambridge: Harvard Univ. Press, 1965.

Flynn, Elizabeth Gurley. *The Rebel Girl*. New York: International Publishers, 1973.

Foner, Philip S. *History of the Labor Movement in the United States*. Vol. 5, *The AFL in the Progressive Era, 1910–1915*. New York: International Publishers, 1980.

Fones-Wolf, Kenneth. *Trade Union Gospel: Christianity and Labor in Industrial Philadelphia, 1865–1915*. Philadelphia: Temple Univ. Press, 1989.

Forney, M. N. *Locomotives and Locomotive Building*. New York, 1886.

Fraser, John Foster. *America at Work*. New York: Cassell, 1903.

Freedley, Edwin T. *Leading Pursuits and Leading Men*. Philadelphia, 1856.

———. *Philadelphia and Its Manufactures*. Philadelphia, 1858.

Galambos, Louis, and Joseph Pratt. *The Rise of the Corporate Commonwealth*. New York: Basic Books, 1988.

Gallman, J. Matthew. *Mastering Wartime: A Social History of Philadelphia during the Civil War*. Cambridge: Cambridge Univ. Press, 1990.

Garner, S. Paul. *Evolution of Cost Accounting to 1925*. University: Univ. of Alabama Press, 1976.

Gibb, George S. *The Saco-Lowell Shops*. Cambridge: Harvard Univ. Press, 1950.

Gordon, David M., et al. *Segmented Work, Divided Workers: The Historical Transformation of Labor in the United States*. Cambridge: Cambridge Univ. Press, 1982.

Gordon, Robert B., and Patrick M. Malone. *The Texture of Industry: An Archaeological View of the Industrialization of North America*. New York: Oxford Univ. Press, 1994.

Graham, Frank D. *Audels Engineers and Mechanics Guide.* Vol. 3. New York: Theo
   Audel, 1921.

Grant Locomotive Works. *A Description of Locomotives Manufactured by the Grant
   Locomotive Works.* New York, 1871.

Grodinsky, Julius. *Transcontinental Railway Strategy, 1869–1893: A Study of Businessmen.*
   Philadelphia: Univ. of Pennsylvania Press, 1962.

Gutman, Herbert G. *Work, Culture, and Society in Industrializing America.* New York:
   Alfred A. Knopf, 1976.

Habakkuk, H. J. *American and British Technology in the Nineteenth Century.* Cambridge:
   Cambridge Univ. Press, 1962.

Harrison, Joseph, Jr. *The Locomotive and Philadelphia's Share in Its Early Improvement.*
   Philadelphia, 1872.

Hartz, Louis. *Economic Policy and Democratic Thought: Pennsylvania, 1776–1860.* Chicago:
   Quadrangle Books, 1968.

Haydu, Jeffrey. *Between Craft and Class: Skilled Workers and Factory Politics in the United
   States and Britain, 1890–1922.* Berkeley: Univ. of California Press, 1988.

Hill, John A. *Jim Skeevers' Object Lessons.* New York, 1899.

Hilton, George W. *American Narrow Gauge Railroads.* Stanford: Stanford Univ.
   Press, 1990.

Hindle, Brooke. *Emulation and Invention.* New York: New York Univ. Press, 1981.

Hirsimaki, Eric. *Lima.* Edmonds, Wash.: Hundman, 1986.

Hoke, Donald R. *Ingenious Yankees: The Rise of the American System of Manufactures in
   the Private Sector.* New York: Columbia Univ. Press, 1990.

Holton, James L. *The Reading Railroad: History of a Coal Age Empire.* Vol. 1, *The
   Nineteenth Century.* Laury's Station, Pa.: Garrigues House, 1989.

Hounshell, David A. *From the American System to Mass Production 1800–1932: The
   Development of Manufacturing Technology in the United States.* Baltimore: Johns
   Hopkins Univ. Press, 1984.

Hughes, Thomas P. *Networks of Power: Electrification in Western Society, 1880–1930.*
   Baltimore: Johns Hopkins Univ. Press, 1983.

Hunter, Louis C. *A History of Industrial Power in the United States.* Vol. 2, *Steam Power.*
   Wilmington: Eleutherian Mills–Hagley Foundation, 1985.

Hunter, Louis C., and Lynwood Bryant. *A History of Industrial Power in the United States.*
   Vol. 3, *The Transmission of Power.* Cambridge: MIT Press, 1991.

Jacoby, Sanford M., ed. *Masters to Managers: Historical and Comparative Perspectives on
   American Employers.* New York: Columbia Univ. Press, 1991.

Kirkland, John F. *The Diesel Builders.* Vol. 3, *Baldwin Locomotive Works.* Pasadena:
   Interurban Press, 1994.

Klein, Maury. *Union Pacific: Birth of a Railroad, 1862–1893.* New York: Doubleday, 1987.

Lamoreaux, Naomi R. *The Great Merger Movement in American Business, 1895–1904.*
   Cambridge: Cambridge Univ. Press, 1985.

Laurie, Bruce. *Artisans into Workers: Labor in Nineteenth Century America.* New York:
   Farrar, Straus & Giroux, 1989.

———. *Working People of Philadelphia, 1800–1850.* Philadelphia: Temple Univ.
   Press, 1980.

Lazonick, William. *Competitive Advantage on the Shop Floor.* Cambridge: Harvard Univ.
   Press, 1990.

Levasseur, E. *The American Workman.* Baltimore: Johns Hopkins Press, 1900.

Lewis, W. David. *Iron and Steel in America*. Wilmington: Eleutherian Mills–Hagley Foundation, 1976.

Licht, Walter. *Getting Work: Philadelphia, 1840–1950*. Cambridge: Harvard Univ. Press, 1992.

————. *Working for the Railroad: The Organization of Work in the Nineteenth Century*. Princeton: Princeton Univ. Press, 1983.

Lindstrom, Diane. *Economic Development in the Philadelphia Region, 1810–1850*. New York: Columbia Univ. Press, 1978.

Martin, Albro. *Enterprise Denied: Origins of the Decline of American Railroads, 1897–1917*. New York: Columbia Univ. Press, 1971.

Mayr, Otto, and Robert C. Post, eds. *Yankee Enterprise: The Rise of the American System of Manufactures*. Washington, D.C.: Smithsonian Institution Press, 1981.

McCraw, Thomas K. *Prophets of Regulation: Charles Francis Adams, Louis D. Brandeis, James M. Landis, Alfred E. Kahn*. Cambridge: Harvard Univ. Press, 1984.

McGaw, Judith. *Most Wonderful Machine: Mechanization and Social Change in Berkshire Paper Making, 1801–1885*. Princeton: Princeton Univ. Press, 1987.

McGuffie, Chris. *Working in Metal: Management and Labour in the Metal Industries of Europe and the USA, 1890–1914*. London: Merlin Press, 1985.

McHugh, Jeanne. *Alexander Holley and the Makers of Steel*. Baltimore: Johns Hopkins Univ. Press, 1980.

Meyer, J. G. A. *Modern Locomotive Construction*. New York, 1904.

Miller, William, ed. *Men in Business: Essays on the Historical Role of the Entrepreneur*. Cambridge: Harvard Univ. Press, 1952.

Mokyr, Joel. *The Lever of Riches: Technological Creativity and Economic Progress*. New York: Oxford Univ. Press, 1990.

Montgomery, David. *The Fall of the House of Labor: The Workplace, the State, and American Labor Activism, 1865–1925*. Cambridge: Cambridge Univ. Press, 1985.

Morison, Elting E. *Men, Machines, and Modern Times*. Cambridge: MIT Press, 1976.

Morrison, John H. *History of New York Shipyards*. New York, 1909.

Mosely Industrial Commission. *Reports of the Delegates of the Mosely Industrial Commission*. New York: Winthrop Press, 1903. Rpt., New York: Arno Press, 1973.

Moss, Michael S., and John R. Hume. *Workshop of the British Empire: Engineering and Shipbuilding in the West of Scotland*. Rutherford, N.J.: Fairleigh Dickinson Univ. Press, 1977.

Navin, Thomas R. *The Whitin Machine Works since 1831*. Cambridge: Harvard Univ. Press, 1950.

Nelson, Daniel. *Frederick W. Taylor and the Rise of Scientific Management*. Madison: Univ. of Wisconsin Press, 1980.

————. *Managers and Workers: Origins of the New Factory System in the United States, 1880–1920*. Madison: Univ. of Wisconsin Press, 1975.

Noble, David F. *America by Design: Science, Technology, and the Rise of Corporate Capitalism*. New York: Alfred A. Knopf, 1977.

Norris, Septimus. *Norris' Handbook for Locomotive Engineers*. Philadelphia, 1853.

Oberholtzer, Ellis P. *Jay Cooke, Financier of the Civil War*. Philadelphia: G. W. Jacobs, 1907.

Pacey, Arnold. *The Culture of Technology*. Cambridge: MIT Press, 1983.

Partington, John E. *Railroad Purchasing and the Business Cycle*. Washington, D.C.: Brookings Institution, 1929.

Perlman, Mark. *The Machinists: A New Study in American Trade Unionism.* Cambridge: Harvard Univ. Press, 1961.

Phenis, Albert. *Yankee Thrift.* Baltimore: Manufacturers' Record, 1905.

Porter, Glenn, and Harold C. Livesay. *Merchants and Manufacturers: Studies in the Changing Structure of Nineteenth-Century Marketing.* Baltimore: Johns Hopkins Press, 1971.

Pursell, Carroll W., Jr. *Early Stationary Steam Engines in America.* Washington, D.C.: Smithsonian Institution Press, 1969.

Pusateri, C. Joseph. *A History of American Business.* Arlington Heights, Ill.: Harlan Davidson, 1984.

Reck, Franklin, M. *On Time: The History of Electro-Motive Division of General Motors Corporation.* N.p., 1948.

Roe, Joseph Wickham. *English and American Tool Builders.* New York: McGraw-Hill, 1926.

Roe, Joseph Wickham, and Charles W. Lytle. *Factory Equipment.* Scranton, Pa.: International Textbook Co., 1935.

Rogers, Daniel T. *The Work Ethic in Industrial America, 1850–1920.* Chicago: Univ. of Chicago Press, 1974.

Rorabaugh, W. J. *The Craft Apprentice: From Franklin to the Machine Age in America.* New York: Oxford Univ. Press, 1986.

Rose, Joshua. *Modern Machine Shop Practice.* 2 vols. New York, 1889.

Rosenberg, Nathan. *Technology and American Economic Growth.* White Plains, N.Y.: M. E. Sharpe, 1972.

Scranton, Philip. *Figured Tapestry: Production, Markets, and Power in Philadelphia Textiles, 1885–1941.* New York: Cambridge Univ. Press, 1989.

———. *Proprietary Capitalism: The Textile Manufacture at Philadelphia, 1880–1885.* New York: Cambridge Univ. Press, 1983.

Scranton, Philip, and Walter Licht. *Work Sights: Industrial Philadelphia, 1890–1950.* Philadelphia: Temple Univ. Press, 1986.

[See, James W.]. *Extracts from Chordal's Letters.* New York, 1880.

Sinclair, Angus. *Development of the Locomotive Engine.* New York: Angus Sinclair, 1907.

Sinclair, Bruce. *Philadelphia's Philosopher Mechanics: A History of the Franklin Institute, 1824–1865.* Baltimore: Johns Hopkins Univ. Press, 1974.

Sloan, Alfred P., Jr. *My Years with General Motors.* New York: MacFadden, 1965.

Smith, J. Russell. *Elements of Industrial Management.* Philadelphia: J. B. Lippincott, 1915.

Smith, Merritt Roe. *Harpers Ferry Armory and the New Technology: The Challenge of Change.* Ithaca: Cornell Univ. Press, 1977.

Stevenson, David. *Sketch of the Civil Engineering of North America.* London, 1838.

Strassmann, W. Paul. *Risk and Technological Innovation: American Manufacturing Methods during the Nineteenth Century.* Ithaca: Cornell Univ. Press, 1959.

Taylor, Frederick W. *The Principles of Scientific Management.* New York: W. W. Norton, 1967.

Taylor, George Rogers. *The Transportation Revolution, 1815–1860.* Armonk, N.Y.: M. E. Sharpe, 1977.

Thompson, Fred, and Patrick Murfin. *The I.W.W.: Its First Seventy Years, 1905–1975.* Chicago: Industrial Workers of the World, 1976.

Tolman, William H. *Social Engineering.* New York: McGraw, 1909.

Trostel, Scott D. *Building a Lima Locomotive.* Fletcher, Ohio: Cam-Tech, 1990.

Tyler, David B. *The American Clyde.* Newark: Univ. of Delaware Press, 1958.

Vansant, J. L. *Royal Road to Wealth*. Philadelphia, 1869.

Vauclain, Samuel M. *Optimism*. Philadelphia, 1924.

Vauclain, Samuel M., and Earl Chapin May. *Steaming Up*. New York: Brewer & Warren, 1930. Rpt., San Marino, Calif.: Golden West Books, 1981.

Wagoner, Harless D. *The U.S. Machine Tool Industry from 1900 to 1950*. Cambridge: MIT Press, 1968.

Wallace, Anthony F. C. *Rockdale: The Growth of an American Village in the Early Industrial Revolution*. New York: W. W. Norton, 1975.

Ward, James A. *J. Edgar Thomson: Master of the Pennsylvania*. Westport, Conn.: Greenwood Press, 1980.

Warner, Paul T. *Motive Power Development on the Pennsylvania Railroad System, 1831–1924*. Philadelphia, 1924.

Warner, Sam Bass. *The Private City: Philadelphia in Three Periods of Its Growth*. Philadelphia: Univ. of Pennsylvania Press, 1968.

Weigley, Russell F., ed. *Philadelphia: A Three Hundred Year History*. New York: W. W. Norton, 1982.

Weitzman, David. *Superpower: The Making of a Steam Locomotive*. Boston: David R. Godine, 1987.

Westing, Fred. *The Locomotives That Baldwin Built*. New York: Bonanza Books, 1966.

Westwood, J. N. *Locomotive Designers in the Age of Steam*. Rutherford, N.J.: Fairleigh Dickinson Univ. Press, 1977.

White, John H., Jr. *American Locomotives: An Engineering History, 1830–1880*. Baltimore: Johns Hopkins Press, 1968.

———. *The American Railroad Passenger Car*. Baltimore: Johns Hopkins Univ. Press, 1978.

———. *Cincinnati Locomotive Builders, 1845–1868*. Washington, D.C.: U.S. Government Printing Office, 1965.

———. *A Short History of American Locomotive Builders in the Steam Era*. Washington, D.C.: Bass, 1982.

Wilson, William B. *History of the Pennsylvania Railroad Company*. 2 vols. Philadelphia, 1899.

Yates, JoAnne. *Control through Communication: The Rise of System in American Management*. Baltimore: Johns Hopkins Univ. Press, 1989.

Zunz, Olivier. *Making America Corporate, 1870–1920*. Chicago: Univ. of Chicago Press, 1990.

ARTICLES

"American Industries No. 90: The Manufacture of Locomotives." *Scientific American* 50 (May 31, 1884): 338–40.

"American Locomotives for Foreign Railways." *Engineering Magazine* 17 (Aug. 1899): 858–59.

"American Locomotives on the Burma Railways." *Board of Trade Journal* (London) 35 (Dec. 5, 1901): 447–49.

Ames, Edward, and Nathan Rosenberg. "The Enfield Arsenal in Theory and History." *Economic Journal* (London) 78 (Dec. 1968): 827–42.

"Appeal to the Machinists and Blacksmiths of America." *Machinists' and Blacksmiths' International Journal* 9 (July 1872): 693.

Arnold, Horace L. "Production Up to the Power Limit." *Engineering Magazine* 9 (Aug. 1895): 916–24.

Arns, Robert. "Embrace of the Past." *Railway and Locomotive Preservation* 11 (Nov.–Dec. 1987): 42–52.

"The Baldwin Locomotive Works." *American Artisan* 4 (June 5, 1867): 482–83.

"The Baldwin Locomotive Works." *Engineering* (London) 22 (Aug. 18, 1876): 138–41.

"The Baldwin Locomotive Works." *Railway Magazine* (London) 33 (Dec. 1913): 457–67.

"The Baldwin Locomotive Works and the Metric System." *Iron Age* 79 (June 20, 1907): 1885–86.

Battison, Edwin. "Eli Whitney and the Milling Machine." *Smithsonian Journal of History* 1 (1966): 9–34.

Bingham, E. A. "The Labor Situation at the Baldwin Works." *Iron Trade Review* (Feb. 12, 1903): 40–41.

Birch, J. Grant. "Some Notes on American Manufactures." *Engineering* (London) 66 (Dec. 23, 1898): 805–9.

Booth, W. H. "Limit Gauges in the Workshop." *Cassier's Magazine* 23 (Nov. 1902): 185–92.

Buffet, Edward P. "The Vail Family and the Speedwell Works." *American Machinist* 32 (July 15, 1909): 117–19.

"Building a Locomotive." *Locomotive Engineering* 11 (Jan. 1898): 1–11.

"Building Locomotives." *Machinery* 2 (July 1896): 206.

Buttrick, John. "The Inside Contract System." *Journal of Economic History* 12 (Summer 1952): 205–21.

Carpenter, Charles. "Jobbing Work and Efficiency." *Industrial Management* 52 (Feb. 1917): 633–39.

Chandler, Alfred D., Jr. "Anthracite Coal and the Beginnings of the Industrial Revolution in the United States." *Business History Review* 46 (Summer 1972): 141–81.

Church, Arthur L. "The Extra Work Department of the Baldwin Locomotive Works." In Burnham, Williams & Co., *Record of Recent Construction #60*. Philadelphia, 1907.

Clark, Malcolm C. "The Birth of an Enterprise: Baldwin Locomotive, 1831–1842." *Pennsylvania Magazine of History and Biography* 90 (Oct. 1966): 423–44.

Converse, John H. "American Locomotives in the Markets of the World." *Independent* 53 (June 13, 1901): 1350–54.

———. "Locomotive Building in the United States." *Cassier's Magazine* 20 (July 1901): 221–37.

Converse, John W. "Some Features of the Labor System and Management at the Baldwin Locomotive Works." *Annals* of the American Academy of Political and Social Science 21 (Jan. 1903): 1–9.

"Cost of Building Locomotives in America." *Engineer* (London) 13 (Jan. 3, 1862): 5–6.

Cotton, M. G. "Baldwin's 'Contract System' Gets Out the Locomotives." *Management* 32 (Mar. 1929): 35–41, 76–84.

Coulson, Thomas. "Some Prominent Members of the Franklin Institute: Part 3, Matthias William Baldwin, 1795–1866." *Journal of the Franklin Institute* 262 (Sept. 1956): 171–84.

Crocker, F. B. "The Electric Distribution of Power in Workshops." *Journal of the Franklin Institute* 151 (Jan. 1901): 1–28.

Dawson, Andrew. "The Paradox of Dynamic Technological Change and the Labor Aristocracy in the United States, 1880–1914." *Labor History* 20 (Summer 1979): 325–51.

Day, Charles. "Metal-Working Plants and Their Machine Tool Equipment." *Engineering Magazine* 39 (June 1910): 364–76.

"The Development of the American Locomotive." *Journal of the Franklin Institute* 164 (Oct. 1907): 233–72.

Devine, Warren. "From Shafts to Wires: Historical Perspective on Electrification." *Journal of Economic History* 43 (June 1983): 347–72.

Drummond, Diane. "Building a Locomotive: Skill and the Work Force in Crewe Locomotive Works, 1843–1914." *Journal of Transport History* 8 (1987): 1–29.

Dubofsky, Melvin. "Technological Change and American Worker Movements, 1870–1970." In *Technology, the Economy, and Society,* ed. Joel Colton and Stuart Bruchey, pp. 162–85. New York: Columbia Univ. Press, 1984.

Durland, Kellogg. "An Industrial Deadlock." Boston *Evening Transcript,* July 1, 1911, sec. 3, p. 3.

Englander, Ernest J. "The Inside Contract System of Production and Organization: A Neglected Aspect of the History of the Firm." *Labor History* 28 (Fall 1987): 429–46.

Ernst, Daniel. "The Yellow Dog Contract and Liberal Reform, 1917–1932." *Labor History* 3 (Spring 1989): 251–74.

"Extension of Electric Driving at the Baldwin Locomotive Works." *American Machinist* 19 (Jan. 30, 1896): 137–38.

Fayant, Frank H. "The Building of an American Locomotive." *Page's Magazine,* Jan. 1903, pp. 13–23.

"Fifty Years at Baldwin's: W. P. Henszey Is Very Ill." Philadelphia *North American,* Mar. 9, 1909, p. 1.

Fincher, Jon. C. "Early History of Our Organization." *Machinists' and Blacksmiths' International Journal* 9 (1872): 520–22, 564–65, 592–93.

Fishlow, Albert. "Productivity and Technological Change in the Railroad Sector." In *Output, Employment, and Productivity in the United States after 1800,* pp. 583–646. New York: National Bureau of Economic Research, 1966.

Fitch, Charles H. "Railroad Shop Tools." *Railway Master Mechanic* 27 and 28 (Apr. 1903–Apr. 1904). Twelve-part series.

Flather, J. J. "The Modern Power Problem." *Cassier's Magazine* 23 (Mar. 1903): 636–49.

Floud, R. C. "The Adolescence of American Engineering Competition, 1860–1900." *Economic History Review,* 2d ser., 27 (Feb. 1974): 57–71.

Fones-Wolf, Ken. "Mass Strikes and Corporate Strategies: The Baldwin Locomotive Works and the Philadelphia General Strike of 1910." *Pennsylvania Magazine of History and Biography* 110 (July 1986): 447–57.

Fries, Russell I. "British Response to the American System: The Case of the Small Arms Industry after 1850." *Technology and Culture* 16 (July 1975): 377–403.

Fry, Lawford H. "The Cost of Locomotives." *Railroad Gazette* 42 (June 7, 1907): 802–4.

"Giants at Eddystone." *Fortune* 2 (July 1930): 58–61.

Gordon, Robert B. "Who Turned the Mechanical Ideal into Mechanical Reality?" *Technology and Culture* 29 (Oct. 1988): 744–78.

Goss, W. F. M. "Tests of the Locomotive at the Laboratory of Purdue University." *Transactions of the American Society of Mechanical Engineers* 14 (1892–93): 826–54.

Henszey, J. Wilmer. "The Organization and Methods of a Modern Industrial Works." *Journal of the Franklin Institute* 158 (Dec. 1904): 401–9.

Houshower, Hans. "We Built the Best." *Locomotive and Railway Preservation,* Mar.–Apr. 1991, pp. 11–36.

Howard, Robert A. "Interchangeable Parts Reexamined: The Private Sector of the American Arms Industry on the Eve of the Civil War." *Technology and Culture* 19 (Oct. 1978): 633–49.

Huddleston, Eugene, L. "Uncle Sam's Locomotives." *Trains* 51 (Mar. 1991): 30–38.

Hughes, L. C. "Trade Unions." *Machinists' and Blacksmiths' International Journal* 8 (Dec. 1870): 47.

"Improvements at the Baldwin Locomotive Works." *Railroad Gazette* 34 (Oct. 31, 1902): 828–29.

James, John A. "Structural Changes in American Manufacturing, 1850–1890." *Journal of Economic History* 43 (June 1983): 433–59.

"The Japanese Visit M. W. Baldwin & Co.'s Locomotive Works." Philadelphia *Inquirer*, June 16, 1860, suppl., p. 1.

Johnson, Alba B. "Efficiency of Modern Railway Equipment." *Manufacturers' Record* 47 (May 25, 1905): 424.

———. "The Market for Locomotives." *Annals* of the American Academy of Political and Social Science 34 (Nov. 1909): 109–13.

———. "The Problem of Motive Power under National Administration." In Baldwin Locomotive Works, *Record of Recent Construction #90*. Philadelphia, 1918.

Kirby, M. W. "Product Proliferation in the British Locomotive Building Industry, 1850–1914: An Engineer's Paradise?" *Business History*, July 1988, pp. 287–305.

Laurie, Bruce, and Mark Schmitz. "Manufacture and Productivity: The Making of an Industrial Base, Philadelphia, 1850–1880." In *Philadelphia*, ed. Theodore Hershberg, pp. 43–92. New York: Oxford Univ. Press, 1981.

"Letter from Professor R. H. Thurston." *Scientific American* 27 (July 20, 1872): 40.

Litterer, Joseph. "Systematic Management: The Search for Order and Integration." *Business History Review* 35 (Winter 1961): 461–76.

Livesay, Harold C. "The Lobdell Car Wheel Co., 1830–1867." *Business History Review* 42 (Summer 1968): 171–94.

"Local Affairs." Philadelphia *Ledger and Transcript*, July 29, 1873, p. 3.

"The Locomotive Builders Meeting." *Van Nostrand's Eclectic Engineering Magazine* 1 (Jan. 1869): 65.

"Locomotive Engines." *New York Tribune*, Oct. 31, 1870. Rpt. in William H. Brown, *The History of the First Locomotives in America*, pp. 229–38. New York, 1871.

"The Locomotive Shops of the Country." *Railroad Advocate* 1 (Nov. 11, 1854): 1.

Luhrsen, Louis. "The Molding of Locomotive Cylinders." *Foundry*, Oct. 1908, pp. 61–63.

Marx, Thomas G. "Technological Change and the Theory of the Firm: The American Locomotive Industry, 1920–1955." *Business History Review* 50 (Spring 1976): 1–24.

McGaw, Judith A. "Accounting for Innovation: Technological Change and Business Practice in the Berkshire Paper Industry." *Technology and Culture* 26 (Oct. 1985): 703–25.

McKelvey, Gerald. "The Whistle Blows No More at the Baldwin Plant." Philadelphia *Inquirer*, May 5, 1972, p. 25.

"Meeting of Locomotive Builders." *American Engineer* 4 (July 18, 1857): 12.

Montgomery, David. "Workers' Control of Machine Production in the Nineteenth Century." *Labor History* 17 (Fall 1976): 485–509.

Moore, Ed. "The Strike at Baldwin's." *International Socialist Review* 12 (Aug. 1911): 90–95.

Moshein, Peter, and Robert R. Rothfus. "Rogers Locomotives: A Brief History and Construction List." *Railroad History*, no. 167 (Autumn 1992): 13–147.

Navin, Thomas R., and Marion V. Sears. "The Rise of a Market for Industrial Securities, 1887–1902." *Business History Review* 24 (June 1955): 105–38.

"Obituary of Charles T. Parry." *Report of the Proceedings of the Twentieth Annual Convention of the American Railway Master Mechanics Association*, pp. 200–207. Chicago, 1887.

Oshima, Harry T. "The Growth of U.S. Factor Productivity: The Significance of New Technologies in the Early Decades of the Twentieth Century." *Journal of Economic History* 44 (Mar. 1984): 161–70.

Outerbridge, A. E., Jr. "Labor Saving Machinery the Secret of Cheap Production." *Engineering Magazine* 12 (Jan. 1897): 650–56.

Park, Donald K. "U.S. Steam Locomotive Builders." *Railroad History Bulletin,* no. 132 (Spring 1975): 47–55.

Phillips, Camillus. "The Reconciliation of Capital and Labor." *The Businessman's Magazine and the Book-Keeper* 27 (May 1905): 915–24.

Pinney, C. G. "Foundry Practice at the Baldwin Locomotive Works." *Santa Fe Magazine* 2 (Jan. 1917): 37–40.

"The Power System of the Baldwin Locomotive Works." *Manufacturers' Gazette* 18 (Jan. 16, 1892): 1.

"A Quarter of a Century's Progress in Locomotive Building at the Baldwin Works." *Railway Master Mechanic* 27 (Feb. 1903): 61–65.

Reed, John. "Back of Billy Sunday." *Metropolitan Magazine,* May 1915, pp. 9–12, 66–72.

Richards, John. "Compensation of Skilled Labor." *Association of Engineering Societies Journal* 28 (Mar. 1902): 101–15.

———. "A Cooperative Contract System." *American Machinist* 17 (Nov. 15, 1894): 1–3.

———. "Economic and Labor Factors in the Distribution of Industries." *Engineering Magazine* 19 (Apr. 1900): 98–106.

Richmond, George. "Operating Machine Tools by Electricity." *Engineering Magazine* 8 (Jan. 1895): 669–86.

Robertson, Ross M. "Changing Production of Metalworking Machinery, 1860–1920." In *Output, Employment, and Productivity in the United States after 1800,* pp. 479–95. New York: National Bureau of Economic Research, 1966.

Roe, Joseph Wickham. "The Industrial Background of 1855." *Iron Age* 75 (Nov. 20, 1930): 1441–47.

Rogers, J. M. "The Greatest Locomotive Works in the World." *Booklover's Magazine* 3 (Jan. 1904): 68–88.

Roland, Henry. "Six Examples of Successful Shop Management." *Engineering Magazine* 12, nos. 1–6 (Oct. 1896–Mar. 1897): 69–85, 270–85, 395–412, 831–37, 994–1000.

Roper, Steven. "The Largest Locomotive Works in the World." *American Machinist* 2 (Apr. 1879): 4.

Rosenberg, Nathan. "Technological Change in the Machine Tool Industry, 1840–1910." *Journal of Economic History* 23 (1963): 414–43.

Sample, N. W. "Apprenticeship System at the Baldwin Locomotive Works." *Annals* of the American Academy of Political and Social Science 33 (1909): 175–77.

Samuel, Raphael. "Workshop of the World: Steam Power and Hand Technology in Mid-Victorian Britain." *History Workshop Journal* 3 (Spring 1977): 6–72.

Sanford, R. H. "A Pioneer Locomotive Builder." *Railway and Locomotive Historical Society Bulletin,* no. 8 (1924): 7–23.

Scranton, Philip. "Diversity in Diversity: Flexible Production and American Industrialization, 1880–1930." *Business History Review* 65 (Spring 1991): 27–90.

———. "The Workplace, Technology, and Theory in American Labor History." *International Labor and Working Class History* 35 (Spring 1989): 3–22.

Sellers, Coleman. "The Progress of the Mechanical Arts in Three-Quarters of a Century." *Journal of the Franklin Institute* 149 (Jan. 1900): 5–25.

"Sensitive Business of Locomotive Making." *New York Times Annalist,* Jan. 20, 1913, p. 10.

Sinclair, Bruce. "At the Turn of a Screw: William Sellers, the Franklin Institute, and a Standard American Thread." *Technology and Culture* 10 (Jan. 1969): 20–34.

Smith, Oberlin. "Economies in Machine Shop Work." *Cassier's Magazine* 18 (May 1900): 243–47.

Sokoloff, Kenneth L. "Investment in Fixed and Working Capital during Early Industrialization: Evidence from U.S. Manufacturing Firms." *Journal of Economic History* 44 (June 1984): 545–56.

"Status of Apprenticeship." *American Machinist* 19 (Dec. 24, 1896): 15–35.

"Status of Employees of the Baldwin Locomotive Works." *Manufacturers' Record* 65 (July 2, 1914): 65.

"Steam on Street Railways." *Journal of the Franklin Institute* 53 (June 1877): 379–90.

Sundstrom, William A. "Internal Labor Markets before World War I: On-the-Job Training and Employee Promotion." *Explorations in Economic History* 25 (1988): 424–45.

Taylor, Frederick W. "A Piece-Rate System." *Transactions of the American Society of Mechanical Engineers* 16 (1895): 856–903.

Usselman, Steven W. "Patents Purloined: Railroads, Inventors, and the Diffusion of Innovation in Nineteenth Century America." *Technology and Culture* 32 (Oct. 1991): 1047–75.

Vauclain, Samuel M. "Is Your Business a Debating Club." *Collier's,* Apr. 2, 1921, pp. 10–11, 18.

———. "The Needs of Industrial Education in America." *Proceedings* of the Engineers' Club of Philadelphia 19 (Jan. 1902): 44–66.

———. "The System of Apprenticeship at the Baldwin Locomotive Works." *Engineering Magazine* 27 (June 1904): 321–33.

———. "Why We Are Always Ready for the Unexpected." *System* 38 (Nov. 1920): 815–17.

"A Visit to the Norris Locomotive Works." *United States Magazine,* Oct. 1855, pp. 151–67.

Waters, Theodore. "Trading in Locomotives." *Frank Leslie's Popular Monthly* 51 (Jan. 1901): 277–91.

Watson, Egbert P. "The Changes of One Lifetime in the Machine Shop." *Engineering Magazine* 30 (Mar. 1906): 883–90.

White, John H., Jr. "Holmes Hinkley and the Boston Locomotive Works." *Railroad History,* no. 142 (Spring 1980): 27–55.

———. "Industrial Locomotives: The Forgotten Servant." *Technology and Culture* 21 (Apr. 1980): 209–16.

———. "Old Ironsides, Baldwin's First Locomotive." *Railway and Locomotive Historical Society Bulletin,* no. 118 (Apr. 1968): 85–87.

———. "Once the Greatest of Builders: The Norris Locomotive Works." *Railroad History,* no. 150 (Spring 1984): 17–56.

———. "Richmond Locomotive Builders." *Railroad History,* no. 130 (Spring 1974): 68–99.

"Workmen." *Engineer* (Philadelphia) 1 (Sept. 27, 1860): 51.

Zeitlin, Jonathan, and Charles Sabel. "Historical Alternatives to Mass Production: Politics, Markets, and Technology in Nineteenth Century Industrialization." *Past and Present* 108 (Aug. 1985): 133–76.

INDEX

References to illustrations and tables are printed in italic type.

Abbott and Ferguson, 16

Accounting, cost, 118, 120–21, 280nn. 85, 86,
  281nn. 97, 98, 100–102

American Federation of Labor, 217, 219

American Locomotive Co. (ALCO), 156,
  216, 227, 230–31; accounting, 281n. 102;
  diversification, 304n. 108; expansion, 205;
  formation, 53–54, 267n. 97; labor organizing,
  218, 301n. 66; liquidation, 305n. 109; losses,
  268n. 106; market share, 268n. 102; Scientific
  Management, 282n. 114

American Locomotive Manufacturers
  Association (ALMA), 51, 53, 266n. 86

American Railway Master Mechanics Association (ARMMA), 57, 85; and standardization,
  71–77

American System, xxv, xxvii, 3, 33, 35, 92, 117,
  180, 183, 291n. 14; described, 172, 252n. 4;
  and machine tools, 164, 169, 189

Apprentices: backgrounds, 137–39, 284nn. 28–
  30; careers, 142, 285n. 41; wages, 141

Apprenticeship, 17, 21, 129, 285nn. 37, 43, 44;
  craft, 136–43, 259nn. 77, 78, 284nn. 22, 23;
  industrial, 211–13, 222

Armory practice, 106, 107, 277n. 48, 295n. 71;
  described, 172, 174–76

Arnold, Horace L., 151

Atchison, Topeka & Santa Fe RR, 48, 76, 150,
  236, 298n. 14

Austin, William L., 84, 95, 100, *124*, 198, 217,
  303n. 89; background, 282n. 108; partner, 123

Baird, M., & Co.: capital, 107; dissolution, 120;
  formation, 106

Baird, Matthew, 22, 25, 40, 76, *100*, 137,
  263n. 48; background, 101, 276n. 26; partner,
  20–21, 99, 102, 277n. 51; retirement, 106–7,
  120, 281nn. 93, 96

Baldwin and Whitney, 12

Baldwin Locomotive Works, 31, 50, 51;
  accounting practices, 118, 120–21; antiunion stance, 23, 146, 219–21; bankruptcy,
  9, 228; capital, *166*, 223, 246; in Civil
  War, 24–25; collaborative strategy, 236–37;
  decline, 224–33; design initiatives, 86–90;
  diversification, 228, 304n. 108; efficiency,
  185–86; in 1837 panic, 9–12; electrification,
  189–94; Employees' Beneficial Association,
  220; employment, *133, 134, 203*; exports,
  44–47, 83–84, 198; gas lighting, 168; in
  Gilded Age, 35–47; growth, *165, 166*, 203–5;
  hostile takeover, 275n. 19; liquidation, 228,
  305n. 111; Loyal Legion, 223; management,
  17, 105, 107–15, 122–25, 170, 176, 182,
  206–7; marketing, 42–47, 77–84, 172,
  182, 264nn. 52, 59; market share, 92, *166*,
  205; origins, 5–7; personnel office, 124;
  product line, 13–14, *15*, 43, 77, 89, 176, 230;
  productivity, *181, 186, 194*, 197, 208, 234;
  profits, 21, 25, 52, 208, 209, 223, 267n. 92,
  299n. 24, 303n. 93; size, 2 (1837), 166–67
  (1850), 28 (1860), 184 (1870), xxv (1872), 33
  (1873), 198 (1902), 215 (1906), xxxi (1907);
  specialization in locos., 21, 37; standard
  locos., 62, 78; standard parts, 103, *170*, 171–
  72, *178;* suppliers, 14–16; test department,
  88; tooling costs, 280n. 82, 297n. 98; trades
  employed, 17, 85, 130–32; Westinghouse

Baldwin Locomotive Works (*continued*)
joint venture, 122; in World War I, 227,
304n. 107. *See also firm's various legal names
and its partners (a list appears on page 97)*
Baldwin, Matthias W., 1, *2*, 17, 25, 32, 39,
61, 66, 92, 99, 255nn. 10–13, 269n. 12;
apprentices, 136–37, 284n. 22; custom locos.,
64–65; death, 25–26; early career, 3–4, 8;
in 1860 strike, 22–23; first engines, 5–7;
management views, 85, 102; "Philadelphia
interests," 40; sole proprietor, 101; workers,
18–19, 20
Baldwin Mutual Relief Fund Association,
152–53, 288n. 83
Baldwin, Vail, and Hufty, 10–11
Baltimore and Ohio RR, 49, 215; custom
locos., 63–65
Banking, 39–40, 263nn. 41, 42
Barry, Job R., 22, 148, 259n. 68
Bement and Dougherty (Bement, Miles &
Co.), 7, 37, 187, 291n. 21; apprentices, 143;
interchangeable parts, 183; "Philadelphia
style" machine tools, 169
Binkerd, Robert, 232
Blacksmiths, xi, 19, *148*, 158, 167, 194, 297n. 99
Boeing Aircraft, 238
Boilermakers, xii, 17, 19, *153*, 159–60, 167, *193*,
220
Brandt, John, 66, 71
Brill Co., J. G., 37
Brooks Locomotive Works, 48, 50, 51, 52, 119
Brown and Sharpe, 147, 177
Brown Brothers, 198, 217
Bryan, Joseph, 57–58
Buick Motor Company, 281n. 102
Building (capital equipment), xxvi, 55, 102–3,
128, 187–89, 231; optimal production, 186,
196, 236–37
Building versus manufacturing, xxvii–xxx, 229–
30, 235, 239–40; apprenticeship, 285n. 44;
capital, 283n. 10; collaboration vs. control,
210, 237; customer relations, 183; design,
custom vs. standard, 58–59; drawings, use of,
292n.26; economies of scale, 49, 54; inside
contracting, 119; interchangeable parts, 172;
machine tools, 297n. 85; managerial control,
195; markets, 28, 253n. 14; production,
optimal vs. efficient, 35–37, 164, 307n. 2;
in recessions, 33; skills, 130. *See also* Capital
equipment
Burnham, George, 17, *94*, 101, 198, 216,
257n. 49, 263n. 44; partner, 106
Burnham, Parry, Williams & Co., 120, 281n. 94

Bush Hill, 7–8, *36*, 37, 56, 135, 189–90, 204, 215,
227

Calvert, Monte, 140, 252n. 6, 300n. 37
Campbell, Henry R., 32, 59–60
Capital equipment, 55, 90, 119, 120, 169,
187–89, 195, 199, 274n. 7; apprenticeships,
137, 143, 211, 285n. 44; decline, 227–29,
254n. 24, 305n. 112, 307n. 4; defined, 251n. 1,
252n. 7; importance, xxvi, 234, 253n. 15;
interchangeable parts, 183; modern, 238–
39, 273n. 91, 307n. 6, 308n. 7; performance
specifications, 273n. 91; production, optimal,
186, 196, 236–37; sector size, xxvi, 253n. 10;
standardized, xxx, 306n. 119. *See also*
Building
Card Books, 110
Central Labor Union, 217
Chandler, Alfred D., Jr., 57, 93, 113, 240,
252n. 5, 253n. 14, 281n. 102, 291n. 14,
305n. 116; and "center firms," 274nn. 6–8
Chief mechanical officer (CMO). *See* Master
mechanics
Chrysler, Walter, 281n. 102
Civil War, 24–25, 260n. 83, 283n. 14
Cleveland, Cincinnati, Chicago, and St. Louis
RR (Big Four), 74
Climax Locomotive Works, 48
Colburn, Zerah, 66
Colvin, Frederick, 74
Commission merchants, 50, 237, 264n. 59,
266n. 83
Converse, John H., *94*, 123, 198, 216, 227;
partner, 281n. 94
Cooke, Jay, & Co., 39
Cooke Locomotive Works, 31, 48, 50, 51, 52, 154
Corliss Engine Works, 252n. 6, 268n. 100
Cost accounting, 120–21, 280n. 86, 281nn. 97,
98, 100–102
Couplers, 56, 271n. 47
Cramps Shipyard, 37, 119
Custom production: height of, 89–90, 202–
3, 209–10, 225–26; market segments, 76–81;
as optimizing strategy, 165, 171, 272n. 62;
origins, 19–20, 63–66, 70. *See also* Building;
Capital equipment; Locomotives

Denver and Rio Grande RR, 78, 263n. 48
Dickson Locomotive Works, 31, 51
Draftsmen, *x*, 85, *86*, 110–12, 131, 158, 178,
273n. 89
Drexel & Co., 39, 198, 220, 263n. 42, 303n. 90
Drexel, Morgan & Co., 39

du Pont, Pierre, 223

Economies of scale, 185, 239, 274n. 6, 296n. 80
Eddystone Ammunition Corporation, 282n. 111
Eight-Hour Movement, 116, 154
Electrification, factory, 189–94, 204
Electro-Motive Corporation. *See* General
    Motors
Emerson, Harrington, 125, 301n. 66
Equalizer, 59–60, *61*
Erecting, xiv–xv, 105, 117, 147, 168, 169, 181, 191,
    204, 205
Ernst, Daniel, 274n. 12
Evans, Oliver, 1, 7, 285n. 39
Expansions, Baldwin plant: first erecting shop,
    169; ca. 1869–73, 183–84; in 1880s, 187; 1890
    erecting shop, 190–91; 1900 machine shop,
    195; 1902 remodeling, 204; Eddystone plant,
    55, 124, 215, 223, 227, *228*

Factory system, 16, 19, 167, 290n. 7; new,
    203–4
Fincher, Jon C., 24, 259n. 77
Fisher brothers, 275n. 19, 304n. 108
Fiske, Pliny, 54
Fitch, Charles, 180, 187
Ford Motor Co., 85
Foremen, 109, 113, 117, 121, 158, 178–79;
    apprentices, training of, 137, 212, powers, 17,
    114, 115, 135, 278n. 66
Founders, xi, 167
Franklin Institute, 1, 255nn. 3, 4, 6; and Sellers
    thread, 183, 271n. 46

Gauges and templates, *173*, *174*, 176–79. *See
    also* Interchangeable parts; Locomotive
    building
General Electric Co., 119, 305n. 109
General Motors, 56, 231–33, 237, 238, 305n. 116,
    306n. 125
Gompers, Samuel, 217
Goss, W. F. M., 88–89
Gotwals, Vernon, 212, 300n. 43
Grant Locomotive Works, 31, 33, 48, 147, 154,
    185, 264n. 54, 295n. 69

Harlan and Hollingsworth, 252n. 6
Harrison, Joseph, Jr., 59–60
Heisler Locomotive Works, 48
Henszey, William P., 77, *88*, *94*, 100, 115, 198;
    death, 216; partner, 106–7; product line
    redesign, 176
Hepburn Act, 215, 224, 300n. 50

Hinkley Locomotive Works, 31, 32, 48, 50, 53,
    119, 185, 261n. 12
Hoadley, Joseph, 53–54, 268n. 100
Hoke, Donald, 295n. 71
Holley, Alexander, 40, 72, 252n. 6, 264n. 49,
    271n. 47
Horizontal combination, 53, 121–22, 267nn. 98,
    99
Hounshell, David, 164, 295n. 71
Hufty, George, 10–11, 17, 101, 256n. 31

Incorporation: BLW in 1909, 216; BLW in 1911,
    223, 303nn. 90, 91; competitors, 33, 261n. 20;
    and managerial capitalism, 123, 275n. 18;
    partners' opposition to, 96, 301n. 55. *See also*
    Partnerships
Inside contracting, 115–16, 119, 278n. 67,
    279n. 81; cost controls, 118, 276nn. 85, 86;
    in depressions, 280n. 90; drawbacks, 119,
    219; in Gilded Age, 117–19, 153–57, 236–37,
    279n. 69; ca. 1900, 205–6, 279nn. 71, 72,
    76, 289n. 90; technical change, 279n. 81,
    280n. 82. *See also* Workers
Interchangeable parts, xiii, 75, 105–6, 164, 172–
    83, 293n. 38; defined, 292n. 35; in locomotive
    repairs, 175, 180–81, 292n. 33, 294nn. 61, 63,
    295n. 65; and marketing, 182–83; origins at
    Baldwin, 37; Sellers thread, 183; wastage, 105,
    294n. 51; and workers, 131–32, 147, 294n. 53
International Association of Machinists, 218,
    219
Interstate Commerce Commission, 215, 224–26

Jervis truck, 6–7, 59
Johnson, Alba B., 96, *124*, 125, 198, 220, 223,
    224–25, 226–27, 304n. 106; background,
    282n. 108; partner, 123; president, 303n. 89

Kansas Pacific RR, 40, 263n. 48
Kuhn, Loeb & Co., 217

Labor policies, Baldwin, 18, 21–22, 237; layoffs,
    33, 133, 145, 150, 215–16, 220; ca. 1900, 210–
    14; promotions, 157–58; recruitment, 134–35,
    143–45; skill dependence, 24, 140. *See also*
    Piecework; Producer Ethos
Laborers (unskilled), 117, 131, 135, 145, 151, 153,
    191–92, 194, 195, 211
Lamoreaux, Naomi, 49
Latrobe, Benjamin H., 64
Laurie, Bruce, 164, 185
Law Books, *110*, 111
Lazonick, William, 283n. 6, 287n. 65

Lehigh Valley RR, 69, 70

Leiter, Joseph, 54

Licht, Walter, 283n. 15

Lighting: electric, 189; gas, 168, 189

Lima Locomotive Works, 48, 268n. 105

List System, 112–13, 203, 224

Locomotive Builders Association (LBA), 50–51, 266n. 81

Locomotive building: batch orders, 175, 182, 293n. 41, 295n. 68; capital-labor ratios, 167, 179, 187, 205; design process, x, 110–12; drawings, 170–72; interchangeable parts, 172–83; labor intensity, 19, 197; production, optimal, 186, 196; standard parts, 171–72; trades in, 17, 130–32. *See also* Locomotives

Locomotive industry: bidding, 34, 42–43, 46, 121, 209, 215, 299n. 27; and business cycle, 31–35, 261n. 16; capital, 30, 53, 261n. 10; competition, 12–13, 19, 25, 33–34, 43, 51; custom production, 50–51; decline, 225–33; diesel, rise of, 231–33; expansion, 33–34, 53, 54, 202–5; exports, 47, 265n. 64; in Great Britain, 30, 44–47, 265nn. 61–63; marketing, 34, 262n. 31; price fixing, 49–53, 230, 266n. 86; profits, 52–53; size, 12, 29, 30–31, 47–49

Locomotive types, 267n. 9; American (4-4-0), 59–60, 66–67, *131;* Atlantic (4-4-2), *201;* Columbia (2-4-2), 86, *88;* Consolidation (2-8-0), 70, 78, *80,* 272n. 66; Decapod (2-10-0), *38, 188;* diesel, 231–33, 306n. 125; English, 29–30; export, *46,* 83–84; flexible beam, 11, 12, 61, *66,* 67, 70; industrial, 43–44, 81–82; Mallet, *201,* 202; mass transit, 43, 82, *83, 84;* metric, 209–10; Mikado (2-8-2), *87;* monorail, *79;* narrow gauge, *42,* 43, 264n. 55; Pacific (4-6-2), *226;* Santa Fe (2-10-2), *202;* Ten-Wheeler (4-6-0), *26,* 67, 74, 84, *200;* turbine, 306n. 124; USRA, 85–86, *87;* Vauclain compound, 87, *88*

Locomotives: blueprint jobs, 73; coal burning, 65, *68;* component specifications, 73; custom, 19–20, 63–65; growth, 53, 67, 69, 76, 189, 195, 202, 224; horsepower, 234; maintenance, 68, 180–81; materials, x, 40, 67, 194; performance guarantees, 62, 70, *161,* 230; performance specifications, 73, 88–89, 202; prices, 9, 32, 46, 50–51, 64, *109,* 121, 209, 256n. 25, 262n. 32, 267n. 89, 299n. 28; quality, 13–14, 182, 205–6, 261n. 24; standard vs. custom, 57–58, 62, 68–69; as symbols, xvii–xix, 251n. 8; tests, 88–89; thermodynamics, 67–68, 270n. 35; tractive effort, 304n. 105. *See also* Master mechanics

Longstreth, Edward, 77, *94,* 142, 183; apprentice, *138;* gauging system, 176–77; partner, 106–7; retirement, 123

Lowell Machine Shop, 119, 147, 256n. 25

Lowell, Ma., textile manufacturers, 2, 28

McCraw, Thomas, 274nn. 6, 8

Machine tools: automated, 187, 297n. 85; Baldwin, 167, 184, 187; costs, 204, 280n. 82; in early industrialization, 3, 252n. 6, 254n. 2; electric power, 192, 297n. 95; interchangeable parts, 179; modern, 237–38, 307nn. 4, 6; "Philadelphia style," 169; Philadelphia's leadership in, 254n. 17, 291n. 22; price fixing, 266n. 76; technological convergence, 164, 187–99, 291n. 21. *See also* Interchangeable parts

Machinery, textile, 3, 255n. 7

Machinists, xiii, 17, 19, 131, 140, 167, 189, 285n. 34; and interchangeable parts, 176, 179

Machinists and Blacksmiths International Union (MBIU), 22–24, 147, 259n. 77

Management: early views of, 94–95, 274nn. 10, 12, 276n. 34; and interchangeable parts, 105–6, 277nn. 47, 48; systematic, 93, 113, 115, 159, 274n. 3. *See also* Scientific Management

Manchester Locomotive Works, 31, 48, 52, 151, 262n. 37

Manufacturing (light machines), xxvii, 85, 172, 187, 231, 305n. 116, 306n. 125; agile, 238; capital in, 283n. 10; problems in, 237–38. *See also* American System; Building versus manufacturing; Mass production

Manufacturing Census, United States, 31, 245, 296n. 77

Martin, Albro, 224, 300n. 50, 303n. 96, 304n. 102

Mason, David, 1, 3–4, 254n. 2, 255nn. 6, 8

Mason Locomotive Works, 31, 53, 151, 295n. 69

Mass production, xxv, 56, 114–15, 123, 146, 229–30, 274n. 6

Master mechanics: and custom locos., 57–58, 63–65, 68–70, 72–73, 91, 298n. 9; empiricism, 69–70, 74; standard locos., 58, 71–72, 85–86; technical virtuosity, 71, 72, 271n. 42. *See also* Railroads

Mechanical drawing, 276n. 40, 292n. 26. *See also* Draftsmen

Mechanical engineering, 71–72, *86,* 140, 252n. 6

Midvale Steel Co., 37, 145, 264n. 50, 282n. 116, 304n. 108

Mitchell, Alexander, 70, 71

Molders, xi, 158, 194

Montgomery, David, 282n. 3

Morgan, J. P., 39–40, 99, *221*

National Civic Federation, 214, 300n. 46
Nelson, Daniel, 196
New York and Erie RR, 66, 180
Norris, Septimus, 61
Norris Locomotive Works, 7, 28, 31, 62, 113, 167, 184, 261n. 12, 285n. 37
Northern Pacific RR, 39, 78, 150, 263n. 11
Novelty Iron Works, 252n. 6

Old Ironsides, 6–7, 59

Panics, financial: in 1837, 9–12; in 1857, 21–22, 30; in 1873, 33, 120; in 1893, 53, 150; in 1907, 215–16
Paris Orléans RR, 209, *210*
Parry, Charles, T., 45, 50, 77, *94*, 100, 115, 129, 183, 238; apprenticeships, 143; contract system, 116, 154; death, 123; and interchangeable parts, 172–76; partner, 106; superintendent, 102–3, 171–72; tooling design, 187
Partnerships, 96–101, 102, 111–12, 260n. 87, 275nn. 17; advantages, 107, 278n. 53, 281nn. 95, 96; contribution shares, 107; decline, 206; dissolutions, 120, 216; limited liability, 121, 275n. 23; specialization in, 106–7. *See also* Incorporation
Patents, 60–61, 87–88, 187, 229
Paterson, N.J., 37, 56, 154, 262n. 36
Peale, Franklin, 5, 255n. 9
Pennsylvania RR, 24, 64, 65, 78, 89, 198, 215, *226*, 306n. 124; and Baldwin, 25, 39, 236, 260n. 81, 263n. 40, 299n. 30; standard locos., 74–75, 272n. 58
Pennsylvania Steel Co., 40
Pettit, William, 17, 101
Philadelphia: efficiency of firms, 185; financial center, 13, 39–40; general strike, 217–18, 301n. 62; industrial center, xxv, 1, 37–39, 164, 251n. 2, 262n. 37; labor pool, 17, 37, 135, 145, 258nn. 51, 52, 283n. 14, 289n. 101; machine tool industry, 169, 251n. 2, 291n. 22; size of firms, 166–67; unions, 214. *See also* Bush Hill
Philadelphia and Columbia RR, 5, 7, 8, 9, 39
Philadelphia and Reading RR, 37–39, 40, 73, 74, 78, 198
"Philadelphia interests," *38*, 40, 263nn. 47, 48. *See also* Thomson, J. Edgar
Philadelphia Rapid Transit Co., strike, 217–18
Piecework, xiii, xxix, 20, 102, 212, 219, 286n. 56; inside contracting, 156–57, 289n. 95; productivity, 20, 146–50, 171, 291n. 15

Pittsburgh Locomotive Works, 31, 51, 151, 185
Porter, H. K., and Co., 48, 51
Portland Locomotive Works, 151
Producer ethos: described, 95, 270n. 12; inside contractors, 119; overhead costs, 106, 125–26, 271n. 13; in partner selection, 100; workers, 128–29, 145–46, 150, 159, 210–11, 221–22
Production schedule, locomotive, x–xvi, 15, 113–14, 167–68, 224, 278n. 62

Railroads: character of American, 6, 30; decline, 225, 233, 303n. 96; early, 5–8; growth, 19, 29–30, 200; innovations in locos., 63–66, 69–71, 202–3; and loco. firms, 30, 42–43, 50, 61–62, 78, 81, 198, 199, 204–5, 209, 224–26, 232, 257nn. 39, 41, 260n. 8; productivity, 29, 58, 75, 200, 202, 268n. 5; regulation, 215, 224–26, 300n. 50; standardization, 57, 74–75, 269n. 6, 270n. 37, 271nn. 46, 47, 273n. 92. *See also* Master mechanics
Rails, 29, 72, 271n. 47
Rhode Island Locomotive Works, 31, 33, 51, 53, 185, 293n. 69
Richmond Locomotive Works, 48, 51, 57, 119, 290n. 105
Rocket, 5, 29
Rogers Locomotive Works, 27, 28, 31, 33, 47, 50, 51, 52, 154, 167, 264n. 57
Roosevelt, Theodore, 215
Rosenberg, Nathan, 164, 252n. 6, 296n. 84

Sample, N. W., 212, 285n. 36
Santa Fe. *See* Atchison, Topeka & Santa Fe RR
Schenectady Locomotive Works, 31, 48, 51, 52, 119, 185
Schmitz, Mark, 164, 185
Scientific Management, 123, 125, 130, 207, 253n. 16, 280n. 116, 283n. 9. *See also* Taylor, Frederick W.
Scranton, Philip, 253n. 9, 254n. 23, 261n. 23, 262nn. 31, 37, 274n. 10
Sellers Co., The William (Sellers and Bancroft), 7, 37, 147, 151, 187, 291n. 21; electric cranes, 191; interchangeable parts, 183; "Philadelphia style" machine tools, 169
Shear, Joseph, 147
Shop Lists, 113, 178
Singer Sewing Machine Co., 35, 93, 154, 183, 262n. 26, 288n. 84
Souther John (Globe Locomotive Works), 31, 121
Southwark Foundry and Machine Co., 37, 304n. 108; apprentices, 143

Specification books, *108,* 109–11
Speedwell Iron Works, 10, 256n. 30
Standard Steel Works, 40, 124, 264n. 50
Standardization, 202, 254n. 24; in diesel locos., 231–32; failures in, 271n. 47, 272n. 58; of loco. parts, 103, 171–72; by manufacturers, 229–30; master mechanics' resistance to, 58, 71–72, 85–86; and patents, 306n. 119; problems, 58, 75, 237–38, 269n. 6; by railroads, 57–58, 74–75; in USRA locos., 85. *See also* Master mechanics
Steel industry, 29, 40, 72, 264nn. 49, 50; capital, 266n. 72; employment, 265n. 66
Steele and Worth, 16
Strikes: in 1860, 20–24, 137, 259n. 70; in 1893, 150, 157; in 1911, 217–23

Taunton Locomotive Works, 31, 48, 53, 119
Taylor, Frederick W., 125, 147, 159, 239, 282n. 116, 286n. 56
Technical virtuosity. *See* Master mechanics
Technological convergence, 189, 296n. 84
Thomson, J. Edgar, 39, 40, 63. *See also* "Philadelphia interests"
Track gauges, 57, 84, 230

Union League, 198–99
Union Pacific RR, 40, 74, 150, *201*
U.S. Military Railroads, 25
U.S. Railway Administration (USRA), 85–86

Vail, George, 10–12
Vail, Stephen, 9, 256n. 30
Vauclain, Samuel M., 87, 100, 150, 198, 204, 216, *222,* 298n. 22; apprenticeships, 211–13; background, 282n. 108; labor unrest, 218–23; partner, 123; plant electrification, 190–93; president, 226–27; tooling design, 187
Vermont Central RR, 64, 65
Vulcan Locomotive Works, 48

War Production Board (WPB), 232, 306n. 123
Westinghouse Electric and Manufacturing Co., 122
Whitin Machine Works, 119, 147, 285n. 37
Whitney, Asa, 12, 17, 37
Williams, Edward, H., 45, 85, *94,* 120; background, 100, 276n. 31; death, 123; partner, 106–7
Wilmarth Locomotive Works, 31, 32
Winans, Ross, 61, 269n. 13
Wootten, John, *68,* 71, 271n. 42, 285n. 36
Workers, 17–18, 20, 21; craft vs. industrial, 129, 155–56, 158–59, *160,* 185, 211–12, 283n. 12, 286n. 58, 299n. 32; divisions among, 152–53, 159–60, 288n. 83; drawings, use of, 111–12, 179; ethnicity, 138, 289n. 101; inside contracting, 117, 153–57; management, 129–30, 148, 157–58; petition of 1848, 18–19; piecework, 145–51, 171; powers, 9, 17–18, 159, 282n. 3; productivity, 151, 197; recruitment, 143–45; relief fund, 152–53; safety, 151–52, 287n. 74; skills, 130–32, 134–35, *160–61,* 167, 211; soldiering, 146, 148–51; technical change, 179, 191–94; trades, 130–32; turnover, 144–45, 223; wages, 33, 147, 150–51, 156–57, 208–9, 219, 287n. 71, 299n. 26. *See also* Strikes; *individual trades*